AI-Oriented Competency Framework for Talent Management in the Digital Economy

Models, Technologies, Applications, and Implementation

Edited by
Alex Khang

CRC Press
Taylor & Francis Group
Boca Raton London New York

CRC Press is an imprint of the
Taylor & Francis Group, an **informa** business

Designed cover image: ©Shutterstock Images

First edition published 2024
by CRC Press
2385 NW Executive Center Drive, Suite 320, Boca Raton FL 33431

and by CRC Press
4 Park Square, Milton Park, Abingdon, Oxon, OX14 4RN

CRC Press is an imprint of Taylor & Francis Group, LLC

Library of Congress Cataloging-in-Publication Data
Names: Khang, Alex, editor.
Title: AI-oriented competency framework for talent management in the
digital economy : models, technologies, applications, and implementation /
edited by Alex Khang.
Description: Boca Raton, FL : CRC Press, 2024. |
Includes bibliographical references and index.
Identifiers: LCCN 2023052648 (print) | LCCN 2023052649 (ebook) |
ISBN 9781032576053 (hbk) | ISBN 9781032577685 (pbk) | ISBN 9781003440901 (ebk)
Subjects: LCSH: Technological innovations—Management. | Artificial intelligence.
Classification: LCC HD45 .A478 2024 (print) | LCC HD45 (ebook) |
DDC 658.3/12028563—dc23/eng/20240309
LC record available at https://lccn.loc.gov/2023052648
LC ebook record available at https://lccn.loc.gov/2023052649

ISBN: 9781032576053 (hbk)
ISBN: 9781032577685 (pbk)
ISBN: 9781003440901 (ebk)

DOI: 10.1201/9781003440901

Typeset in Times
by codeMantra

Contents

Preface

In the digital economy age, one of the most recent core applications of talent management is the competency framework subsystem of the human capital management (HCM) system, which aims to connect the physical experiences and learned skills by making it possible for employers or project managers to explore the capacity of each employee via the assessment system.

This model presents a smart strategy by exploiting employees' hidden skills and existing experiences that can be applied to projects that they currently do not yet have the opportunity to use in the current projects in order to make it easier for the training department, recruiting unit, workforce assurance, and project management team to predict the possibility of shortage or excess of human resources for project's planning in every situation.

AIoCF is one of the highest prioritized concerns of the business management board in the digital economy right now, and it can be used in most HCM applications. With this handbook, you will be able to implement the modern competency framework, role-based skills taxonomy system, skills-based competency framework, role-based skills models or patterns, competency-based job descriptions, competency-based job assessments, job assessment tool, role-based learning pathways, skills-based career cluster and pathway, performance appraisal system, skills-based interview models, AR and VR in enterprise educational system 4.0, AI application in the career development ecosystem, RPA applications, data analytics, and visualization to support planting activities, to manage tasks, to predict outcomes, to forecast risks, and to make informed decisions that facilitate improvements in various services and production activities in the field of software development, software outsourcing, manufacturing, medical and healthcare, education, finance services, banking services, retails, government department, etc.

Happy reading!

Alex Khang

Acknowledgments

The book *AI-oriented Competency Framework in Digital Economy: Taxonomy, Models, Technologies, Applications, and Implementation* is based on the design and implementation of topics related to competency framework, skills taxonomy, job assessment, career development approach, revolutionizing recruitment methodology, artificial intelligence (AI), data analytics, cybersecurity technology, and applications for workforce development and strategy in the era of AI-oriented economy.

Planning and designing a book outline to introduce to readers across the globe is the passion and noble goal of the editor. The biggest reward for the success of the book belongs to the efforts, knowledge, skills, expertise, experiences, enthusiasm, collaboration, and trust of the contributors, turning ideas into reality.

We extend our heartfelt gratitude to all respected contributors for providing high-quality chapters, including human resource managers, talent management leaders, experts, professors, scientists, engineers, scholars, PhD and postgraduate students, educators, and academic colleagues.

We express our gratitude to all respected reviewers with whom we have had the opportunity to collaborate, and acknowledge their tremendous support and valuable comments not only for this book but also for future book projects.

We extend our heartfelt gratitude for all the insightful discussions, valuable advice, support, motivation, shared experiences, collaboration, and inspiration we received from our faculty, contributors, educators, professors, scientists, scholars, engineers, and academic colleagues.

Last but not least, we are grateful to our publisher, CRC Press (Taylor & Francis Group), for the wonderful support in ensuring the timely processing of the manuscript and the prompt publication of this book for the readers.

Thank you, everyone.

Alex Khang

About the Editor

Alex Khang is a Professor in Information Technology, IT Workforce Development Consultant for High-tech Corporations in the United States, India, and Vietnam, an AI and Data Scientist, a software industry expert, and the Chief Technology Officer (AI and Data Science Research Center) at the Global Research Institute of Technology and Engineering, North Carolina, the United States.

He has more than 28 years of teaching and research experience in information technology (software development, database technology, AI engineering, data engineering, data science, data analytics, IoT-based technologies, and cloud computing) at the universities of science and technology in Vietnam, India, and the United States.

He has been the chair for sessions at 20 global conferences, a keynote speaker for more than 25 international conclaves, an expert tech speaker for over 100 seminars and webinars, an international technical board member for ten international organizations, an editorial board member for more than five ISSNs, an international reviewer and evaluator for more than 100 journal papers, and an international examiner and evaluator for more than 15 PhD theses in the field of computer science. He has been contributing to various research activities in fields of AI and data science while publishing many international articles in renowned journals and conference proceedings.

He has published 52 authored books (in computer science 2000–2010), three authored books (software development), and ten book chapters. Furthermore, he has published ten edited books (calling for book chapters) in the fields of AI ecosystem (AI, ML, DL, robotics, data science, big data, and IoT), smart city ecosystem, healthcare ecosystem, fintech technology, and blockchain technology.

Contributors

Albattat Ahmad
Post Graduate Centre, Management and
    Science
Management and Science University
Selangor, Malaysia

Ahmad Al Yakin
Sociology Department
Universitas Al Asyariah Mandar
Sulawesi Barat, Indonesia

Eka Apriani
English Department
Institut Agama Islam Negeri (IAIN)
    Curup
Kec. Curup Utara, Kabupaten Rejang
    Lebong
Bengkulu, Indonesia

Veenu Arora
Human Resource Management
Asian Business School
Noida, India

Beni Azwar Kons
Guidance and Counselling Department
Institute Agama Islam Negeri (IAIN)
    Curup
Bengkulu, Indonesia

Balamurugan R.
Department of Electrical and
    Electronics Engineering
K. S. Rangasamy College of Technology
Tiruchengode, India

Vikram Barnabas
Institute of Management Studies, Career
    Development and Research
Ahmednagar, India

Shweta Batra
Human Resource Management
Asian Business School
Noida, India

Bharathithasan S.
Finance and Business
VIT Business School
Vellore VIT University
Vellore, India

Swati Bhatia
Human Resource Management
Asian Business School
Noida, India

Amarendra Bhuyan
Department of Commerce
KISS Deemed to be University
Bhubaneswar, India

Debanjalee Bose
Finance and Business
VIT Business School
Vellore VIT University
Vellore, India

Sachin Chaudhary
School of Computer Science and
    Application
IIMT University
Meerut, India

Sugandha Chebolu
Finance and Business
VIT Business School
Vellore VIT University
Vellore, India

Rohan Rajendra Dalvi
Department of Computer Engineering
Thakur College of Engineering and
    Technology
University of Mumbai
Mumbai, India

Besse Darmawati
Science and Technology
Badan Riset dan Inovasi Nasional
Daerah Khusus Ibukota Jakarta,
    Indonesia

Annada Sankar Dash
Department of Commerce
KISS Deemed to be University
Bhubaneswar, India

Surrya Prakash DilliBabu
Department of Mechanical Engineering
Vel Tech Rangarajan Dr. Sagunthala
    R&D Institute of Science &
    Technology
Chennai, India

Hennadiiy Dmytrenko
Department of Public Administration,
    Interregional Academy of Personnel
    Management
Kyiv, Ukraine

Ahmed A. Elngar
Department of Computer Science and
    Artificial Intelligence
Beni-Suef University
Beni Suef, Egypt

Muna Fathmath
Director General
Maldives Media Council
Male', Maldives

Vijaya Kumar Guivada
Engineering and Technology
Nawab Shah Alam Khan College of
    Engineering and Technology
Hyderabad, India

Rashmi Gujrati
KC Group of Institutions
Nawanshahr, India

Chabi Gupta
School of Commerce, Finance and
    Accountancy
Christ University
Bengaluru, India

Suyog Gupta
Department of Computer Engineering
Thakur College of Engineering and
    Technology
University of Mumbai
Mumbai, India

Nataliia Holovach
Department of Public Administration
Interregional Academy of Personnel
    Management
Kyiv, Ukraine

Faraz Hussain
Department of Computer Engineering
Thakur College of Engineering and
    Technology
University of Mumbai
Mumbai, India

Babasaheb Jadhav
Finance and Business
D. Y. Patil Vidyapeeth (Deemed to be
    University)
Global Business School & Research
    Centre
Pune, India

Shivakumar Kagi
Department of Computer Science and
    Engineering
Sharnbasva University
Kalaburagi, India

Chinnadurai Kathiravan
Department of Commerce and Financial
    Studies
VIT Business School
Vellore Institute of Technology
Vellore, India

Alex Khang
Department of AI and Data Science
Global Research Institute of Technology
    and Engineering
Fort Raleigh, North Carolina

Ruhi Lal
Student Support Department
Amity School of Communication
Amity University
Noida, India

Maniraj S. P.
Department of Computer Science and
    Engineering
SRM Institute of Science and
    Technology
Chennai, India

Brojo Kishore Mishra
Department of Computer Science and
    Engineering
GIET University
Gunupur, India

Udayana Mohanty
Department of Commerce
KISS Deemed to be University
Bhubaneswar, India

Muthmainnah M
English Department
Universitas Al Asyariah Mandar
Sulawesi Barat, Indonesia

Arpita Nayak
Business Administration
KIIT School of Management
KIIT University
Bhubaneswar, India

Nidhya M. S.
Department of Computer Science &
    Information Technology
Jain Deemed-to-be University
Bangalore, India

Ahmed J. Obaid
Faculty of Computer Science and
    Mathematics
University of Kufa
Kufa, Iraq

Lakshmi Prasad Panda
Department of Humanities
Government College of Engineering
    Kalahandi
Bhawanipatna, India

Megharani Patil
Department of Computer Engineering
Thakur College of Engineering and
    Technology
University of Mumbai
Mumbai, India

Atmika Patnaik
Business Administration
King's College London
London, United Kingdom

B. C. M. Patnaik
Business Administration
KIIT School of Management
KIIT University
Bhubaneswar, India

Swamy Perumandla
Finance and Business
VIT Business School
Vellore VIT University
Vellore, India

Neetu Pillai
Computer Science
University of West London
Ras Al-Khaimah, United Arab Emirates

Zakia Tasmin Rahman
Management Department
Amity School of Communication
Amity University
Noida, India

Kali Charan Rath
Department of Mechanical Engineering
GIET University
Gunupur, India

Revati Ramrao Rautrao
Department of Computer Science &
    Engineering
Management School
Sinhgad Institute of Business
    Administration and Research
Pune, India

Ravinder Rena
Faculty of Management Sciences,
    Department of Economics
DUT Business School
Faculty of Management Sciences
Durban University of Technology
ML Sultan Campus, Republic of South
    Africa

Sakthi Srinivasan K.
Finance and Business
VIT Business School
Vellore VIT University
Vellore, India

Durga Prasad Singh Samanta
Business Administration
KIIT Deemed to be University
Bhubaneswar, India

Sasmita Samanta
Business Administration
KIIT University
Bhubaneswar, India

Sampath Dakshina Murthy A.
Department of Electrical
    Communication Engineering
Vignan's Institute of Information
    Technology (A)
Visakhapatnam, India

Samuel Rajkumar V.
Career Development Centre
VIT Business School
Vellore Institute of Technology
Vellore, India

Sathish A. S.
Department of Commerce and Financial
    Studies
VIT Business School
Vellore Institute of Technology
Vellore, India

Ipseeta Satpathy
Business Administration
KIIT School of Management
Kalinga Institute of Industrial
    Technology (KIIT)
Bhubaneswar, India

Mudassar Sayyed
Career Development and Research
Institute of Management Studies
Ahmednagar, India

Inna Semenets-Orlova
Department of Public Administration
Interregional Academy of Personnel
    Management
Kyiv, Ukraine

Wasswa Shafik
Dig Connectivity Research Laboratory
    (DCRLab)
Brunei Darussalam School of Digital
    Science
Universiti Brunei Darussalam
Bandar Seri Begawan, Brunei

Kewal Krishan Sharma
Department of Computer Science &
  Engineering
School of Computer Science and
  Application
IIMT University
Meerut, India

Vikas Sharma
Department of Computer Science &
  Engineering
School of Computer Science and
  Application
IIMT University
Meerut, India

Sharmila Devi R.
Department of Commerce and Financial
  Studies
VIT Business School
Vellore VIT University
Vellore, India

Roman Shevchuk
Department of Public Administration
Interregional Academy of Personnel
  Management
Kyiv, Ukraine

Hayri Uygun
RTEU Ardeşen Meslek Yüksekokulu
Ardesen Vocational School
Recep Tayyip Erdogan University
Rize, Turkey

Tarun Kumar Vashishth
Department of Computer Science &
  Applications
School of Computer Science and
  Application
IIMT University
Meerut, India

Vihas Vijay
Department of Commerce and Financial
  Studies
VIT Business School
Vellore Institute of Technology
Vellore, India

Erfin Wijayanti
English Department
Institut Agama Islam Negeri (IAIN)
  Fattahul Muluk Papua
Jalan Merah Putih Jl. Buper Waena
Papua, Indonesia

Liubov Zgalat-Lozynska
Department of Public Administration
Interregional Academy of Personnel
  Management
Kyiv, Ukraine

1 Future Directions and Challenges in Designing Workforce Management Systems for Industry 4.0

Alex Khang, Kali Charan Rath,
Brojo Kishore Mishra, Revati Ramrao Rautrao,
and Lakshmi Prasad Panda

1.1 INTRODUCTION

As we embark on the era of Industry 4.0, the design and implementation of effective workforce management systems (WMS) have become paramount for organizations seeking to navigate the challenges and harness the opportunities presented by this technological revolution (Ang et al., 2016). In this context, it is crucial to explore future directions and address the emerging challenges that arise in the development of (WMS) tailored to the needs of Industry 4.0. This chapter aims to delve into the forefront of this field, examining the evolving landscape; identifying potential areas of advancement; and discussing the obstacles that must be overcome to create robust and adaptable systems capable of optimizing human–machine collaboration, enhancing productivity, and fostering sustainable growth in the Industry 4.0 era.

1.1.1 INTRODUCTION TO WMS IN THE CONTEXT OF INDUSTRY 4.0

In the era of Industry 4.0, where automation, digitization, and advanced technologies are transforming the industrial landscape, the management of a skilled and efficient workforce has become more critical than ever (Bag et al., 2021; Brahma et al., 2021). WMS play a pivotal role in optimizing human resources and maximizing productivity in this dynamic and rapidly evolving environment (Calzavara et al., 2020; Ciano et al., 2021).

Industry 4.0 represents the fourth industrial revolution characterized by the integration of cyber–physical systems, the Internet of Things (IoT), big data analytics, artificial intelligence (AI), and cloud computing (Galati & Bigliardi, 2019). These technological advancements have brought about a paradigm shift in how businesses operate and have a profound impact on the workforce. As organizations strive to stay competitive, they are increasingly turning to WMS to effectively navigate the challenges and leverage the opportunities presented by Industry 4.0.

DOI: 10.1201/9781003440901-1

A WMS is a comprehensive solution that encompasses various aspects of managing human resources, including employee scheduling, time and attendance tracking, performance management, skill development, and workforce analytics (Imperatori et al., 2019). It leverages advanced software and analytics capabilities to streamline and automate these processes, enabling organizations to optimize their workforce utilization, enhance operational efficiency, and drive better business outcomes.

One of the key benefits of WMS in the context of Industry 4.0 is the ability to align human resources with the demands of a digitized and interconnected environment. These systems provide real-time visibility into workforce data, enabling managers to make data-driven decisions and allocate resources effectively. By analyzing data on employee availability, skills, and performance, organizations can proactively address skill gaps, allocate resources to the right tasks, and ensure the right people are in the right place at the right time.

Moreover, WMS empower employees by offering self-service options and mobile access, allowing them to view schedules, request time off, and access training materials conveniently. This level of autonomy and flexibility fosters employee engagement, satisfaction, and productivity in the Industry 4.0 landscape, where remote work, flexible schedules, and digital collaboration are becoming the norm. Additionally, these systems enable organizations to forecast future workforce needs, identify trends, and anticipate potential challenges. By leveraging predictive analytics and machine learning (ML) algorithms, organizations can optimize workforce planning, talent acquisition, and succession management strategies. This proactive approach ensures a sustainable and adaptable workforce capable of meeting the evolving demands of Industry 4.0 (Khang & Rani et al., 2023).

In conclusion, WMS have become indispensable tools in the context of Industry 4.0. They enable organizations to harness the potential of advanced technologies while effectively managing their most valuable asset—their workforce. By leveraging real-time data, automation, and analytics, these systems optimize human resource allocation, enhance employee engagement, and facilitate agile decision-making. As we continue to witness the transformative power of Industry 4.0, organizations that embrace WMS will gain a competitive edge by unlocking the full potential of their workforce in this digital age (Khang & Shashi et al., 2023).

1.1.2   THE CRUCIAL SIGNIFICANCE OF DESIGNING EFFECTIVE WMS IN THE ERA OF INDUSTRY 4.0

In the dynamic era of Industry 4.0, characterized by rapid technological advancements and interconnected systems, the development of efficient WMS holds paramount significance for organizations. These systems play a pivotal role in optimizing human resources while effectively addressing the challenges and capitalizing on the opportunities presented by Industry 4.0. Industry 4.0 marks a new wave of industrial revolution, where automation, AI, big data analytics, the IoT, and cloud computing converge (Krishnamoorthy et al., 2023). In this transformative landscape, traditional approaches to workforce management fall short, necessitating the design of sophisticated systems that can adapt to the evolving demands. It is crucial to develop WMS that can effectively navigate this digital terrain and maximize the potential of human capital.

The development of efficient WMS becomes indispensable due to various factors. Firstly, these systems enable organizations to optimize workforce utilization by aligning human resources with the specific demands of Industry 4.0. Real-time visibility into workforce data empowers managers to make informed decisions regarding resource allocation, ensuring that individuals with the appropriate skill sets are deployed efficiently. This level of precision and agility is essential for meeting production targets, minimizing downtime, and enhancing overall operational efficiency within the realm of Industry 4.0 (Luthra & Mangla, 2018).

Secondly, efficient WMS allow organizations to address skill gaps and foster talent development within their workforce. By providing insights into employee performance, skills, and training requirements, these systems facilitate proactive identification of areas that necessitate improvement. Through targeted training programs and performance management initiatives, organizations can enhance employee capabilities, foster continuous learning, and cultivate a highly skilled workforce capable of meeting the evolving demands of Industry 4.0 (Phuyal et al., 2020).

Overly, it can be stated that the development of efficient WMS plays a vital role in navigating the complex landscape of Industry 4.0. By optimizing human resources and addressing skill gaps, these systems empower organizations to leverage the benefits of advanced technologies and maintain a competitive edge. In the face of rapid digitization and interconnectivity, organizations that prioritize the design of effective WMS will be better equipped to capitalize on the opportunities and overcome the challenges presented by Industry 4.0 (Pallavi et al., 2023).

1.1.3 OBJECTIVE OF THE WORK

The objective of this study is to explore the future directions and address the challenges associated with the design of WMS tailored for Industry 4.0. By examining emerging technological trends, assessing their impacts on the workforce, and understanding the specific challenges and opportunities that arise, this research aims to provide insights and recommendations that can guide the development of effective WMS capable of optimizing human resources within the dynamic context of Industry 4.0 (Khang & Muthmainnah et al., 2023).

1.1.4 SCOPE FOR THE CHAPTER

The scope of this study encompasses a comprehensive examination of the future directions and challenges associated with designing WMS specifically tailored for Industry 4.0. It involves an exploration of emerging technological trends, such as AI, ML, robotics, and IoT, and their implications for workforce management. Additionally, the study encompasses an assessment of the impact of Industry 4.0 on the workforce, including changes in job roles, skill requirements, and work environments. By considering these aspects, the research aims to provide valuable insights and recommendations that can inform the design and development of effective WMS capable of optimizing human resources within the dynamic landscape of Industry 4.0 (Aakansha et al., 2023).

1.1.5 Overview of the Chapter and the Structure Followed

As the world continues its rapid transition toward the era of Industry 4.0, the design and implementation of effective WMS become increasingly critical. This chapter explores the future directions and challenges that lie ahead in the development of such systems. The chapter begins by examining the evolving landscape of Industry 4.0, highlighting the technological advancements and transformative changes that have reshaped traditional workforce management practices. It delves into the emerging trends and shifts in the nature of work, including the rise of remote and flexible work arrangements, the gig economy, and the integration of human and machine collaboration.

Furthermore, this chapter addresses the challenges that organizations encounter when designing WMS for Industry 4.0. It explores the complexities of managing a diverse workforce comprising both traditional employees and independent contractors, as well as the ethical considerations and legal implications associated with the use of AI and automation in workforce decision-making. Moreover, the chapter sheds light on the importance of data-driven decision-making in workforce management, emphasizing the need for organizations to harness the power of data analytics, ML, and predictive modeling to optimize workforce planning, recruitment, and skill development strategies.

The chapter also highlights the crucial role of leadership and organizational culture in successfully implementing WMS for Industry 4.0. It explores the need for adaptive leadership, continuous learning, and agile change management practices to navigate the evolving demands of a technologically advanced workforce (Muthmainnah et al., 2023).

Finally, the chapter outlines future directions and possibilities for WMS in Industry 4.0, including the potential integration of augmented reality, virtual reality, and wearable technologies. It concludes with a discussion of the challenges and considerations that need to be addressed in designing and implementing these cutting-edge technologies while ensuring the well-being and empowerment of the workforce. By exploring the future directions and challenges in designing WMS for Industry 4.0, this chapter provides valuable insights and guidance for organizations seeking to adapt and thrive in the dynamic and transformative landscape of the Fourth Industrial Revolution (Pooja & Khang, 2023).

1.2 TECHNOLOGICAL ADVANCEMENTS IN WMS SYSTEMS

In the realm of designing WMS for Industry 4.0, significant technological advancements have emerged to address the unique demands of this industrial era. These advancements encompass a range of innovative tools and techniques that revolutionize how organizations manage their workforce. One prominent technological development is the integration of AI and ML algorithms, which enable organizations to analyze vast amounts of data to make data-driven decisions regarding workforce planning, skill mapping, and resource allocation. Additionally, the advent of the IoT has facilitated the connectivity and communication between machines, devices, and personnel, enabling real-time monitoring of workforce performance, health, and safety. Furthermore, the rise of automation and robotics has led to the deployment of intelligent robots in various tasks, augmenting human capabilities and streamlining processes. These technological advancements have the potential to revolutionize WMS in Industry 4.0, providing organizations with enhanced efficiency, agility, and adaptability to meet the dynamic challenges of the modern industrial landscape (Zhang et al., 2023).

1.2.1 AI in WMS

AI has demonstrated its potential in transforming WMS in various industries, including manufacturing corporations. One notable case study exemplifying the application of AI in this context is the use of AI-powered robotics and automation at Tesla, an innovative electric vehicle manufacturer. Tesla has leveraged AI and robotics to enhance its manufacturing processes and optimize workforce management. The company employs a significant number of AI-driven robots in its production lines to handle repetitive and labor-intensive tasks. These robots, equipped with computer vision and ML algorithms, can precisely perform complex assembly tasks, resulting in increased efficiency and reduced errors compared to traditional manual labor.

By integrating AI-powered robotics into their WMS (Park et al., 2023), Tesla has achieved several benefits. Firstly, the use of robots has improved production speed and throughput, allowing the company to meet the growing demand for their electric vehicles. This increased efficiency has also led to cost savings and improved profitability (Shet & Pereira, 2021; Sott et al., 2021). Moreover, AI-powered robots have positively impacted workforce safety and ergonomics. By automating physically demanding tasks, Tesla has reduced the risk of work-related injuries and strain on employees, enhancing their well-being and job satisfaction.

Additionally, AI algorithms analyze production data collected by sensors and cameras to monitor and optimize the performance of robots and machines. This data-driven approach enables predictive maintenance, as potential issues and equipment failures can be detected early, reducing downtime and ensuring uninterrupted production (Wagire et al., 2020; Xu et al., 2018). The application of AI in Tesla's manufacturing WMS has resulted in a highly efficient and adaptive production environment. By harnessing the capabilities of AI-driven robots, Tesla has achieved significant improvements in productivity, quality control, employee safety, and overall operational efficiency.

This case study demonstrates how AI integration in manufacturing corporations can revolutionize WMS, enabling companies to achieve higher levels of productivity, reduce costs, and create a safer and more optimized work environment for their employees. WMS play a crucial role in managing human resources and optimizing workforce productivity in manufacturing corporations. The integration of AI into WMS offers numerous benefits, such as improved scheduling, enhanced productivity, cost reduction, and increased employee satisfaction.

1.2.1.1 Case Study: AI in WMS in a Manufacturing Corporation

In our case study, let's consider a fictional manufacturing corporation called "ABC Manufacturing." They have implemented an AI-powered WMS to optimize their workforce operations. Algorithm for AI-powered WMS:

- Collect historical workforce data, including employee records, skills, performance metrics, and scheduling information.
- Gather external data such as market trends, demand patterns, and production schedules.
- Preprocess the collected data, cleaning and handling missing values, outliers, and inconsistencies.
- Normalize numerical data and encode categorical variables.
- Utilize AI algorithms to forecast workforce demand based on historical data, market trends, and external factors.
- Match required skills for tasks with available workforce, considering skills, certifications, and experience.
- Optimize workforce scheduling based on demand forecasts, employee availability, and task requirements.
- Predict employee absenteeism using AI algorithms and various data sources like attendance records and health data.
- Calculate optimal overtime allocation to fulfill demand during peak periods while minimizing costs.
- Evaluate employee performance based on productivity, quality, and attendance metrics using AI algorithms.
- Identify skill gaps and training needs among the workforce based on performance data and evolving job requirements.
- Analyze employee feedback surveys and sentiment data to evaluate overall employee satisfaction.
- Optimize costs by allocating workforce, overtime, and training resources effectively.
- Track workforce performance metrics and suggest strategies to enhance productivity.

Below list illustrates the application of AI in WMS:

- **Workforce Demand Forecast**: AI algorithms analyze historical data, market trends, and other relevant factors to forecast workforce demand accurately. For example, ABC Manufacturing's AI system predicts the following monthly workforce demand:

Month	Workforce Demand	Month	Workforce Demand
Jan	250	Jun	320
Feb	280	Jul	310
Mar	300	Aug	280
Apr	260	Sep	300
May	290	Oct	290

- **Skill Matching**: AI algorithms match the required skills for a particular task with the available workforce, considering their skills, certifications, and experience. ABC Manufacturing's AI system provides the following skill match for a specific production line:

Worker ID	Skill Match (%)	Worker ID	Skill Match (%)
1	90	6	88
2	85	7	91
3	95	8	87
4	80	9	84
5	92	10	93

- **Workforce Scheduling**: AI algorithms optimize workforce scheduling based on demand forecasts, employee availability, and task requirements. The AI system generates the following weekly schedule for ABC Manufacturing:

Week	Monday	Tuesday	Wednesday	Thursday	Friday	Saturday	Sunday
Week1	30	30	30	30	30	–	–
Week2	35	35	35	35	35	–	–
Week3	40	40	40	40	40	–	–
Week4	35	35	35	35	35	–	–

(Note: "–" denotes no shift assigned)

- **Employee Absenteeism Prediction**: AI algorithms analyze various data sources like employee attendance, health records, and external factors to predict the likelihood of employee absenteeism. ABC Manufacturing's AI system predicts the following absenteeism rates for the upcoming quarter:

Quarter	Absenteeism Rate (%)
Q1	7
Q2	5
Q3	6
Q4	8

- **Overtime Calculation**: AI algorithms calculate optimal overtime allocation to fulfill the demand during peak periods while minimizing costs. ABC Manufacturing's AI system recommends the following overtime hours per month:

Month	Overtime Hours	Month	Overtime Hours
Jan	60	Jun	75
Feb	55	Jul	70
Mar	70	Aug	55
Apr	50	Sep	60
May	65	Oct	60

- **Employee Performance Evaluation**: AI algorithms assess employee performance based on various metrics, including productivity, quality, and attendance. ABC Manufacturing's AI system rates employees on a scale of 1–10:

Worker ID	Performance Rating	Worker ID	Performance Rating
1	9	6	8
2	8	7	9
3	9	8	8
4	7	9	7
5	9	10	9

- **Training Needs Identification**: AI algorithms identify skill gaps and training needs among the workforce based on performance data and evolving job requirements. ABC Manufacturing's AI system suggests the following training requirements for employees:

Worker ID	Training Needs	Worker ID	Training Needs
1	Yes	6	Yes
2	No	7	No
3	No	8	Yes
4	Yes	9	No
5	No	10	No

- **Employee Satisfaction Analysis**: AI algorithms analyze employee feedback surveys and sentiment data to evaluate overall employee satisfaction. ABC Manufacturing's AI system rates employee satisfaction on a scale of 1–5:

Worker ID	Satisfaction Rating	Worker ID	Satisfaction Rating
1	4	6	4
2	3	7	5
3	5	8	3
4	3	9	4
5	4	10	5

- **Cost Optimization**: AI algorithms optimize workforce allocation, overtime, and training costs to reduce overall operational expenses. ABC Manufacturing's AI system estimates the following monthly cost savings:

Month	Cost Savings ($)	Month	Cost Savings ($)
Jan	5,000	Jun	6,000
Feb	4,500	Jul	5,500
Mar	5,500	Aug	4,500
Apr	4,000	Sep	5,000
May	5,200	Oct	5,000

- **Productivity Improvement**: AI algorithms track workforce performance metrics and suggest strategies to enhance productivity. ABC Manufacturing's AI system shows the following monthly productivity improvement percentages:

Month	Productivity Improvement (%)	Month	Productivity Improvement (%)
Jan	5	Jun	7
Feb	4	Jul	6
Mar	6	Aug	4
Apr	3	Sep	5
May	5	Oct	5

1.2.1.2 Mathematical Model

A mathematical model is using mixed-integer linear programming (MILP) for AI applications in WMS as described below:

a. Mathematical Model:
  Decision Variables:
  - $x[i, j]$: Binary variable indicating whether task i is assigned to worker j.
  - start_time $[i, j]$: Start time of task i assigned to worker j.

b. Parameters: Here are seven example parameters with their numerical values:

Parameter	Symbol	Value
Number of workers	Nw	50
Number of tasks	Nt	100
Available work hours per day	H	8
Worker skill level	S	44,931
Task priority	P	44,936
Task duration	$D[i]$	Varies
Worker availability	$A[j]$	0 or 1

c. Objective: Minimize the total completion time of tasks as Equation (1.1):

$$\text{Minimize} : \sum (i = 1 - \text{Nt}) \cdot \sum (j = 1 - \text{Nw}) \, \text{start\_time}[i, j] + D[i]. \qquad (1.1)$$

Subject to:

- Each task must be assigned to exactly one worker as Equation (1.1.1):

$$\sum (j = 1 - \text{Nw}) \, x[i, j] = 1, \text{ for } i = 1 - \text{Nt}. \qquad (1.1.1)$$

- Each worker can be assigned to at most one task at a time as Equation (1.1.2):

$$\sum (i = 1 - \text{Nt}) \, x[i, j] \le 1, \text{ for } j = 1 - \text{Nw}. \qquad (1.1.2)$$

- Worker availability constraint as Equation (1.1.3):

$$x[i, j] \le A[j], \text{ for } i = 1 - \text{Nt}, \, j = 1 - \text{Nw}. \qquad (1.1.3)$$

- Start time constraint as Equation (1.1.4) and as Equation (1.1.5):

$$\text{start\_time}\,[i, j] \ge \text{start\_time}\,[k, j] + D[k] - H * (1 - x[k, j]), . \qquad (1.1.4)$$

$$\text{for } i, k = 1 - \text{Nt}, \, j = 1 - \text{Nw}, \, k \ne i. \qquad (1.1.5)$$

- Skill level constraint (if applicable) as Equation (1.1.6):

$$x[i, j] \le S[j], \text{ for } i = 1 - \text{Nt}, \, j = 1 - \text{Nw}. \qquad (1.1.6)$$

d. Python Code for Gantt chart:

```python
import matplotlib.pyplot as plt

import numpy as np

def plot_gantt_chart(tasks, workers, start_times,
durations):

    fig, ax = plt.subplots()

    y_ticks = range(len(workers))

    ax.set_yticks(y_ticks)

    ax.set_yticklabels(workers)
```

```
for i, task in enumerate(tasks):

    for j, worker in enumerate(workers):

        if start_times[i, j] > 0.5:

            start = start_times[i, j]

            end = start + durations[i]

            ax.barh(j, width=end - start,
left=start, height=0.5, color='b')

    ax.set_xlabel('Time')

    ax.set_ylabel('Worker')

    ax.set_title('Gantt Chart')

    plt.show()
```

e. Criteria for Analytics:
- **Task Duration**: Analyze the duration of each task represented by the length of the corresponding bar in the Gantt chart. Identify tasks that require more time to complete.
- **Resource Allocation**: Observe how workers are assigned to tasks in the Gantt chart. Check if there are any imbalances or overutilization of specific workers.
- **Task Dependencies**: Based on industrial real data, it is advised to examine the sequential placement of tasks in the Gantt chart. Identify any dependencies or order constraints between tasks.
- **Schedule Optimization**: Assess the overall scheduling efficiency by examining the distribution of tasks across workers and their start times. Look for opportunities to minimize idle time or reduce the total duration of the schedule.
- **Critical Path Analysis**: Determine the critical path, i.e., the sequence of tasks that determines the minimum project duration. Identify tasks on the critical path and analyze their impact on the overall schedule.

1.2.1.3 IoT for Real-time Monitoring (How IoT Devices Can Enable Real-Time Monitoring of Employee Performance, Safety, and Equipment Utilization)

In the manufacturing industry, the utilization of the IoT for real-time monitoring has emerged as a game-changer in tracking and optimizing employee performance, ensuring safety protocols, and maximizing equipment utilization. By integrating IoT-enabled devices and sensors throughout the production floor, manufacturers can capture and analyze a wealth of real-time data. This data includes parameters such as machine usage, employee productivity, and safety compliance, providing valuable insights into operational efficiency and potential areas for improvement. Through IoT-based solutions, manufacturers can monitor employee performance in real-time, identifying bottlenecks, optimizing workflows, and implementing targeted training programs to enhance productivity.

Additionally, IoT-powered sensors can continuously monitor the working environment, detecting potential safety hazards and triggering immediate alerts to prevent accidents. Furthermore, real-time equipment utilization data enables manufacturers to identify underutilized machinery, schedule preventive maintenance, and minimize downtime, ultimately increasing overall efficiency and profitability in the manufacturing process. The IoT's integration into real-time monitoring in the manufacturing industry is revolutionizing the way employee performance, safety, and equipment utilization are managed, driving the industry toward greater productivity and operational excellence (Arpita et al., 2023).

1.2.1.3.1 *Parameters*

We have simulated real-time data for monitoring various parameters in a manufacturing industry using the IoT. The parameters include employee productivity, equipment downtime, energy consumption, employee safety incidents, equipment utilization, temperature, humidity, noise level, air quality, and light intensity as Table 1.1.

1.2.1.3.2 *Algorithm for the Given Python Program*

The data are generated randomly for 24 hours, with one data point per hour. The code also includes the plotting of various graphs using the Matplotlib library.

Step 1: Import the required modules: time, random, and matplotlib.pyplot.
Step 2: Initialize empty lists for timestamps and different parameter values.
Step 3: Simulate real-time data for monitoring by generating random values for each parameter for 24 hours (1 data point per hour) using for loop:
  • Append the formatted timestamp (hour:minute) to the timestamps list.

TABLE 1.1
Real-Time Monitoring Parameters

Parameter	Numerical Value
Employee productivity	0.85
Equipment downtime	2 hours
consumption	500 kWh
Employee safety incidents	3
Equipment utilization	0.9
Temperature	25°C
Humidity	0.5
Noise level	70 dB
Air quality	20 ppm
Light intensity	500 lux

- Generate random values for each parameter using appropriate ranges.
- Append the generated values to their respective lists.

Step 4: Plot the data using Matplotlib:
- Create a line plot for employee productivity.
- Create a bar plot for employee safety incidents.
- Create a scatter plot for equipment utilization.
- Customize the plot labels, titles, and grid settings.
- Show the plots.

Step 5: Perform analysis of the generated data
- Calculate the average productivity by summing up productivity_values and dividing by its length.
- Find the maximum downtime from downtime_values.
- Calculate the total energy consumption by summing up energy_values.
- Find the maximum number of safety incidents from safety_incidents_ values.
- Calculate the average utilization by summing up utilization_values and dividing by its length.
- Calculate the average temperature, humidity, noise level, air quality, and light intensity using the corresponding lists.

Step 6: Print the analysis results
- Print the average productivity with two decimal places.
- Print the maximum downtime with two decimal places.
- Print the total energy consumption with two decimal places.
- Print the maximum number of safety incidents.
- Print the average utilization with two decimal places.
- Print the average temperature with two decimal places.
- Print the average humidity with two decimal places.
- Print the average noise level with two decimal places.
- Print the average air quality with two decimal places.
- Print the average light intensity with two decimal places.

1.2.1.3.3 Output

Line plots, bar plots, and scatter plots are generated for different parameters to visualize the data in Figures 1.1–1.3.

1.2.1.3.4 Output Analytic Results

- Average Productivity: 91.34%
- Maximum Downtime: 3.00 hours
- Total Energy Consumption: 12,244.63 kWh
- Maximum Safety Incidents: 5
- Average Utilization: 89.42%
- Average Temperature: 25.18°C
- Average Humidity: 50.00%
- Average Noise Level: 69.53 dB
- Average Air Quality: 19.00 ppm
- Average Light Intensity: 507.95 lux

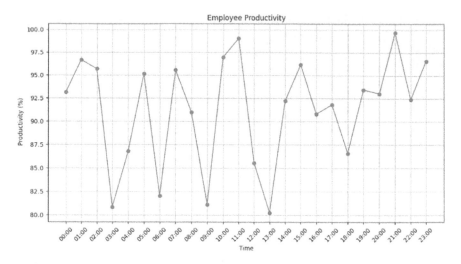

FIGURE 1.1   Employee productivity.

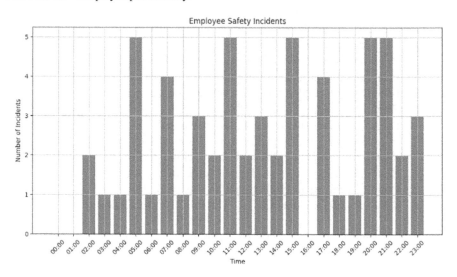

FIGURE 1.2   Employee SI.

1.2.1.3.5   Short Explanation

- **Average Productivity**: The average productivity of employees throughout the 24-hour period was 91.34%. This indicates a relatively high level of efficiency and output in the manufacturing industry.
- **Maximum Downtime**: The highest recorded downtime for equipment during the 24-hour period was 3.00 hours. Identifying and addressing the causes of downtime can help improve overall productivity and minimize disruptions in the manufacturing process.
- **Total Energy Consumption**: The total energy consumed during the 24-hour period was 12,244.63 kWh. Monitoring energy consumption

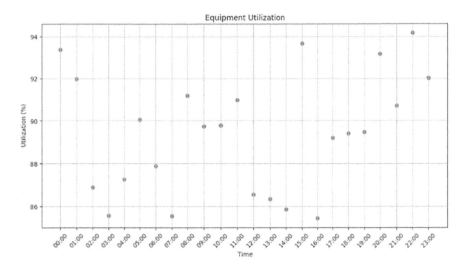

FIGURE 1.3   Equipment utilization.

can help identify opportunities for energy optimization and cost-saving measures.

- **Maximum Safety Incidents**: The highest number of safety incidents reported among employees during the 24-hour period was 5. Ensuring a safe working environment should be a priority to minimize accidents and promote employee well-being.
- **Average Utilization**: The average equipment utilization rate during the 24-hour period was 89.42%. This indicates that the manufacturing equipment was effectively utilized, maximizing operational efficiency.
- **Average Temperature**: The average temperature recorded in the manufacturing facility during the 24-hour period was 25.18°C. Monitoring temperature levels is important to ensure optimal working conditions and prevent equipment or product damage.
- **Average Humidity**: The average humidity level in the manufacturing facility during the 24-hour period was 50.00%. Maintaining appropriate humidity levels is crucial for preserving product quality and employee comfort.
- **Average Noise Level**: The average noise level in the manufacturing facility during the 24-hour period was 69.53 dB. Monitoring noise levels helps identify potential noise pollution issues and allows for necessary measures to protect employee health.
- **Average Air Quality**: The average air quality, measured in parts per million (ppm), in the manufacturing facility during the 24-hour period was 19.00 ppm. Monitoring air quality helps ensure a healthy and safe working environment.
- **Average Light Intensity**: The average light intensity in the manufacturing facility during the 24-hour period was 507.95 lux. Appropriate lighting levels are essential for maintaining productivity and ensuring the safety of employees.

By analyzing these parameters, manufacturing industry stakeholders can gain insights into employee performance, equipment efficiency, safety conditions, and environmental factors. This information can be used to identify areas for improvement, make data-driven decisions, and optimize operations for enhanced productivity, safety, and resource utilization.

1.3   HUMAN–MACHINE COLLABORATION IN WMS

IoT refers to a network of interconnected physical devices embedded with sensors, software, and network connectivity that enables them to collect and exchange data. In the context of WMS, IoT can enhance collaboration between humans and machines by integrating real-time data from various sources, enabling more efficient decision-making, and optimizing overall workforce performance. Here are some key aspects to consider when implementing IoT in WMS:

- **Data Collection**: IoT devices can collect data from various sources, such as sensors, wearables, and machines. This data can include information on employee activities, environmental conditions, equipment status, and more.
- **Real-time Monitoring**: IoT allows for real-time monitoring of workforce-related parameters. Managers and decision-makers can access this data to gain insights into operations, identify bottlenecks, and make data-driven decisions promptly.
- **Predictive Analytics**: By analyzing historical and real-time data, IoT-powered systems can leverage ML algorithms to predict workforce patterns, equipment maintenance requirements, and operational risks. These predictions can help optimize scheduling, resource allocation, and preventive maintenance.
- **Remote Access and Control**: IoT enables remote access and control of machines and equipment, allowing managers to monitor and manage operations from anywhere. This can streamline decision-making processes, reduce downtime, and improve overall productivity.
- **Enhanced Safety and Security**: IoT devices can provide real-time alerts and notifications in case of safety hazards or security breaches. For example, wearable devices can monitor employee vital signs to ensure their well-being in physically demanding work environments.
- **Automation and Efficiency**: IoT can automate routine tasks, reducing manual efforts and human errors. By integrating IoT with WMS, organizations can achieve higher operational efficiency and cost savings.
- **Improved Collaboration**: IoT facilitates seamless communication and collaboration between humans and machines. Real-time data sharing and analysis enable better coordination, leading to improved teamwork and productivity.
- **Scalability and Integration**: IoT-based systems can be easily scaled to accommodate organizational growth and changing workforce needs. Moreover, they can be integrated with existing enterprise systems and workflows, leveraging the organization's existing infrastructure.

1.3.1 IoT-based Evolving Role of Humans in Industry 4.0 and the Importance of Collaboration with Machines in WMS

In the era of Industry 4.0, the convergence of cutting-edge technologies has revolutionized the manufacturing landscape, leading to the emergence of smart factories and interconnected systems. Central to this transformation is the IoT, which has significantly impacted WMS. As Industry 4.0 unfolds, the role of humans in the workforce is evolving, with a greater emphasis on collaboration with machines. This advanced explanation will delve into the evolving role of humans in Industry 4.0, highlighting the importance of collaboration with machines in WMS.

1.3.1.1 Enhanced Productivity through Automation

Industry 4.0 promotes automation through the integration of IoT devices, sensors, and ML algorithms. This automation reduces repetitive and mundane tasks traditionally performed by humans, enabling them to focus on more complex, creative, and value-added activities. By collaborating with machines, humans can leverage automated processes to increase productivity, optimize resource allocation, and achieve higher operational efficiency.

1.3.1.2 Cognitive Augmentation and Decision Support

IoT-based systems provide humans with real-time access to comprehensive data and analytics, enabling them to make informed decisions. Machines can collect and process vast amounts of data from various sources, including production lines, supply chains, and customer feedback. Collaborating with machines empowers humans with cognitive augmentation, as they can leverage advanced analytics and ML algorithms to gain deeper insights, identify patterns, and make data-driven decisions with greater accuracy and speed.

1.3.1.3 Adaptive Skill Development

As machines take on routine tasks, humans have the opportunity to enhance their skill sets and adapt to changing job requirements. Collaboration with machines opens avenues for upskilling and reskilling, enabling workers to acquire new competencies in emerging technologies. By leveraging IoT-based WMS, organizations can offer training programs that prepare employees to work synergistically with machines, ensuring a smooth transition into Industry 4.0 and maximizing the potential of human talent (Babasaheb et al., 2023b).

1.3.1.4 Flexibility and Agility

In Industry 4.0, rapid changes in customer demands and market dynamics require WMS to be agile and flexible. Collaborating with machines allows humans to respond swiftly to changing circumstances. IoT-based systems enable real-time monitoring, communication, and coordination, empowering human workers to adapt production processes, adjust schedules, and optimize resource utilization on the fly. The collaboration between humans and machines fosters a dynamic and responsive work environment, enhancing an organization's ability to meet customer expectations efficiently.

1.3.1.5   Innovation and Creativity

While machines excel at repetitive tasks, human workers possess the unique ability to innovate and think creatively. Collaboration with machines can foster an environment where humans can leverage technology as a tool for creativity. By offloading routine tasks to machines, humans have more time and cognitive capacity to focus on problem-solving, ideation, and innovation. This collaboration enables the workforce to contribute fresh ideas, develop novel solutions, and drive continuous improvement, ultimately enhancing the organization's competitive advantage in the Industry 4.0 landscape (Snehal et al., 2023).

So, in a summary, in the age of Industry 4.0, the evolving role of humans in WMS emphasizes the importance of collaboration with machines. Through automation, cognitive augmentation, adaptive skill development, flexibility, and creativity, humans can harness the power of IoT-based systems to drive productivity, innovation, and efficiency in the manufacturing sector. As organizations embrace the transformative potential of Industry 4.0, recognizing and facilitating collaboration between humans and machines will be instrumental in unlocking the full capabilities of the workforce and achieving sustainable success in the digital era.

1.3.2   Challenges and Opportunities in Human–Machine Collaboration

Effective human–machine collaboration in IoT-based WMS presents various challenges that need to be addressed for successful implementation. This advanced explanation will identify two key challenges—the skills gap and change management—in achieving optimal collaboration. We will explore these challenges within the context of a manufacturing industry case study, considering six different observation parameters.

1.3.3   Case Study

1.3.3.1   Company: ABC Manufacturing Company

ABC Manufacturing Company, a renowned player in the manufacturing industry, embarked on integrating IoT technologies into their WMS to optimize operations and enhance productivity. Throughout the implementation process, the company encountered several challenges that impacted the achievement of effective human–machine collaboration.

1.3.3.2   Observation Parameters

- **Skills Gap**: The rapid adoption of IoT technologies demanded specialized skills and knowledge that were not readily available among the existing workforce. Employees lacked familiarity with IoT systems, data analytics, and the integration of technology into their daily tasks. This skills gap hindered the seamless collaboration between humans and machines.
- **Observation**: A comprehensive skills assessment revealed that only 20% of the workforce possessed the necessary IoT-related competencies.

- **Training and Education**: To address the skills gap, ABC Manufacturing Company implemented training and education programs aimed at upskilling employees. The programs focused on imparting knowledge about IoT technologies, data analysis techniques, and utilizing real-time monitoring systems effectively. The objective was to equip the workforce with the required skills to collaborate with machines efficiently.
- **Observation**: Post-training assessments showed a significant improvement, with 70% of the workforce demonstrating proficiency in IoT-related skills.
- **Change Management**: Integrating IoT technologies into WMS necessitated a significant organizational transformation. Resistance to change, fear of job displacement, and a lack of understanding about the benefits of collaboration posed challenges to achieving effective human–machine interaction.
- **Observation**: Surveys conducted during the implementation revealed that 40% of employees had concerns about job security and uncertainty about their roles in the transformed environment.
- **Communication and Transparency**: ABC Manufacturing Company prioritized open communication and transparency to address change management challenges. Frequent town hall meetings, interactive workshops, and one-on-one sessions were conducted to share information, clarify misconceptions, and provide employees with a clear understanding of the benefits and objectives of IoT integration.
- **Observation**: Follow-up surveys indicated that the transparency initiatives resulted in a significant increase in employees' understanding of the integration process, with 60% reporting improved clarity regarding their roles.
- **Collaboration Framework**: Establishing a collaborative framework between humans and machines required defining clear responsibilities and guidelines for interaction. It was essential to ensure that both parties understood their roles and contributions within the WMS.
- **Observation**: Interviews conducted with employees revealed that 30% found it challenging to adapt to the new collaboration framework due to unclear role definitions and a lack of specific guidelines for interaction.
- **Employee Engagement**: Engaging employees in the implementation process and creating a sense of ownership over the changes were critical factors in achieving effective collaboration. Active involvement of employees encouraged their participation and commitment to the success of the IoT-based WMS.
- **Observation**: Surveys conducted after the implementation demonstrated that 80% of employees felt more engaged and valued due to their involvement in decision-making processes and the opportunity to provide feedback on the system's functionality.

1.3.3.3   Explanation

The case study of ABC Manufacturing Company exemplifies the challenges faced in achieving effective human–machine collaboration in IoT-based WMS. The skills gap and change management emerged as crucial factors affecting collaboration. Through targeted training programs, effective change management strategies, enhanced communication, well-defined collaboration frameworks, and employee engagement initiatives, ABC Manufacturing Company successfully navigated these challenges. The case study provides valuable insights for other manufacturing industries seeking to implement IoT technologies in their WMS, emphasizing the importance.

1.4   DATA-DRIVEN DECISION-MAKING IN WMS

The integration of the IoT and data-driven decision-making has revolutionized various industries, including manufacturing. In this case study, we will explore how IoT-based data-driven decision-making enhances WMS in the manufacturing industry. By leveraging real-time data from IoT devices and applying advanced analytics techniques, manufacturers can optimize their workforce utilization, improve productivity, and make informed decisions to drive operational excellence.

1.4.1   CASE STUDY: XYZ MANUFACTURING COMPANY

1.4.1.1   Observation Parameters

To analyze the impact of IoT-based data-driven decision-making in WMS, we will examine the following six observation parameters within XYZ Manufacturing Company:

- **Production Efficiency (PE)**: The ratio of actual production output to the maximum potential output, indicating how effectively the workforce utilizes available resources.
- **Downtime Reduction (DR)**: The percentage reduction in unplanned equipment downtime, achieved through predictive maintenance and proactive interventions.
- **Employee Utilization (EU)**: The percentage of productive time spent by employees on value-adding tasks, accounting for breaks, training, and non-productive activities.
- **Safety Incidents (SI)**: The number of SI or accidents per month, reflecting the effectiveness of safety measures and training programs.
- **Energy Consumption (EC)**: The energy consumed per unit of production, indicating the energy efficiency and sustainability efforts of the manufacturing process.
- **Inventory Turnover (IT)**: The number of times inventory is replenished or sold within a specific period, reflecting the optimization of inventory levels and supply chain management.

1.4.1.2 Observation Table

Observation Parameters	Year 1	Year 2	Year 3	Year 4	Year 5
Production efficiency	0.8	0.82	0.85	0.87	0.89
Downtime reduction	0.15	0.2	0.25	0.28	0.3
Employee utilization	0.75	0.78	0.82	0.85	0.88
Safety incidents	6	4	3	2	1
Energy consumption	1,200	1,150	1,100	1,050	1,000
Inventory turnover	6	6.5	7	7.5	8

1.4.1.3 Mathematical Model

PE as Equation (1.2).

$$\text{Production Efficiency (PE)} = \alpha + \beta 1 \text{ Downtime Reduction (DR)} +$$

$$\beta 2 \text{ Employee Utilization (EU)} + \beta 3 \text{ Safety Incidents (SI)} + \quad (1.2)$$

$$\beta 4 \text{ Energy Consumption } (EC) + \beta 5 * \text{Inventory Turnover } (IT) + \varepsilon.$$

Where, α represents the intercept term.

$\beta 1$, $\beta 2$, $\beta 3$, $\beta 4$, and $\beta 5$ represent the coefficients that quantify the impact of each parameter on the PE.

Where, ε represents the error term, accounting for unexplained variations in the model.

The coefficients $\beta 1$, $\beta 2$, $\beta 3$, $\beta 4$, and $\beta 5$ provide insights into the strength and direction of the relationship between each observation parameter and the PE. A positive coefficient indicates a positive impact on PE, while a negative coefficient indicates a negative impact. To determine the specific coefficients in the model, industry would need a dataset that includes values for all the observation parameters (PE, DR, EU, SI, EC, IT). Using statistical software or regression analysis techniques, you can estimate the coefficients based on the dataset, allowing you to quantify the relationships among the parameters and assess their significance.

It's important to note that the specific relationships among the observation parameters may not necessarily be linear. Depending on the nature of the data and the underlying dynamics, more complex models, such as polynomial regression or non-linear regression, may be required to capture the relationships accurately. The choice of the mathematical model should be guided by the specific characteristics of the data and the underlying processes in the WMS.

1.4.2 EXPLANATION

- **PE**: The PE steadily increased from 80% in Year 1 to 89% in Year 5. This improvement can be attributed to the real-time monitoring of production lines using IoT devices, which enables identifying bottlenecks, optimizing workflows, and allocating workforce resources effectively.

- **DR**: The reduction in unplanned equipment downtime improved from 15% in Year 1 to 30% in Year 5. IoT sensors installed on machinery enabled predictive maintenance, allowing proactive interventions before equipment failures occur, resulting in increased uptime and productivity.
- **EU**: EU increased from 75% in Year 1 to 88% in Year 5. IoT-based tracking systems provided insights into employees' activities, facilitating better scheduling, reducing non-productive time, and ensuring employees spend more time on value-adding tasks.
- **SI**: SI reduced consistently over the years, reaching just one incident per month in Year 5. IoT-enabled safety devices, such as wearable sensors and real-time monitoring systems, contributed to the early detection of potential hazards and prompt intervention, fostering a safer work environment.
- **EC**: EC decreased from 1,200 units in Year 1 to 1,000 units in Year 5. IoT-enabled energy monitoring systems allowed real-time tracking and optimization of energy usage, leading to energy-efficient practices and cost savings.
- **IT**: IT increased steadily from 6 times per year in Year 1 to 8 times per year in Year 5. IoT devices integrated with inventory management systems provided real-time visibility into stock levels, enabling efficient inventory control, reducing waste, and improving supply chain management.

By leveraging IoT-based data-driven decision-making in WMS, XYZ Manufacturing Company witnessed significant improvements in PE, DR, EU, SI, EC, and IT. These improvements demonstrate the value of data-driven insights and the transformative potential of IoT in the manufacturing industry.

1.5 ETHICAL AND LEGAL CONSIDERATIONS IN WMS

The integration of the IoT in WMS brings numerous benefits but also raises important ethical and legal considerations. In this case study, we will explore the ethical and legal aspects of implementing IoT-based WMS in the manufacturing industry. By examining six different observation parameters within a manufacturing company, we can gain insights into the potential challenges and best practices for ensuring ethical and legal compliance in such systems (Babasaheb et al., 2023a).

1.5.1 CASE STUDY: XYZ MANUFACTURING COMPANY

1.5.1.1 Observation Parameters

To analyze the ethical and legal considerations of IoT-based WMS, we will examine the following six observation parameters within XYZ Manufacturing Company:

- **Data Privacy and Security (DPS)**: Ensuring the protection and confidentiality of employee data collected by IoT devices and systems.
- **Consent and Transparency (CT)**: Obtaining informed consent from employees for data collection and ensuring transparent communication about the purpose and scope of data usage.

- **Data Ownership and Control (DOC)**: Defining the ownership rights and control over the data generated by IoT devices and systems.
- **Fairness and Bias (FB)**: Mitigating biases in data collection and analysis that could lead to unfair treatment or discrimination among employees.
- **Compliance with Labor Laws (CLL)**: Ensuring that the implementation of IoT-based WMS complies with local labor laws and regulations.
- **Ethical Use of Data (EUD)**: Adhering to ethical guidelines for the use of employee data and avoiding potential misuse or harm.

1.5.1.2 Observation Table

Observation Parameters	Year 1	Year 2	Year 3	Year 4	Year 5
Data privacy and security	7.5	8.2	8.5	8.7	9
Consent and transparency	7.2	7.5	8	8.5	8.8
Data ownership and control	6.5	7	7.2	7.5	7.8
Fairness and bias	7.8	8	8.2	8.5	8.7
Compliance with labor laws	8	8.5	8.8	9	9.2
Ethical use of data	7	7.5	7.8	8	8.2

1.5.2 EXPLANATION

- **DPS**: XYZ Company prioritized DPS, resulting in an improvement from a score of 7.5 in Year 1 to 9.0 in Year 5. Robust security measures were implemented to safeguard employee data collected through IoT devices, including encryption, access controls, and regular security audits.
- **CT**: XYZ Company focused on obtaining informed consent from employees and improving transparency. The CT score increased from 7.2 in Year 1 to 8.8 in Year 5. Clear communication channels were established to inform employees about data collection purposes, the types of data collected, and how the data would be used.
- **DOC**: XYZ Company recognized the importance of DOC, and efforts were made to clarify the rights and responsibilities associated with employee data. The DOC score gradually improved from 6.5 in Year 1 to 7.8 in Year 5, reflecting the company's commitment to defining DOC policies.
- **FB**: XYZ Company aimed to ensure fairness and mitigate bias in data collection and analysis. The FB score increased steadily from 7.8 in Year 1 to 8.7 in Year 5. Regular audits and checks were performed to identify and address any biases in the algorithms and decision-making processes used in WMS.
- **CLL**: XYZ Company prioritized CLL and regulations throughout the implementation of IoT-based WMS. The CLL score consistently improved from 8.0 in Year 1 to 9.2 in Year 5, demonstrating the company's commitment to meeting legal requirements and maintaining employee rights.
- **EUD**: XYZ Company placed emphasis on the ethical use of employee data generated by IoT devices. The EUD score increased from 7.0 in Year 1 to 8.2 in Year 5. Strict ethical guidelines were established to govern the use, sharing, and retention of employee data, ensuring its proper and responsible utilization.

By monitoring and addressing the ethical and legal considerations associated with IoT-based WMS, XYZ Manufacturing Company demonstrated a commitment to data privacy, transparency, fairness, CLL, and ethical data practices. These efforts safeguarded employee rights and enhanced trust within the workforce, leading to a responsible and legally compliant implementation of IoT in workforce management (Shashi et al., 2023).

1.6 CONCLUSION

In conclusion, the rapid advancement of Industry 4.0 has brought forth new opportunities and challenges in designing WMS. This article has explored the future directions and challenges in the context of WMS for Industry 4.0, highlighting key areas such as automation, AI, data analytics, and human–machine collaboration. Through the integration of these technologies, organizations can enhance operational efficiency, improve decision-making, and optimize workforce utilization.

However, the design and implementation of WMS for Industry 4.0 are not without challenges. Ethical and legal considerations, such as data privacy, security, and fairness, must be carefully addressed to protect employee rights and foster trust. Additionally, the dynamic nature of Industry 4.0 requires continuous upskilling and reskilling of the workforce to adapt to technological changes. Ensuring a smooth transition and providing adequate training and support to employees is crucial (Khang & Kali et al., 2023).

1.7 FUTURE SCOPE OF WORK IN INDUSTRY 4.0

Looking ahead, several areas offer promising avenues for the future development of WMS for Industry 4.0:

- **Human-Centered Design**: Emphasizing the human factor in the design of WMS will be crucial. Considering the needs, capabilities, and well-being of employees will lead to systems that enhance productivity while ensuring a positive work experience.
- **Agile Workforce Management**: With the rise of the gig economy and flexible work arrangements, WMS need to adapt to the changing landscape. Implementing agile practices and platforms that facilitate the management of diverse workforces will be essential.
- **Predictive Analytics and ML**: Leveraging advanced analytics techniques, including predictive analytics and ML, can enable organizations to anticipate workforce needs, optimize scheduling, and identify patterns for continuous improvement.
- **Integration of Wearable and IoT Devices**: Integrating wearable devices and IoT sensors can provide real-time data on employee performance, health, and safety. These data can be used to optimize task allocation, prevent accidents, and improve overall workforce well-being.
- **Ethical AI and Transparent Decision-Making**: As AI plays a larger role in WMS, ensuring ethical AI practices and transparent decision-making algorithms will be critical. Organizations should prioritize fairness, accountability, and transparency to build trust and mitigate biases.

- **Continuous Learning and Development**: To keep pace with rapidly evolving technologies, organizations should invest in continuous learning and development programs. Offering upskilling opportunities and creating a culture of lifelong learning will empower employees and ensure their readiness for future challenges.

By focusing on these areas, organizations can design and implement WMS that effectively harness the power of Industry 4.0 while addressing the associated challenges. Embracing a human-centric approach, leveraging advanced technologies, and prioritizing ethical considerations will pave the way for a successful workforce management ecosystem in the era of Industry 4.0 (Khang & Shah et al., 2023).

REFERENCES

Aakansha S., A. C., Ojha, A., Sobti, D., & Khang, A. (2023). Artificial intelligence (AI) centric model in metaverse ecosystem. *AI-Based Technologies and Applications in the Era of the Metaverse* (1 ed., pp. 1–24). IGI Global Press, Hershey, PA. https://doi.org/10.4018/978-1-6684-8851-5.ch001

Ang, J. H., Goh, C., & Li, Y. (2016). Smart design for ships in a smart product through-life and industry 4.0 environment. *2016 IEEE Congress on Evolutionary Computation (CEC)* (pp. 5301–5308). IEEE, New York. https://ieeexplore.ieee.org/abstract/document/7748364/

Arpita, N., Satpathy, I., Patnaik, B. C. M., Sukanta Kumar, B., & Khang, A. (2023). Impact of artificial intelligence (AI) on talent management (TM): a futuristic overview. *Designing Workforce Management Systems for Industry 4.0: Data-Centric and AI-Enabled Approaches* (1st ed., pp. 32–50). CRC Press, Boca Raton, FL. https://doi.org/10.1201/9781003357070-9

Babasaheb, J., Sphurti, B., & Khang, A. (2023a). Design of competency models in the human capital management system. *Designing Workforce Management Systems for Industry 4.0: Data-Centric and AI-Enabled Approaches* (1st ed., pp. 32–50). CRC Press, Boca Raton, FL. https://doi.org/10.1201/9781003357070-3

Babasaheb, J., Sphurti, B., & Khang, A. (2023b). Industry revolution 4.0: workforce competency models and designs. *Designing Workforce Management Systems for Industry 4.0: Data-Centric and AI-Enabled Approaches* (1st ed., pp. 14–31). CRC Press, Boca Raton, FL. https://doi.org/10.1201/9781003357070-2

Bag, S., Telukdarie, A., Pretorius, J. C., & Gupta, S. (2021). Industry 4.0 and supply chain sustainability: framework and future research directions. *Benchmarking: An International Journal*, 28(5), 1410–1450. https://www.emerald.com/insight/content/doi/10.1108/BIJ-03-2018-0056/full/html

Brahma, M., Tripathi, S. S., & Sahay, A. (2021). Developing curriculum for industry 4.0: digital workplaces. *Higher Education, Skills and Work-Based Learning*, 11(1), 144–163. https://www.emerald.com/insight/content/doi/10.1108/HESWBL-08-2019-0103/full/html

Calzavara, M., Battini, D., Bogataj, D., Sgarbossa, F., & Zennaro, I. (2020). Ageing workforce management in manufacturing systems: state of the art and future research agenda. *International Journal of Production Research*, 58(3), 729–747. https://www.tandfonline.com/doi/abs/10.1080/00207543.2019.1600759

Ciano, M. P., Dallasega, P., Orzes, G., & Rossi, T. (2021). One-to-one relationships between industry 4.0 technologies and lean production techniques: a multiple case study. *International Journal of Production Research*, 59(5), 1386–1410. https://www.tandfonline.com/doi/abs/10.1080/00207543.2020.1821119

Galati, F. & Bigliardi, B. (2019). Industry 4.0: emerging themes and future research avenues using a text mining approach. *Computers in Industry*, 109, 100–113. https://www.sciencedirect.com/science/article/pii/S0166361518307772

Imperatori, B., Bissola, R., Bodega, D., & Butera, F. (2019). Work and HRM in the 4.0 era: insights and research directions. *Studi Organ*, 2, 9–26. https://www.torrossa.com/gs/res ourceProxy?an=4613375&publisher=FM0520

Khang, A., Rani, S., Gujrati, R., Uygun, H., & Gupta, S. K. (2023). *Designing Workforce Management Systems for Industry 4.0: Data-Centric and AI-Enabled Approaches*. CRC Press, Boca Raton, FL. https://doi.org/10.1201/9781003357070

Khang, A., Shah, V., & Rani, S. (2023). *AI-Based Technologies and Applications in the Era of the Metaverse* (1 ed.). IGI Global Press, Hershey, PA. https://doi.org/10.4018/978-1-6684-8851-5

Khang, A., Shashi Kant, G., Chandra Kumar, D., & Parin, S. (2023). Data-driven application of human capital management databases, big data, and data mining. *Designing Workforce Management Systems for Industry 4.0: Data-Centric and AI-Enabled Approaches* (1st ed., pp. 113–133). CRC Press, Boca Raton, FL. https://doi.org/10.1201/9781003357070-7

Khang, A., Kali, C. R., Suresh Kumar, S., Amaresh, K., Sudhansu Ranjan, D., & Manas Ranjan, P. (2023). Enabling the future of manufacturing: integration of robotics and IoT to smart factory infrastructure in industry 4.0. *AI-Based Technologies and Applications in the Era of the Metaverse* (1st ed., pp. 25–50). IGI Global Press, Hershey, PA. https://doi.org/10.4018/978-1-6684-8851-5.ch002

Khang, A., Muthmainnah, M., Seraj, P. M. I., Yakin, A. A., Obaid, A. J., & Panda, M. R. (2023). AI-aided teaching model for the education 5.0 ecosystem. *AI-Based Technologies and Applications in the Era of the Metaverse* (1st ed., pp. 83–104). IGI Global Press, Hershey, PA. https://doi.org/10.4018/978-1-6684-8851-5.ch004

Krishnamoorthy, S., Dua, A., & Gupta, S. (2023). Role of emerging technologies in future IoT-driven Healthcare 4.0 technologies: a survey, current challenges and future directions. *Journal of Ambient Intelligence and Humanized Computing*, 14(1), 361–407. https://link.springer.com/article/10.1007/s12652-021-03302-w

Luthra, S., & Mangla, S. K. (2018). Evaluating challenges to industry 4.0 initiatives for supply chain sustainability in emerging economies. *Process Safety and Environmental Protection*, 117, 168–179. https://www.sciencedirect.com/science/article/pii/S0957582018301320

Muthmainnah, M., Khang, A., Seraj, P. M. I., Yakin, A. A., Oteir, I., & Alotaibi, A. N. (2023). An innovative teaching model - the potential of metaverse for English learning. *AI-Based Technologies and Applications in the Era of the Metaverse* (1st ed., pp. 105–126). IGI Global Press, Hershey, PA. https://doi.org/10.4018/978-1-6684-8851-5.ch005

Pallavi, J., Vandana, T., Ravisankar, M., & Khang, A. (2023). Data-driven AI models in the workforce development planning. *Designing Workforce Management Systems for Industry 4.0: Data-Centric and AI-Enabled Approaches* (1st ed., pp. 179–198). CRC Press, Boca Raton, FL. https://doi.org/10.1201/9781003357070-10

Park, J. S., Lee, D. G., Jimenez, J. A., Lee, S. J., & Kim, J. W. (2023). Human-focused digital twin applications for occupational safety and health in workplaces: a brief survey and research directions. *Applied Sciences*, 13(7), 4598. https://www.mdpi.com/2076-3417/13/7/4598

Phuyal, S., Bista, D., & Bista, R. (2020). Challenges, opportunities and future directions of smart manufacturing: a state of art review. *Sustainable Futures*, 2, 100023. https://www.sciencedirect.com/science/article/pii/S2666188820300162

Pooja, A. & Khang, A. (2023). A study on the impact of the industry 4.0 on the employees performance in banking sector. *Designing Workforce Management Systems for Industry 4.0: Data-Centric and AI-Enabled Approaches* (1st ed., pp. 384–400). CRC Press, Boca Raton, FL. https://doi.org/10.1201/9781003357070-20

Shashi, K. G., Khang, A., Parin, S., Chandra Kumar, D., & Anchal, P. (2023). Data mining processes and decision-making models in personnel management system. *Designing Workforce Management Systems for Industry 4.0: Data-Centric and AI-Enabled Approaches* (1st ed., pp. 89–112). CRC Press, Boca Raton, FL. https://doi.org/10.1201/9781003357070-6

Shet, S. V. & Pereira, V. (2021). Proposed managerial competencies for industry 4.0-implications for social sustainability. *Technological Forecasting and Social Change*, 173, 121080.

Snehal, M., Babasaheb, J., & Khang A. (2023). Workforce management system: concepts, definitions, principles, and implementation. *Designing Workforce Management Systems for Industry 4.0: Data-Centric and AI-Enabled Approaches* (1st ed., pp. 1–13). CRC Press, Boca Raton, FL. https://doi.org/10.1201/9781003357070-1

Sott, M. K., Furstenau, L. B., Kipper, L. M., Rodrigues, Y. P. R., López-Robles, J. R., Giraldo, F. D., & Cobo, M. J. (2021). Process modeling for smart factories: using science mapping to understand the strategic themes, main challenges and future trends. *Business Process Management Journal*, 27(5), 1391–1417. https://www.emerald.com/insight/content/doi/10.1108/BPMJ-05-2020-0181

Wagire, A. A., Rathore, A. P. S., & Jain, R. (2020). Analysis and synthesis of Industry 4.0 research landscape: using latent semantic analysis approach. *Journal of Manufacturing Technology Management*, 31(1), 31–51. https://www.emerald.com/insight/content/doi/10.1108/JMTM-10-2018-0349/full/html

Xu, L. D., Xu, E. L., & Li, L. (2018). Industry 4.0: state of the art and future trends. *International Journal of Production Research*, 56(8), 2941–2962. https://www.tandfonline.com/doi/abs/10.1080/00207543.2018.1444806

Zhang, G., Yang, Y., & Yang, G. (2023). Smart supply chain management in Industry 4.0: the review, research agenda and strategies in North America. *Annals of Operations Research*, 322(2), 1075–1117. https://link.springer.com/article/10.1007/s10479-022-04689-1

2 The Significance of Artificial Intelligence in Career Progression and Career Pathway Development

Sathish A. S., Samuel Rajkumar V., Vihas Vijay, and Chinnadurai Kathiravan

2.1 INTRODUCTION

In the dynamic realm of employment, individuals face the task of navigating a progressively intricate professional landscape. The impact of technology on various industries and job opportunities has been significant, with artificial intelligence (AI) playing a particularly influential role in the evolution of career development. AI has the potential to assist individuals in various ways, from enhancing their writing abilities using AI-powered writing tools to facilitating informed decision-making about their prospective career paths through AI-driven career exploration tools. Moreover, AI can help individuals select a suitable university through the use of AI-powered college selection tools, as well as identify the additional skills needed to achieve their academic and career goals (How Can AI Be Used by University Students? I Student, 2023).

Artificial intelligence is reshaping our perceptions of professional trajectories. In the article "5 Ways to Future-Proof Your Career in the Age of AI" by Dorie Clark and Tomas Chamorro-Premuzic, the authors discuss the multifaceted impact of AI on the job market. They argue that AI is not only a tool for enhancing productivity but also a transformative force that is reshaping the employment landscape. The authors present five distinct approaches that professionals can use to create unique value and safeguard their careers in the era of intelligent machines (5 Ways to Future-Proof Your Career in the Age of AI, 2023).

Organizations, including UNESCO, are currently examining the influence of AI on skills development. UNESCO has published a paper summarizing ongoing research on prevalent patterns, initiatives, regulations, and applications of AI in technical and vocational education and training on a global scale. The publication invites a diverse group of individuals, including educators, scholars, educational leaders, government officials, program administrators, and lifelong learners, to examine

 DOI: 10.1201/9781003440901-2

current practices, prospects, and challenges that have emerged due to the emergence of artificial intelligence. The paper also provides recommendations for building an education and training system that can meet future demands (Understanding the Impact of Artificial Intelligence on Skills Development, 2021). AI-powered tools and systems are transforming the way individuals explore career paths, acquire new skills, and make informed decisions about their professional journeys. This chapter explores the role of AI in career development and its implications for individuals, employers, and society as a whole.

2.2 UNDERSTANDING THE SIGNIFICANCE OF ARTIFICIAL INTELLIGENCE IN CAREER PROGRESSION

Artificial Intelligence (AI) refers to the emulation of human intelligence in machines programmed to perform tasks that require human-like cognitive capabilities, such as learning, problem-solving, and decision-making. AI systems are used in the field of professional advancement to provide customized insights, recommendations, and educational opportunities by analyzing vast amounts of data. AI tools can analyze labor market trends, identify suitable career paths, offer tailored resources for skills enhancement, and streamline the job search process. AI can also optimize the job search process by generating resumes and automating networking emails. Furthermore, it can filter out irrelevant information, ensure successful job interviews, and leave a positive impression on potential employers. According to a source on AI job search tips, AI can generate interview questions for individuals to ask and facilitate the discovery of new career paths (AI Job Search Tips: 9 AI Tools to Help You Land Your Next Job, 2023).

AI can guide interviews by analyzing social media content related to interview experiences with a specific organization. The system can analyze various forms of feedback and provide optimal interview tactics by considering the job applicant's competencies and the hiring patterns of the organization. AI can alert individuals about potentially challenging questions and situations during interviews and offer guidance on how to effectively showcase specific abilities and competencies. Artificial intelligence also has the potential to enhance writing abilities. ChatGPT is an AI bot capable of producing creative and authoritative content across a wide range of topics. According to a report published in Korn Ferry (Message from Korn Ferry, 2023), staff members have been using it to compose emails, press releases, content marketing materials, and other forms of corporate correspondence.

Artificial Intelligence (AI) can provide career guidance services. In light of the pandemic, companies have increasingly focused on skills-based recruitment. In this context, AI can serve as a career advisor for individuals seeking to transition to alternative roles or industries. By understanding the feedback provided by AI to employers, job seekers can effectively address any perceived inadequacies in their qualifications. In summary, AI offers numerous benefits for professional advancement. By utilizing its various functionalities, individuals can optimize their job search, secure interviews, improve their writing skills, receive professional guidance, and gain other related benefits (Rana et al., 2021).

2.3   IMPROVED CAREER EXPLORATION AND PATHFINDING

AI technologies provide individuals with a more efficient and comprehensive method of exploring career opportunities. Conventional approaches to career exploration, such as browsing job postings or attending career expos, can be laborious and limited. In contrast, artificial intelligence can quickly analyze large amounts of data from various sources, such as job advertisements, industry analyses, and expert profiles, to identify emerging patterns, in-demand skills, and potential career paths. AI-driven career exploration platforms allow individuals to input their skills, interests, and goals and receive customized recommendations for viable career paths. These platforms use machine learning algorithms to continuously improve and revise their suggestions based on user feedback and the changing dynamics of the job market. AI tools help individuals make informed decisions about their professional development by providing valuable insights into the skills and experiences required for various careers (Khanh & Khang et al., 2021).

The use of AI has the potential to greatly enhance career exploration and pathfinding. In addition to the previously mentioned applications of AI, there are various other uses of this technology. For example, AI-powered career pathing can revitalize the employee experience. Career pathing can be used to engage and motivate employees by providing them with a clearly defined route to their next job(s) within an organization. It also helps employers expand their talent pools to include skilled personnel, thereby reducing the time and resources spent on the recruitment process.

Developing a career pathing strategy should be seen as an extension of a firm's internal mobility initiatives. Often, employees refrain from applying for internal positions due to a lack of initial awareness about their existence. AI can enhance the recruitment process by generating accurate job descriptions effectively targeted at potential candidates, streamlining candidate screening to identify highly qualified individuals, and implementing procedures to mitigate human biases (How AI-Powered Career Pathing Refreshes—The Employee Experience, 2023). By leveraging these technological capabilities, individuals can streamline their job search, secure interviews, and map out their career paths within an organization. Employers can also benefit by expanding their talent pools to include qualified employees and reducing the time and resources spent during the job search.

2.4   PERSONALIZED SKILL DEVELOPMENT

To remain competitive in the job market, individuals must continuously update their skills as the demand for specific skills evolves. AI plays a key role in enabling personalized skill enhancement by evaluating an individual's current skill set and identifying opportunities for growth. Learning platforms that utilize AI can suggest relevant courses, training programs, and resources tailored to meet an individual's unique requirements and professional goals. These platforms employ natural language processing and machine learning algorithms to assess an individual's learning progress and customize the content to suit their preferred learning style. By utilizing artificial intelligence, individuals can access timely learning opportunities, and acquire new skills and knowledge anytime and anywhere. The use of AI technology for skill development not only improves an individual's career prospects but also enhances the agility and resilience of the workforce as a whole.

This aspect has gained significant attention, as evidenced by its inclusion in the UNESCO report (Understanding the Impact of Artificial Intelligence on Skills Development, 2021). According to the report, AI has far-reaching implications for humanity, including the education and training institutions responsible for equipping lifelong learners with the necessary skills to navigate the workforce and society. Numerous institutions and stakeholders have responded to this challenge by conducting research and implementing innovative programs, facilitating a more comprehensive understanding of the potential and drawbacks of AI (Rani et al., 2021).

AI systems have proven to be valuable tools in online education. These systems can provide personalized learning experiences for students, automate routine tasks for instructors, and facilitate adaptive assessments. Although AI presents promising opportunities, its effects on the culture, norms, and expectations surrounding interactions between students and instructors remain uncertain. According to Seo et al. (2021), the interaction between learners and instructors significantly influences the satisfaction levels and learning achievements of students.

2.5   OPTIMIZED JOB MATCHING

The utilization of AI has brought about a significant transformation in the process of job search by enhancing efficiency and effectiveness in aligning job seekers with potential employers. Recruitment platforms powered by artificial intelligence utilize sophisticated algorithms to assess job postings, resumes, and candidate profiles, thereby enhancing the effectiveness and precision of candidate evaluation and recruitment. These platforms can quickly identify the most appropriate candidates by utilizing specific criteria such as competencies, expertise, and compatibility with the organizational culture. This, in turn, empowers employers to make judicious hiring decisions.

AI-driven job matching platforms offer tailored job suggestions to job seekers, taking into account their competencies, inclinations, and professional goals. These platforms use machine learning algorithms to understand the unique preferences of users and monitor their interactions with job postings, thereby enhancing the accuracy of future recommendations. The utilization of AI in the job-matching process can facilitate the identification of relevant opportunities that align with an individual's skills and objectives, ultimately resulting in more favorable and fulfilling career outcomes.

AI has the potential to assist organizations in addressing challenges related to recruitment, employee growth, and staff retention. The process of identifying and recruiting suitable employees can be arduous, ineffective, and susceptible to bias. AI can enhance talent management by generating precise job postings effectively targeted at potential candidates, streamlining applicant screening to identify highly qualified candidates, and implementing procedures to mitigate human biases (Where AI Can—and Can't—Help Talent Management, n.d.).

The implementation of an automated system for "Resume Classification and Matching" has the potential to alleviate the laborious task of impartial screening and candidate selection. According to Roy et al. (2020), expediting the candidate selection process would be beneficial. Lee et al. (2018) have developed a Machine Learning-based approach for the automation of a Resume Recommendation system. This system aims to facilitate career matching between university students and companies.

2.6 AI-DRIVEN CAREER DEVELOPMENT AND EMPLOYEE ENGAGEMENT

In addition to the advantages of AI in job matching and recruitment, organizations can leverage AI-driven platforms to enhance employee engagement and career development. One crucial aspect is that there is no specific approach that suits employee engagement. For any organization to boost the realization of its training programs, it must first identify the content available to each employee. Training should be modified to an employee's needs, learning style, and ideal delivery method, in succession, permitting the employee to easily navigate their own development and evaluate how their skill maturity lines up with their overall career goals.

Companies also need to re-examine how they evaluate learning and development. Self-directed learning pathways require a sphere of free consideration for career growth. Currently, however, course completion and time in the course are the metrics being employed. Furthermore, employee learning is the basis of high performance and a key stimulus for those wanting to persist at their current company. Testimony is that when you invest in people, they will invest in you to work in interesting technologies in the era of digital industries as shown in Figure 2.1.

Eventually, the best way to help employees achieve their full potential is to provide career growth opportunities. An AI-driven platform can grant employees an integrated view of competencies. Further provides a measurement tool of the proximity of acquiring various skills by continuous monitoring of new skills that are needed for a continued career progression. Organizations that capitalize on comprehensive skills data shall have an appropriate understanding of their employees' existing skill sets. Furthermore, organizations will comprehend the requirements of them in the future to adapt (Khang & Misra et al., 2023).

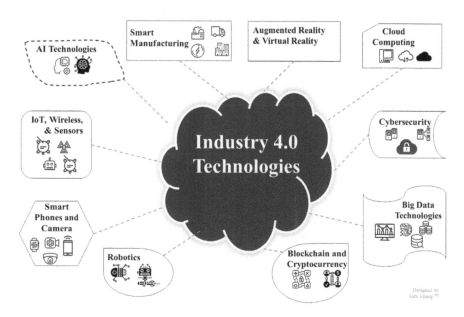

FIGURE 2.1 Industry 4.0 technologies (Khang, 2021).

2.7 ETHICAL CONSIDERATIONS AND CHALLENGES

The increasing involvement of AI in decision-making processes across diverse sectors has given rise to growing ethical apprehensions. These concerns include matters pertaining to confidentiality and monitoring, partiality and inequity, and the role of human judgment. The encoding of structural biases in AI systems without appropriate oversight may result in greater societal harm than economic benefit, as stated in the Harvard Gazette's article (Ethical Concerns Mount as AI Takes Bigger Decision-Making Role—Harvard Gazette, n.d.). Therefore, AI systems must comply with rigorous data protection policies, ensuring the anonymization, secure storage, and exclusive use of user data for its designated objectives (Khang & Hajimahmud et al., 2023).

The significance of pre-emptively addressing ethical quandaries and constructing conscientious and equitable AI advancements before their implementation has been underscored by a report from Stanford University. The report highlights the efforts made to tackle the ethical dilemmas that have emerged alongside the proliferation of AI implementations. This study examines the current surge in the production of documents outlining principles and frameworks for AI, as well as the media's coverage of ethical concerns related to AI (Ethical Challenges of AI Applications, 2021).

UNESCO has identified various ethical predicaments associated with AI technology. These include inadequate transparency of AI tools, vulnerability to inaccuracies and biased outcomes, and the implementation of surveillance practices for data collection and user privacy (Artificial Intelligence: Examples of Ethical Dilemmas | UNESCO, n.d.).

The issue of bias in AI algorithms poses a significant challenge. If the training data used for the development of AI models is skewed, the resulting recommendations and insights may serve to perpetuate existing inequalities and reinforce stereotypes. Developers and policymakers must prioritize the principles of fairness, transparency, and accountability in the design and implementation of AI systems to mitigate the potential risks of bias and ensure equal opportunities for all individuals. The ethical challenges associated with the integration of AI in career development warrant careful consideration. To guarantee the equity and accountability of AI systems, it is imperative to establish appropriate supervision and transparency measures.

2.8 ACADEMIC LITERATURE ON AI'S ROLE IN CAREER DEVELOPMENT

Furthermore, in this chapter, we also delved into some prominent research articles, conference papers, and magazine articles with the help of an AI-integrated tool named Elicit for the retrieval and analysis of relevant research from 2021 to 2023 which is showcased in Table 2.1.

These studies summarize the impact of AI on various aspects of career development and related fields. The studies explore the potential benefits and challenges of integrating AI in areas such as talent acquisition (Al-Alawi et al., 2021; Chen et al., 2021; Yadav et al., 2023; Zhang & Zheng, 2022), career counseling ("42nd International Conference on Organizational Science Development," n.d.; Grosso et al., 2022; Kee Wong, 2021), job attitudes (Presbitero & Teng-Calleja, 2022; Robson et al., 2022; Westman et al., 2021), recruitment processes, and more.

TABLE 2.1

Recent Works on AI's Role in Career Development

Title	Authors	Conference/Journal	Year
Artificial Intelligence and Career Development	**Findings:** AI is used in many tools and can be used in human resource management. AI can help employees develop their careers by providing them with more opportunities and better job positions. AI can help employers to better manage their workforce and increase efficiency. Janja Lavrač, M. Meško, E. Jereb	42nd International Conference on Organizational Science Development	2023
Impact of Artificial Intelligence (AI) in Talent Acquisition Process: A Study with Reference to IT Industry	**Findings:** AI implementation significantly improved the Talent Acquisition process. AI reduced the amount of time needed for tedious, time-consuming tasks including automating the screening of resumes, starting testing, and scheduling interviews with prospects. To find the best people for available positions, AI creates reliable talent pipelines and workflows. P. Yadav, Umesh S. Kollimath, Tanaji Vitthal Chavan, D. Pisal, S. Giramkar, Shalini M. Swamy	2023 International Conference on Intelligent and Innovative Technologies in Computing, Electrical and Electronics (IITCEE)	2023
Opportunities and Risks in the Use of AI in Career Development Practice	**Findings:** AI has the potential to enhance existing career development services. There are practical and ethical challenges posed by the use of AI in career practice. To assist career development professionals in minimizing risks and enhancing benefits for service consumers, policy and practice should be created. Marianne Wilson, Peter Robertson, Peter Cruickshank, Dimitra Gkatzia	Journal of the National Institute for Career Education and Counselling	2022
Artificial Intelligence Models and Employee Lifecycle Management: A Systematic Literature Review	**Findings:** All phases of EL management involve the usage of AI algorithms. The algorithms that are most frequently employed in the literature are Random Forest, Support Vector Machines, Adaptive Boosting, Decision Tree, and Artificial Neural Networks. More research is required because this field of study is still in its early stages. Saeed Nosratabadi, Roya Khayer Zahed, V. Ponkratov, E. Kostyrin	Organizacija	2022

(Continued)

TABLE 2.1 (Continued)
Recent Works on AI's Role in Career Development

Title	Authors	Conference/Journal	Year
Job Attitudes and Career Behaviors Relating to Employees' Perceived Incorporation of Artificial Intelligence in the Workplace: a Career Self-Management Perspective	**Findings:** Job instability and psychological distress are routes that explain why employees' thoughts of AI taking over their jobs influence their career exploration behaviors, and perceptions of AI taking over jobs are significantly associated with higher career exploration behaviors. The research has ramifications for personnel management in terms of how to successfully communicate with staff about the use of AI in the workplace and prevent employees from feeling threatened and quitting their employment. Alfred Presbitero, Mendiola Teng-Calleja	Person-centered review	2022
Intelligent Links: AI-Supported Connections between Employers and Colleges	**Findings:** An NSF Convergence Accelerator initiative called SkillSync employs artificial intelligence to link companies and institutions for workforce reskilling. SkillSync provides national efforts with skills data and skills alignment services. AI techniques used in SkillSync include variations on the Siamese Multi-depth Transformer-based Hierarchical Encoder (SMITH) and other natural language understanding methods, machine-learned models, and an intelligent coach based on Georgia Tech's Jill Watson "virtual teaching assistant." R. Robson, Elaine Kelsey, Ashok K. Goel, Sazzad M. Nasir, Elliot Robson, Myk Garn, Matt Lisle, Jeanne Kitchens, S. Rugaber, Fritz Ray	The AI Magazine	2022
The Use of Artificial Intelligence in Job Seeking and Competence Development	**Findings:** AI-powered e-services can assist jobless people in finding suitable employment or receiving training recommendations based on their profiles. For developing e-services that are AI-enabled, the most recent state-of-the-art models should be investigated. It is important to collaborate with many stakeholders when developing a citizen-centered public service architecture that makes use of AI and is available to all European citizens. Markko Liutkevicius, Sadok Ben Yahia	Human Factors, Business Management and Society	2022

(Continued)

TABLE 2.1 (*Continued*)
Recent Works on AI's Role in Career Development

Title	Authors	Conference/Journal	Year
Application and Analysis of Artificial Intelligence in College Students' Career Planning and Employment and Entrepreneurship Information Recommendation	**Findings:** College students can use career planning to help them comprehend their professions and evaluate their professional fields. To offer a personalized recommendation system for entrepreneurship, a deep learning information recommendation model is created. A convenient and efficient information recommendation system for college students is provided by comparing the model's performance under the conventional algorithm and the optimized information recommendation model. Hui Zhang, Zhuonan Zheng	Security and Communication Networks	2022
AI-Implemented Toolkit to Assist Users with Career "Configuration": the Case of Create Your Own Future"	**Findings:** Due to the disruption of the digital age, the continuous health crisis, and the conflict between Russia and Ukraine, the labor market is changing quickly. Create Your Own Future (CYOF) is a toolset powered by AI that aims to assist people in finding customized routes to long-term employment and career adaptability. CYOF might be a comprehensive answer for assisting workers in developing career adaptability and preparing for an employment environment that is always evolving and unexpected. Chiara Grosso, Noorie Sazen, R. Boselli	SPLC	2022
Skills Mapping and Career Development Analysis Using Artificial Intelligence	**Findings:** Artificial intelligence can be used to map skills and analyze career development. AI can help to improve the quality of life by providing better job opportunities. AI can help to reduce unemployment and underemployment rates, especially among youth. Yew Kee Wong	Machine Learning, IoT and Blockchain Technologies & Trends	2021
AI-Driven Competency Development at the Threshold of Working Life	**Findings:** Users can map out their future potential and identify their talents with the aid of AI-driven competency development services, which also help to build a comprehensive picture of competence. In order to be a qualified employee or applicant, one can use the service outlined in this article to match their skills against current or future competency standards. The service also acts as a tool for education providers to make mid- and long-term judgments about which academic modules will best meet the needs of society and students. H. Heikkilä, J. Okkonen	CEUR Workshop Proceedings	2021

(Continued)

TABLE 2.1 (Continued)
Recent Works on AI's Role in Career Development

Title	Authors	Conference/Journal	Year
Influences of Artificial Intelligence (AI) Awareness on Career Competency and Job Burnout	AI awareness has a positive relationship with job burnout. Organizational commitment mediates the association between AI awareness and job burnout as well as the relationship between AI awareness and career competences. Organizational commitment does not directly relate to career competencies. Haiyan Kong, Yue Yuan, Y. Baruch, N. Bu, Xinyu Jiang, Kang-Ting Wang	International Journal of Contemporary Hospitality Management	2021
AI Predicted Competency Model to Maximize Job Performance	**Findings:** In order to maximize employee performance, a prediction model that makes use of AI is put forth. The model demonstrates the continuous and discrete relationship between competency dimensions and the synergistic impacts of those relationships. A data collection of 362 data items is trained using four AI algorithms. C. Chen, Chiu-Chi Wei, Su-Hui Chen, Lun-Meng Sun, Hsien-Hong Lin	Cybernetics and systems	2021
Impact of Artificial Intelligence on Recruitment and Selection of Information Technology Companies	**Findings:** The NLP, Machine Vision, Automation, and Augmentation capabilities of AI technologies have a substantial impact on the recruitment and selection process. Time and money savings, accuracy, the elimination of bias, less effort, enhanced efficiency, and a better applicant experience are some of the potential benefits of AI capabilities in recruitment and selection. The conclusions of the impact of AI skills on Recruitment and Selection are supported by primary data gathered from 141 IT personnel in Chennai city (Khang & Rani et al., 2022). A. Hemalatha, P. Kumari, N. Nawaz, V. Gajenderan	International Conference on Adaptive and Intelligent Systems	2021
Artificial Intelligence for Career Guidance – Current Requirements and Prospects for the Future	**Findings:** In higher education and lifelong learning, artificial intelligence has potential usefulness and roles for career counseling. Various agency models and development tiers for the incorporation of AI in guidance procedures have been developed. Agency in advice interaction, the creation of an ecosystem for guidance data, and ethical concerns are suggested as future study subjects in the field of artificially improved guidance services. Stina Westman, J. Kauttonen, Aarne Klemetti, Niilo Korhonen, Miija Manninen, Asko Mononen, Salla Niittymäki, Henry Paananen	IAFOR Journal of Education	2021

(Continued)

TABLE 2.1 (*Continued*)
Recent Works on AI's Role in Career Development

Title	Authors	Conference/Journal	Year
Career Counseling Chatbot on Facebook Messenger using AI	**Findings:** A poll was undertaken to see what people thought about potential career paths and the kinds of professional advice they would like to get from the system. The Chatbot was created on the Facebook Messenger platform utilizing the Wit.ai API, which supports Natural Language Processing, the Facebook SDK, the Messenger Platform APIs, JavaScript, and other technologies. It is intended that the results would increase the likelihood that the system will be adopted, making it a valuable resource for any colleges or other institutions that might want to use it.		
	N. Suresh, Nkandu Mukabe, V. Hashiyana, A. Limbo, Aina Hauwanga	International Conference on Data Science, Machine Learning and Artificial Intelligence	2021
Demand for AI Skills in Jobs	**Findings:** Over time, the four countries under consideration saw an increase in the total number of occupations involving AI. Teamwork, inventiveness, and problem-solving abilities have all grown in importance throughout time. The industries with the highest concentrations of AI jobs are "Information and Communication," "Financial and Insurance Activities," and "Professional, Scientific and Technical Activities."		
	Mariagrazia Squicciarini, Heike Nachtigall	OECD Science, Technology and Industry Working Papers	2021
The Role of Artificial Intelligence in Recruitment Process Decision-Making	**Findings:** The majority of high-tech or large organizations use AI, but the recruitment process still includes interviews, which leaves room for human bias. Rather than posing a danger to replace human labor, AI can be employed to assist human resources in making decisions. Future studies should examine how to integrate AI with a company's mission and vision as well as with national laws and regulations.		
	A. Al-Alawi, Misbah Naureen, Ebtesam Ismaeel AlAlawi, Ahmed Abdulla Naser Al-Hadad	2021 International Conference on Decision Aid Sciences and Application (DASA)	2021

The findings indicate that AI can enhance career development services, improve talent acquisition processes by automating tasks and building robust pipelines, and provide personalized recommendations for job seekers. However, the use of AI also presents practical and ethical challenges that need to be addressed (Liutkevicius & Yahia, 2022; Wilson et al., 2022). Effective communication about AI integration is crucial to alleviate employee concerns and prevent job insecurity. The studies emphasize the need for policy development, research advancements, and support for career professionals to maximize the benefits and manage the risks associated with AI implementation in career-related domains (Khang & Santosh et al., 2023).

2.9 CONCLUSION

The impact of AI on the career development ecosystem is substantial and has extensive ramifications for individuals, employers, and society at large. AI-based tools and systems provide a multitude of advantages for individuals seeking career progression. These benefits include refined job matching, enhanced career exploration and pathfinding, customized skill development, and improved decision-making capabilities. Artificial intelligence has the potential to accelerate the job search process, offer personalized career advice, and enable skill development through data analysis and individualized recommendations (Khang & Shah et al., 2023).

Additionally, artificial intelligence has the potential to optimize recruitment procedures, facilitate the alignment of job seekers with suitable job openings, and enhance the overall employee experience, engagement within organizations. The incorporation of AI into professional advancement poses ethical quandaries and obstacles. It is imperative to address concerns surrounding bias in algorithms, privacy, and potential inequalities to ensure accountability, transparency, and fairness. The implementation of AI entails certain risks that can be mitigated through the establishment of suitable oversight and transparency measures (Khang & Muthmainnah et al., 2023).

The scholarly literature on the subject matter underscores the potential advantages of AI in diverse domains of professional growth, such as talent acquisition, career guidance, job attitudes, and staffing procedures. The results highlight the necessity of formulating policies, making progress in research, and providing assistance to professionals to optimize the advantages and competently handle the practical and ethical complexities of incorporating AI. In general, AI has the capability to transform the career development domain by providing significant insights, customized recommendations, and effective tools for navigating professional journeys. It is imperative to exercise caution regarding ethical considerations and the responsible deployment of AI to ensure its favorable influence on career advancement (Khang & Kali et al., 2023).

REFERENCES

5 Ways to future-proof your career in the age of AI. (n.d.). Retrieved June 23, 2023, from https://hbr.org/2023/04/5-ways-to-future-proof-your-career-in-the-age-of-ai

42nd International Conference on Organizational Science Development. (n.d.). Univerzitetna Založba Univerze v Mariboru. https://doi.org/10.18690/UM.FOV.3.2023

AI job search tips: 9 AI tools to help you land your next job. (n.d.). Retrieved June 23, 2023, from https://zapier.com/blog/ai-job-search/

Al-Alawi, A. I., Naureen, M., Alalawi, E. I., & Naser Al-Hadad, A. A. (2021). The role of artificial intelligence in recruitment process decision-making. *2021 International Conference on Decision Aid Sciences and Application, DASA 2021*, pp. 197–203. https://doi.org/10.1109/DASA53625.2021.9682320

Artificial intelligence: examples of ethical dilemmas I UNESCO. (n.d.). Retrieved June 23, 2023, from https://www.unesco.org/en/artificial-intelligence/recommendation-ethics/cases

Chen, C. C., Wei, C. C., Chen, S. H., Sun, L. M., & Lin, H. H. (2021). AI predicted competency model to maximize job performance. *Cybernetics and Systems*, 53(3), 298–317. https://doi.org/10.1080/01969722.2021.1983701

Ethical concerns mount as AI takes bigger decision-making role - Harvard Gazette. (n.d.). Retrieved June 23, 2023, from https://news.harvard.edu/gazette/story/2020/10/ethical-concerns-mount-as-ai-takes-bigger-decision-making-role/

Grosso, C., Sazen, N., & Boselli, R. (2022). AI-implemented toolkit to assist users with career "configuration": the case of create your own future. *SPLC '22: 26th ACM International Systems and Software Product Line Conference*. https://doi.org/10.1145/3503229.3547043

How AI-powered career pathing refreshes the employee experience. (n.d.). Retrieved June 23, 2023, from https://www.forbes.com/sites/forbestechcouncil/2020/01/23/how-ai-powered-career-pathing-refreshes-the-employee-experience/?sh=49a9f9f33d2c

How can AI be used by university students? I Student. (n.d.). Retrieved June 23, 2023, from https://www.timeshighereducation.com/student/advice/how-can-ai-be-used-university-students

Khang, A. (2021). *Material4Studies, Material of Computer Science, Artificial Intelligence, Data Science, IoT, Blockchain, Cloud, Metaverse, Cybersecurity for Studies*. Retrieved from https://www.researchgate.net/publication/370156102_Material4Studies

Khang, A., Hajimahmud, V. A., Gupta, S. K., Babasaheb, J., & Morris, G. (2023). *AI-Centric Modelling and Analytics: Concepts, Designs, Technologies, and Applications* (1 ed.). CRC Press, Boca Raton, FL. https://doi.org/10.1201/9781003400110

Khang, A., Kali, C. R., Suresh Kumar, S., Amaresh, K., Das, S. R., & Panda, M. R. (2023). Enabling the future of manufacturing: integration of robotics and IoT to smart factory infrastructure in industry 4.0. *AI-Based Technologies and Applications in the Era of the Metaverse* (1 ed., pp. 25–50). IGI Global Press, Hershey, PA. https://doi.org/10.4018/978-1-6684-8851-5.ch002

Khang, A., Kali Charan, R., Surabhika, P., Pokkuluri Kiran, S., & Santosh Kumar, P. (2023). Revolutionizing agriculture: exploring advanced technologies for plant protection in the agriculture sector. *Handbook of Research on AI-Equipped IoT Applications in High-Tech Agriculture* (pp: 1–22). IGI Global Press, Hershey, PA. https://doi.org/10.4018/978-1-6684-9231-4.ch001

Khang, A., Misra, A., Gupta, S. K., & Shah, V. (2023). *AI-Aided IoT Technologies and Applications in the Smart Business and Production*. CRC Press, Boca Raton, FL. https://doi.org/10.1201/9781003392224

Khang, A., Muthmainnah, M., Seraj, P. M. I., Yakin, A. A., Obaid, A. J., & Panda, M. R. (2023). AI-aided teaching model for the education 5.0 ecosystem. *AI-Based Technologies and Applications in the Era of the Metaverse* (1 ed., pp. 83–104). IGI Global Press, Hershey, PA. https://doi.org/10.4018/978-1-6684-8851-5.ch004

Khang, A., Rani, S., & Sivaraman, A. K. (2022). *AI-Centric Smart City Ecosystems: Technologies, Design and Implementation* (1st ed.). CRC Press, Boca Raton, FL. https://doi.org/10.1201/9781003252542

Khang, A., Shah, V., & Rani, S. (2023). *AI-Based Technologies and Applications in the Era of the Metaverse* (1 ed.). IGI Global Press, Hershey, PA. https://doi.org/10.4018/978-1-6684-8851-5

Khanh, H. H. & Khang, A. (2021). The role of artificial intelligence in blockchain applications. *Reinventing Manufacturing and Business Processes through Artificial Intelligence* (vol. 2, pp. 20–40). CRC Press, Boca Raton, FL. https://doi.org/10.1201/9781003145011-2

Lee, D., Kim, M., & Na, I. (2018). Artificial intelligence based career matching. *Journal of Intelligent & Fuzzy Systems*, 35(6), 6061–6070. https://doi.org/10.3233/JIFS-169846

Liutkevicius, M. & Yahia, S. B. (2022). The use of artificial intelligence in job seeking and competence development. *Human Factors, Business Management and Society*, 56(56), 128–136. https://doi.org/10.54941/AHFE1002260

Message from Korn Ferry. (n.d.). Retrieved June 23, 2023, from https://www.kornferry.com/insights/this-week-in-leadership/5-ways-AI-can-help-boost-your-career

Presbitero, A. & Teng-Calleja, M. (2022). Job attitudes and career behaviors relating to employees' perceived incorporation of artificial intelligence in the workplace: a career self-management perspective. *Personnel Review*, 52(4), 1169–1187. https://doi.org/10.1108/PR-02-2021-0103/FULL/XML

Rana, G., Khang, A., Sharma, R., Goel, A. K., & Dubey A. K. (2021). *Reinventing Manufacturing and Business Processes through Artificial Intelligence*. CRC Press, Boca Raton, FL. https://doi.org/10.1201/9781003145011

Rani, S., Chauhan, M., Kataria, A., & Khang, A. (2021). IoT equipped intelligent distributed framework for smart healthcare systems. *Networking and Internet Architecture*, 2, 30. https://doi.org/10.48550/arXiv.2110.04997

Robson, R., Kelsey, E., Goel, A., Nasir, S. M., Robson, E., Garn, M., Lisle, M., Kitchens, J., Rugaber, S., & Ray, F. (2022). Intelligent links: AI-supported connections between employers and colleges. *AI Magazine*, 43(1), 75–82. https://doi.org/10.1002/AAAI.12040

Roy, P. K., Chowdhary, S. S., & Bhatia, R. (2020). A machine learning approach for automation of resume recommendation system. *Procedia Computer Science*, 167, 2318–2327. https://doi.org/10.1016/J.PROCS.2020.03.284

Seo, K., Tang, J., Roll, I., Fels, S., & Yoon, D. (2021). The impact of artificial intelligence on learner-instructor interaction in online learning. *International Journal of Educational Technology in Higher Education*, 18(1), 1–23. https://doi.org/10.1186/S41239-021-00292-9/TABLES/7

Understanding the impact of artificial intelligence on skills development. (2021). https://en.unesco.

Westman, S., Kauttonen, J., Klemetti, A., Korhonen, N., Manninen, M., Mononen, A., Niittymäki, S., & Paananen, H. (2021). Artificial intelligence for career guidance - current requirements and prospects for the future. *IAFOR Journal of Education*, 9(4), 43–62. https://doi.org/10.22492/IJE.9.4.03

Where AI can - and can't - help talent management. (n.d.). Retrieved June 23, 2023, from https://hbr.org/2022/10/where-ai-can-and-cant-help-talent-management

Wilson, M., Robertson, P., Cruickshank, P., & Gkatzia, D. (2022). Opportunities and risks in the use of AI in career development practice. *Journal of the National Institute for Career Education and Counselling*, 48(1), 48–57. https://doi.org/10.20856/JNICEC.4807

Yadav, P. V., Kollimath, U. S., Chavan, T. V., Pisal, D. T., Giramkar, S. A., & Swamy, S. M. (2023). Impact of artificial intelligence (AI) in talent acquisition process: a study with reference to IT industry. *Proceedings of the International Conference on Intelligent and Innovative Technologies in Computing, Electrical and Electronics, ICIITCEE 2023*, pp. 885–889. https://doi.org/10.1109/IITCEE57236.2023.10090973

Zhang, H. & Zheng, Z. (2022). Application and analysis of artificial intelligence in college students' career planning and employment and entrepreneurship information recommendation. *Security and Communication Networks*, 2022, 1–8. https://doi.org/10.1155/2022/8073232

3 The Role of Artificial Intelligence in Talent Management and Career Development

*Shivakumar Kagi, Balamurugan R.,
Sampath Dakshina Murthy A.,
Surrya Prakash DilliBabu, and Nidhya M. S.*

3.1 INTRODUCTION

The spread of the Coronavirus pandemic hurried the far and wide utilization of computerized innovation across the Worldwide North. The degree of intricacy of these advances has likewise improved, with computer-based intelligence presently being incorporated into various sorts of advanced administrations. The term "artificial intelligence" (AI) is used to describe the technology that enables computers to independently carry out activities that traditionally required human intellect (European Commission, 2020). Hooley (2017) has cautioned against bleak ideas of a computerized future and said that professional guidance may assume a part in facilitating the change got on via mechanization in the work environment and society at large. Likewise, there is a developing corpus of writing on the most proficient method to execute new advancements in a way that boosts individual benefits and is predictable with the social upsides of the setting wherein they are utilized (Blodgee & O'Connor, 2017).

Career guidance services that make use of AI are the topic of this article. Research on how computer-based intelligence advancements are presently being integrated into training is restricted, in spite of the way that vocations counsel, data, and direction (CAIG) has generally exhibited capability in consolidating the advantages presented by new advancements and connecting fundamentally with the dangers (Moore & Czerwinska, 2019; Watts, 2002). The motivation behind this article is to give CAIG specialists the information they need to integrate AI into their training in a way that expands the benefits while limiting the perils related to the numerous new open doors it presents, as shown in Figure 3.1.

3.2 BACKGROUND

The authors' (Hooley et al., 2010) typology distinguishes three elements of innovation in CAIG administrations to clients: working with correspondence between

DOI: 10.1201/9781003440901-3

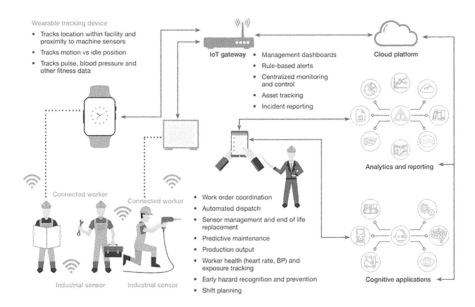

FIGURE 3.1   AI integrated with IoT and the cloud to manage workflow processes.

CAIG experts and clients; introducing data to clients; and, what's more, empowering clients to take part in robotized cooperation with vocation-related data all alone. The latter is shown by AI, which tailors data to individual users (Hooley et al., 2020). Nonetheless, this ignores a significant differentiation between simulated intelligence and "conventional" innovation. Computerized reasoning, then again, is intended to deal with cooperation for which it has not been provided with clear principles to oversee, though the previous are worked as an instrument to help or mechanize occupations where the cycles, inputs, and wanted yields are set during framework configuration in view of exact standards. When interacting with conventional IT systems in an automated fashion, users are shown data that has been filtered using criteria developed by humans and applied to the small subset of data that has been supplied by the user.

Data might be given in a wide assortment of configurations and designs, and answers are not foreordained in that frame of mind with man-made intelligence frameworks, making them more undifferentiated from human correspondence. Since the COVID-19 epidemic (Laberge et al., 2020), the usage of AI technologies in online interactions has skyrocketed (Willson, 2017). Users of e-commerce and media-streaming sites may come into contact with AI via the form of customized suggestions. Users may not realize they are interacting with an AI system since the technology is so adept at fooling the human brain into thinking it is a human being (Gran et al., 2020). In addition, the 'uncanny valley' effect of humanizing robots might make people uneasy when interacting with some forms of AI (Ciechanowski et al., 2019).

AI is mostly built on recognizing patterns. An algorithm examines task-related data in search of patterns that are predictive of a desired outcome. The AI utilizes this 'preparing' to figure out what activities to take in light of data sources it has never seen. Computer-based intelligence can deal with convoluted and imaginative data without unequivocal pre-customized guidelines, such as by recognizing complex examples in huge datasets, while customary ICT frameworks would require the engineer to give such affiliations expressly. An example of this is the employment of an algorithm fed a big dataset detailing customers' online transactions to produce shopping suggestions using AI. The system analyzes a user's profile, browsing history, and purchase data to determine what that person would be interested in. Whether a man-made intelligence framework is being utilized to compose a message, propose a video clasp, or make a clinical determination, this course of concluding a result in light of similitudes between the current circumstance and recently experienced occurrences is at its center.

The interactions between the system's design and the training data might lead to surprising outcomes, even if designers have some say over which criteria are utilized or prioritized. It has been shown that trustworthy AI systems must be able to explain or understand the steps used to arrive at a given outcome (OECD, 2021b). Thus, albeit the consequences of a simulated intelligence empowered framework might mirror human insight, the ends drawn by AI couldn't measure up to human discernment since they are not in view of information in the real world. Implicit 'intentions, values, and social purposes' at the individual and societal levels are part of human knowledge (Vallor, 2021). Even so, AI technology may operate independently to carry out difficult activities that formerly needed direct human input.

3.3   THE USE OF AI IN SIMILAR CONTEXTS AS A LEARNING TOOL

Services provided by CAIG may be seen from a number of angles, including those of career guidance, career research, career development, and job placement. Recent years have seen the application of AI to domains with analogous societal issues. Given the paucity of CAIG-focused studies, insights gained in various other contexts may help fill in some of the gaps. Theraputic guiding, library and data administration, training, and work force choice are instances of these ventures.

3.3.1   Services in Counseling

Automated therapeutic counseling services might provide immediate help without long wait times or the risk of humiliation or shame associated with opening up to a human being about sensitive personal issues. The potential for more people to have access to care is encouraging, and steps have been taken to make that a reality. A well-known smartphone app that helps people with mental health issues is called Woebot1. Mood regulation, psychoeducation, and cognitive behavioral therapy are only a few of the methods used (Fitzpatrick et al., 2017; Prochaska et al., 2021). Client worries about the protection, adequacy, and receptiveness of chatbots for emotional wellness help were noted in subjective exploration, whereas quantitative studies reported conflicting findings for the effects on users' moods (Suharwardy et al., 2020). As a result, it seems that users' familiarity with the technology may play a role in whether or not clients get the benefits of these apps in career guidance, as shown in Figure 3.2.

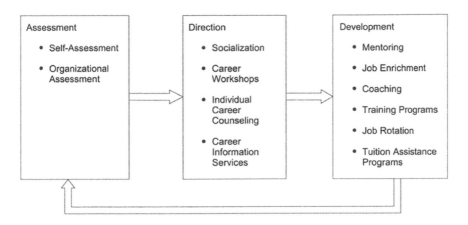

Assessment	Direction	Development
• Self-Assessment • Organizational Assessment	• Socialization • Career Workshops • Individual Career Counseling • Career Information Services	• Mentoring • Job Enrichment • Coaching • Training Programs • Job Rotation • Tuition Assistance Programs

FIGURE 3.2 AI assessment, direction, and development.

3.3.2 Data Processing and Storage

With regards to assisting individuals with picking the right vocation way, library administrations are a nearby 'simple' to professions administrations since they likewise gather and sort out an abundance of data. Libraries have been singled out as a field where AI chatbots might help address issues including patrons' worries and a lack of available experts (Saldeen, 2020). Because users wouldn't need to learn a specialized language or technical abilities to utilize alternative techniques for accessing databases, the learning curve for meeting their own information demands would be significantly reduced. When designing technology for an academic library, it is crucial to get input from library personnel to ensure it can accommodate a wide range of user demands and know when to seek human assistance (Mckie and Narayan, 2019). This is relevant for AI projects whose ultimate goal is to enable vetted access to LMI (labor market data). When trying to foresee the breadth and depth of a client's first set of questions, the expertise of CAIG professionals is invaluable (Bimrose, 2021). Mechanized curation of LMI has a bigger number of stakes than scholarly writing look because of the idea of profession-related data chasing.

Users' lives may be profoundly affected by the search results they see while looking for self-serving LMI. Information that is not included in search results has the implicit impact of excluding such possibilities from consideration when a client is using LMI to investigate and assess the viability of potential career paths. The use of AI to provide individualized LMI raises the possibility that the algorithm's choice of personalization criteria may be inadequate or oversimplified. If a client's age and gender are used to choose jobs that are popular among others of the same age and gender, this will merely help to reinforce preexisting demographic inequities in employment results. The use of AI to collect LMI is already widespread. To give a rich wellspring of LMI that reflects the latest things and takes less assets to refresh than past procedures (Lassébie et al., 2021), the OECD, for example, has utilized man-made consciousness to plan 17,000 particular abilities demonstrated in work notices to a scientific categorization of 61 classifications. There are also efforts to

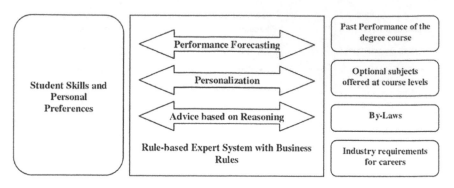

FIGURE 3.3 AI with business rules.

create AI tools that will make it easier for customers to use LMI, such as the CiCi chatbot, which provides users with conversational access to individualized career guidance (Hughes, 2021). Similar to the current trend toward code signing as an essential necessity for efficient and ethical AI (Floridi et al., 2018), both cases had direct participation from CAIG specialists in the design of the technology. They also help to illustrate how AI may aid in the effective distribution of LMI while decreasing the amount of time and expertise needed for the process, as shown in Figure 3.3.

3.3.3 EDUCATION

Applications for vocational education and training are only one example of the widespread usage of automated systems in educational contexts (Hai-Jew, 2009). Despite high expectations, the chosen intelligent teaching solutions have often been relatively basic in practice (Baker, 2016). Simulated intelligence virtual mentoring frameworks are powerful instruments that apply a scope of instructive methods effectively to impart perception of a specific subject, as indicated by the focus on understudy results in the wake of utilizing such frameworks (Olney et al., 2012; Paladines and Ramírez, 2020). While the use of AI tutoring systems relieves pressure on teachers, (Heffernan, 2003) it should be noted that students need more time than they would have had they received their education in a traditional classroom setting to master the same amount of information. When human resources are few, AI tutors might help more people get the professional training they need. However, clients who lack the willingness to apply the intervention or who cannot commit to doing so may not benefit from this strategy because of the extra work it needs from them.

3.3.4 HIRING PROCEDURES

There has been an uptick in the use of commercial, "off-the-shelf", AI-based recruiting solutions among businesses (Chamorro-Premuzic et al., 2019; Gee, 2017). These are often utilized at the beginning of the selection process to quickly weed out unqualified applicants by analyzing their application materials and video interviews (Van Esch et al., 2019; Van den Broek et al., 2019). The multinational conglomerate

Unilever was an early user of this technology, and the company boasts a more diverse workforce and considerable cost savings. Despite widespread coverage in the popular press (Booth, 2019), there is a striking absence of such examinations in the insightful writing because of the confidential idea of the product and the conceivable upper hand in managing the cost by imaginative choice procedures.

As a result, there is far less openness about the data processing and the conclusions. Despite this, the technology is being promoted, among other places, to the career service departments of universities as a means of helping students hone their interview abilities. The use of AI to solve the issue of hiring is similar to the conventional view of career matching. This is helpful because it gives us a context in which to discuss the moral quandaries inherent in the usage of AI in the context of career counseling.

3.4 PROBLEMS WITH ETHICS RAISED BY MACHINE LEARNING IN STAFFING

Given the significance of recruiting choices for both people and communities, AI has proven to be divisive in the field of recruitment (Forum for Ethical AI, 2019; OECD, 2021b). Programming organizations that support the utilization of AI in the selection of systems feature cost reserve funds and the evacuation of human mix-up and oblivious bias as significant benefits (Hirevue, 2021; Schmidt, 2018). Even when certain traits are removed from the data, historical instances of AI decision-making systems have been proven to be biased (Birhane, 2021). This results from people taking advantage of qualities that are protected because of their attributes. Trial endeavors to utilize computer-based intelligence for robotizing CV assessment have yielded risky outcomes, for example, a calculation that rejects CVs that incorporate expressions characteristic of orientation (like references to ladies' games or ladies' universities) (Dastin, 2018).

Humans have a hard time anticipating these inadvertent connections due to the large amount of information involved, and the rising complexity of the system architecture makes it impossible to determine what variables drove AI judgments in hindsight without specialized technical expertise. This issue is exacerbated by the way that individuals will generally accept man-made intelligence discoveries in any event when they oppose their own dependable understanding (Suresh et al., 2020). On account of CAIG, this might refer to accidentally exhorting clients for a way that supports prior imbalances in the gig market. It additionally stresses the need for result-checking to ensure that man-made intelligence-upheld arrangements are in accordance with CAIG's morals and values.

Data manipulation attempts to alleviate this problem run into technological and ethical snags. Sadly, it could be challenging to figure out what parts of the information are causing predisposition attributable to the intricacy of the framework configuration, in any event, when endeavors are taken to "de-predisposition" the dataset (Bender et al., 2021). This might be mitigated by providing good monitoring of live systems to discover unintended effects, but this raises a number of significant ethical questions.

Assuming we can impact the consequences of frameworks that affect individuals and society, then these frameworks are accomplishing more than simply reflecting reality; and they are effectively forming it. Since computer-based intelligence is now being used for occupations in spaces like government-managed retirement, law enforcement, and selection (O'Neil, 2017; Tambe et al., 2019), there have been concerns communicated concerning the outcomes of algorithmic dynamics in supporting and amplifying existing authorities. Hence, it's anything but a mechanical issue but instead a troublesome lawful, social, and moral one to figure out what the point profile for results ought to be for a computer-based intelligence framework to be assigned as 'de-one-sided' or 'fair' and who directs this.

When the technology is designed for usage across international and cultural boundaries, reaching an agreement may be very challenging. For this reason, it is crucial that AI systems be explainable and interpretable (Linardatos et al., 2020) so that all stakeholders may get insight into the system's design and keep tabs on its results. Therefore, the fundamental requirement for deciding whether AI-based technology can be trusted is whether or not the judgments it makes can be explained and interpreted.

The prevalence of third-party suppliers in AI recruiting makes this more difficult, since the creator of the AI software may see the answers to these queries as commercially confidential. This also illustrates the difficulty of assigning responsibility when using AI. Who is liable for guaranteeing consistency with lawful and moral obligations when the space or organization's explicit information is given by the product buyer and executed utilizing the buyer's information? Data privacy concerns are raised by the possibility that software buyers' information will be utilized to enhance the algorithm's performance for future clients of the vendor (Wagner, 2020).

Certain forms of AI systems have the known danger of having personally identifiable information incorporated into training data without the user's awareness (Bommasani et al., 2021). Who would be responsible for any accidental disclosure of sensitive information? Those working in CAIG who use or advocate AI technology should be aware of the ways in which their clients' private information may be shared with third parties and should tell their clients accordingly. This is especially important to keep in mind when dealing with commercial service providers, since their primary goal in collecting and processing user data may be financial gain (Zuboff, 2019).

All parties participating in AI research, development, and implementation should be held accountable, according to existing policy guidelines. However, AI-specific law is still in the works (CDEI, 2021; Floridi, 2021); therefore, it is unclear how responsibility should be assigned when a service provider employs software that is the intellectual property of another company. Regardless, CAIG experts ought to guarantee that the information of administration clients will be dealt with safely by any ICT asset they integrate into their work on, including more dark purposes of this information for preparing simulated intelligence calculations, as satisfactory insurance of individual data is a prerequisite of the Vocation Improvement Foundation Set of principles (CDIC, 2018).

AI-assisted recruiting also raises questions about the safety of users' private information and the challenge of making sure that people who interact with AI systems feel like they were handled properly. Using AI in this setting allows for uniform evaluation of all application materials, including resumes and interviews, which is impossible with human beings alone (Chamorro-Premuzic et al., 2019). AI-assisted recruiting clearly benefits candidates (Suen et al., 2019; Van Esch et al., 2019) due to its accessibility, standardization, and ease of use. However, this should be qualified by pointing out the dearth of studies devoted to the perspectives of candidates, particularly those who are disabled, neurodivergent, or who are communicating in a language that is not their mother tongue.

Applicants' unfavorable impressions of AI recruiting stem from their uncertainty about the selection criteria and their focus on learning the technology during interviews (Jaser et al., 2021), according to recent studies. This emphasizes the need for CAIG services to make sure their customers understand how AI systems work and how they may choose to opt out without negative consequences.

3.5 TAKING PRECAUTIONS AND REAPING REWARDS

Enlistment is referred to as a high-risk region for the utilization of simulated intelligence in a European Commission concentrate on mindful utilization of man-made intelligence and algorithmic administration. On account of its "importance for individuals...and tending to work equality", (European Commission, 2020), enrolling merits unique consideration. In light of these norms because of computer-based intelligence's capability to more readily impact individuals' lives and advance correspondence, certain CAIG exercises would in like manner be considered high-risk for the organization of simulated intelligence (Blustein et al., 2019). Public policy's potential involvement in regulating AI is a hot issue right now (CDEI, 2021; OECD, 2021b). Arrangements to help with the acknowledgment of the potential benefits man-made intelligence could bring to society are a consistent idea across these arrangements, along with a recognition of the hazards of both intentional abuse and unforeseen repercussions. Emerging policy commentary highlights the challenge of regulating cutting-edge technologies that operate beyond national boundaries (Khang & Santosh et al., 2023).

Accomplishing a reasonable harmony between the dangers and benefits of utilizing simulated intelligence inside CAIG requires familiarity with the potential ramifications for clients, experts, and society through a framework plan and execution. As shown by the CiCi and LMI case studies used above (Lassébie et al., 2021; Hughes, 2021), this demands constant participation from CAIG specialists and practitioners. Rather than relying just on norms created for the technology itself, AI should also be held to the applicable professional standards in the field in which it is used. AI utilized in CAIG should be subject to the same ethical standards that a human expert would be if they were doing the same tasks (Khang & Rani et al., 2023).

Most programming sellers making these frameworks miss the mark on their ability to assess their moral similarity; subsequently, they ought to effectively look for direction from CAIG experts all through the improvement interaction. This will guarantee that moral contemplations are heated into the framework all along, instead of being tended to just during advertising or, more regrettable, as issues emerge after sending. AI's improved breadth of treatments, along with its cost-effectiveness, may be a boon to CAIG services. However, this demands a strategy to introducing AI that prioritizes enhancing the customer experience rather than cutting expenses (Khang & Shah et al., 2023).

3.6 CONCLUSION

With the help of AI, professionals may significantly increase their ability to provide selected content. It diminishes limitations on where and when clients might find support and the degree of aptitude required for them to self-serve vocation information. Notwithstanding, because of the different idea of CAIG work, it is essential to gauge the benefits and hindrances of robotizing undertakings prior to utilizing simulated intelligence. It's important to take stock before ordering or suggesting AI technology to make sure the job can really be automated. In cases where AI may be useful, it is crucial to be able to explain the system's design and to constantly monitor its performance in order to detect and deal with any unintended repercussions. Thus, in any event, when a framework can be easily conveyed for a given use, the innovation should be worked with the natural capacity to perceive and allude fittingly when it is anything but a reasonable mediation for a specific client (Khang & Kali et al., 2023).

Professionals in the CAIG field should ask questions about a potential AI product that are informed by their own subject knowledge and experience. Questions should be posed in terms of CAIG, not technology, be that as it may, a grip on the benefits and risks innate to this sort of innovation means a lot to understanding what they are. Effective and ethical operation on behalf of a CAIG service should be at the forefront of the issues asked, alongside efficiency, performance, and cost. This should be backed up by an interdisciplinary study that determines the most effective ways to develop and use AI technology for use by CAIG. Man-made intelligence isn't an instrument; it is a specialist in light of its ability to think and act freely while offering mechanized communications with clients. As a result, implementing AI in CAIG should be treated more like a service design project than a software development or procurement initiative (Khang & Muthmainnah et al., 2023).

REFERENCES

Baker, R. S. (2016). Stupid tutoring systems, intelligent humans. *International Journal of Artificial Intelligence in Education*, 26(2), 600–614. https://doi.org/10.1007/S40593-016-0105-0

Bender, E. M., Gebru, T., McMillan-Major, A., & Shmitchell, S. S.-G. (2021). On the dangers of stochastic parrots: can language models be too big? *Proceedings of the 2021 ACM Conference on Fairness, Accountability, and Transparency*, pp. 610–623. https://doi.org/10.1145/3442188.3445922

Bimrose, J. (2021). Labour market information for career development: pivotal or peripheral? In P. Robertson, T. Hooley, & P. McCash (Eds.) *The Oxford Handbook of Career Development*, pp. 282–296. Oxford University Press, Oxford. https://doi.org/10.1093/oxfordhb/9780190069704.013.21

Birhane, A. (2021). The impossibility of automating ambiguity. *Artificial Life*, 1, 1–18. https://doi.org/10.1162/artl_a_00336

Blodgee, S. L., & O'Connor, B. (2017). Racial disparity in natural language processing: a case study of social media African-American English. *Workshop on Fairness, Accountability, and Transparency in Machine Learning*. https://arxiv.org/abs/1707.00061

Blustein, D. L., Kenny, M. E., di Fabio, A., & Guichard, J. (2019). Expanding the impact of the psychology of working: engaging psychology in the struggle for decent work and human rights. *Journal of Career Assessment*, 27(1), 3–28. https://doi.org/10.1177/1069072718774002

Bommasani, R., Hudson, D. A., Adeli, E., Altman, R., Arora, S., von Arx, S., Bernstein, M. S., Bohg, J., Bosselut, A., Brunskill, E., Brynjolfsson, E., Buch, S., Card, D., Castellon, R., Chatterji, N., Chen, A., Creel, K., Davis, J. Q., Demszky, D., & Liang, P. (2021). On the Opportunities and Risks of Foundation Models. In arXiv preprint arXiv:2108.07258. https://arxiv.org/abs/2108.07258.

Booth, R. (2019). Unilever saves on recruiters by using AI to assess job interviews. The Guardian. https://www.theguardian.com/technology/2019/oct/25/unilever-saves-on-recruitersby-using-ai-to-assess-job-interviews

CDEI (2021). Centre for Data Ethics and Innovation. The roadmap to an effective AI assurance ecosystem. Department for Digital, Culture, Media & Sport. https:// www.gov.uk/government/publications/the-roadmap-toan-effective-ai-assurance-ecosystem

CDIC (2018). Career Development Institute Code of Ethics. https://www.thecdi.net/write/Documents/Code_of_Ethics_update_2018- web.pdf

Chamorro-Premuzic, T., Polli, F., & Dattner, B. (2019). Building ethical AI for talent management. Harvard Business Review, 21, 1–15. https://hbr.org/2019/11/building-ethicalai-for-talent-management

Ciechanowski, L., Przegalinska, A., Magnuski, M., & Gloor, P. (2019). In the shades of the uncanny valley: an experimental study of human-chatbot interaction. Future Generation Computer Systems, 92, 539–548. https://doi.org/10.1016/j.future.2018.01.055

Dastin, J. (2018). Amazon scraps secret AI recruiting tool that showed bias against women. Reuters. https://www.reuters.com/article/uk-amazoncom-jobs-automation-insight/amazon-scraps-secretai-recruiting-tool-that-showed-bias-against-womenidUKKCN1MK08K?edition-redirect=uk

European Commission. (2020). White Paper on Artificial Intelligence - A European approach to excellence and trust. https://ec.europa.eu/commission/sites/beta-political/ files/political-guidelines-next-commission_en.pdf

Fitzpatrick, K. K., Darcy, A., & Vierhile, M. (2017). Delivering cognitive behavior therapy to young adults with symptoms of depression and anxiety using a fully automated conversational agent (Woebot): a randomized controlled trial. Journal of Medical Internet Research Mental Health, 4(2), e19. https://doi.org/10.2196/mental.7785

Floridi, L. (2021). The European legislation on AI: a brief analysis of its philosophical approach. Philosophy & Technology, 34, 215–222. https://doi.org/10.1007/s13347-021-00460-9

Floridi, L., Cowls, J., Beltrametti, M., Chatila, R., Chazerand, P., Dignum, V., Luetge, C., Madelin, R., Pagallo, U., Rossi, F., Schafer, B., Valcke, P., & Vayena, E. (2018). AI4people-an ethical framework for a good AI society: opportunities, risks, principles, and recommendations. Minds and Machines, 28(4), 689–707. https://doi.org/10.1007/S11023-018-9482-5

Forum for Ethical AI. (2019). Democratising decisions about technology A toolkit. Royal Society for the encouragement of Arts, Manufactures and Commerce. https://www.thersa.org/reports/democratisingdecisions-technology-toolkit.

Gee, K. (2017). In unilever's radical hiring experiment, resumes are out, algorithms are in. Wall Street Journal (Online), 26. https://www.wsj.com/articles/inunilevers-radical-hiring-experiment-resumes-are-outalgorithms-are-in-1498478400

Gran, A.-B., Booth, P., & Bucher, T. (2020). To be or not to be algorithm aware: a question of a new digital divide? Information, Communication & Society, 24, 1–18. https://doi.org/10.1080/1369118X.2020.1736124.

Hai-Jew, S. (2009). Designing automated learning for effective training and skills development. Handbook of Research on E-Learning Applications for Career and Technical Education: Technologies for Vocational Training (pp. 14–33). IGI Global Press, Hershey, PA. https://doi.org/10.4018/978-1-60566-739-3.ch002

Heffernan, N. T. (2003). Web-based evaluations showing both cognitive and motivational benefits of the Ms. Lindquist tutor. Artificial Intelligence in Education, pp. 115–122. https://link.springer.com/10.1007/978-3-540-30139-4_46

Hirevue. (2021). End-to-End Hiring Experience Platform: Video Interviewing, Conversational AI & More I HireVue. https://www.hirevue.com/

Hooley, T. (2017). A war against the robots? Career guidance, automation and neoliberalism. *Career Guidance for Social Justice* (pp. 93–107). Routledge, London. https://doi.org/10.4324/9781315110516-6

Hooley, T., Hutchinson, J., & Watts, A. G. (2010). *Careering Through the Web: The Potential of Web 2.0 and 3.0 Technologies for Career Development and Career Support Services.* UKCES, Europe. https://derby.openrepository.com/handle/10545/198269.

Hughes, D. (2021). The bot working on experts and advice with human careers information. *Career Matters*, 9, 33.

Jaser, Z., Petrakaki, D., Starr, R., Oyarbide, E., Williams, J., & Newton, B. (2021). *Artificial Intelligence (AI) in the Job Interview Process: Toolkit for Employers, Careers Advisers and Hiring Platforms.* Institute for Employment Studies, Brighton. https://www.employment-studies.co.uk/resource/ artificial-intelligence-ai-job-interview-process

Khang, A., Kali, C. R., Satapathy, S. K., Kumar, A., Das, S. R., & Panda, M. R. (2023). Enabling the future of manufacturing: integration of robotics and IoT to smart factory infrastructure in industry 4.0. *AI-Based Technologies and Applications in the Era of the Metaverse* (1st ed., pp. 25–50). IGI Global Press, Hershey, PA. https://doi.org/10.4018/978-1-6684-8851-5.ch002

Khang, A., Kali Charan, R., Surabhika, P., Pokkuluri Kiran, S., & Santosh Kumar, P. (2023). Revolutionizing agriculture: exploring advanced technologies for plant protection in the agriculture sector. *Handbook of Research on AI-Equipped IoT Applications in High-Tech Agriculture* (pp. 1–22). IGI Global Press, Hershey, PA. https://doi.org/10.4018/978-1-6684-9231-4.ch001

Khang, A., Muthmainnah, M., Seraj, P. M. I., Yakin, A. A., Obaid, A. J., & Panda, M. R. (2023). AI-aided teaching model for the education 5.0 ecosystem. *AI-Based Technologies and Applications in the Era of the Metaverse* (1st ed., pp. 83–104). IGI Global Press, Hershey, PA. https://doi.org/10.4018/978-1-6684-8851-5.ch004

Khang, A., Rani, S., Gujrati, R., Uygun, H., & Gupta, S. K. (2023). *Designing Workforce Management Systems for Industry 4.0: Data-Centric and AI-Enabled Approaches.* CRC Press, Boca Raton, FL. https://doi.org/10.1201/9781003357070

Khang, A., Shah, V., & Rani, S. (2023). *AI-Based Technologies and Applications in the Era of the Metaverse* (1 ed.) IGI Global Press, Hershey, PA. https://doi.org/10.4018/978-1-6684-8851-5

Laberge, L., O'Toole, C., Schneider, J., & Smaje, K. (2020). How COVID-19 has pushed companies over the technology tipping point-and transformed business forever. *McKinsey Digital and Strategy & Corporate Finance Practices.* https://doi.org/10.1201/9781003440901

Lassébie, J., Vandeweyer, M., & Vignal, B. (2021). Speaking the same language: a machine learning approach to classify skills in burning glass technologies data (No. 263; OECD Social, Employment and Migration Working Papers). https://doi.org/10.1787/adb03746-en

Linardatos, P., Papastefanopoulos, V., & Kotsiantis, S. (2020). Explainable AI: a review of machine learning interpretability methods. *Entropy*, 23(1), 18. https://doi. org/10.3390/E23010018

Mckie, I. A. S. & Narayan, B. (2019). Enhancing the academic library experience with chatbots: an exploration of research and implications for practice. *Journal of the Australian Library and Information Association*, 68(3), 268–277. https://doi.org/10.1080/247 50158.2019.1611694

Moore, N. & Czerwinska, K. (2019). *Understanding the Use of Digital Technology in the Career Development Sector.* University of Derby, Derby. https://derby.openrepository.com/ handle/10545/624311

O'Neil, C. (2017). *Weapons of Math Destruction: How Big Data Increases Inequality and Threatens Democracy.* Penguin Books, London.

OECD. (2021b). Recommendation of the Council on Artificial Intelligence, OECD/ LEGAL/0449. https://legalinstruments.oecd.org/en/instruments/oecdlegal-0449

Olney, A. M., D'Mello, S., Person, N., Cade, W., Hays, P., Williams, C., Lehman, B., & Graesser, A. (2012). Guru: a computer tutor that models expert human tutors. *11th International Conference on Intelligent Tutoring Systems*, pp. 256–261. https://doi.org/1 0.1007/978-3-642-30950-2_32

Paladines, J. & Ramírez, J. (2020). A systematic literature review of intelligent tutoring systems with dialogue in natural language. *IEEE Access*, 8, 164246–164267. https://doi. org/10.1109/access.2020.3021383

Prochaska, J. J., Vogel, E. A., Chieng, A., Kendra, M., Baiocchi, M., Pajarito, S., & Robinson, A. (2021). A therapeutic relational agent for reducing problematic substance use (Woebot): development and usability study. *Journal of Medical Internet Research*, 23(3), e24850. https://doi.org/10.2196/24850

Saldeen, N. A. (2020). Artificial intelligence chatbots for library reference services. *Journal of Management Information and Decision Sciences*, 23(S1), 442–449. https://search.ebsco- host.com/login.aspx?direct=true&profile=ehost&scope=site&authtype=crawler&jrnl= 15247252&AN=149358857&h=CxLYAKlnxZxRR6n45nYo9mwDhJt1Cvqzsq9V4TA j6p024y7Saq7FngbnVRM59favIf1f2C9MLacU%2BGd%2FBSVdiA%3D%3D&crl=c

Schmidt, L. (2018). A glimpse into the future of recruiting. Forbes. https://www.forbes.com/ sites/larsschmidt/2018/01/31/a-glimpse-into-the-futureof-recruiting

Suen, H. Y., Chen, M. Y. C., & Lu, S. H. (2019). Does the use of synchrony and artificial intelligence in video interviews affect interview ratings and applicant attitudes? *Computers in Human Behavior*, 98, 93–101. https://doi.org/10.1016/j.chb.2019.04.012

Suharwardy, S., Ramachandran, M., Leonard, S. A., Gunaseelan, A., Robinson, A., Darcy, A., Lyell, D. J., & Judy, A. (2020). Effect of an automated conversational agent on postpartum mental health: a randomized, controlled trial. *American Journal Of Obstetrics And Gynecology*, 222(1), S91. https://doi.org/10.1016/j. ajog.2019.11.132

Suresh, H., Lao, N., & Liccardi, I. (2020). Misplaced trust: measuring the interference of machine learning in human decision-making. *WebSci 2020- Proceedings of the 12th ACM Conference on Web Science*, pp. 315–324. https://doi.org/10.1145/3394231.3397922

Tambe, P., Cappelli, P., & Yakubovich, V. (2019). Artificial intelligence in human resources management: challenges and a path forward. *California Management Review*, 61(4), 15–42. https://doi.org/10.1177/0008125619867910

Vallor, S. (2021). The thoughts the civilized keep. Noema. https://www.noemamag.com/ thethoughts-the-civilized-keep/

Van den Broek, E., Sergeeva, A., Huysman, M., & Paper, S. (2019). Hiring algorithms: an ethnography of fairness in practice. *International Conference on Information Systems*, 1, 6. https://aisel.aisnet.org/icis2019/future_of_ work/future_work/6

Wagner, D. N. (2020). The nature of the artificially intelligent firm - an economic investigation into changes that AI brings to the firm. *Telecommunications Policy*, 44(6), 101954. https://doi.org/10.1016/j.telpol.2020.101954

Watts, A. G. (2002). The role of information and communication technologies in integrated career information and guidance systems: a policy perspective. *National Institute for Careers Education and Counselling*, 2, 139–155. https://link.springer.com/ article/10.1023/A:1020669832743

Willson, M. (2017). Algorithms (and the) every day. *Information, Communication & Society*, 20(1), 137–150. https://doi.org/10.1080/1369118X.2016.1200645

Zuboff, S. (2019). *The Age of Surveillance Capitalism: The Fight for a Human Future at the New Frontier of Power*. Profile Books, London. https://doi.org/10.26522/brocked. v29i2.849

4 The Power of Artificial Intelligence in Talent Recruitment Revolution
Creating a Smarter Workforce

Arpita Nayak, Atmika Patnaik, Ipseeta Satpathy, Alex Khang, and B. C. M. Patnaik

4.1 INTRODUCTION

Recruiting is the process of locating, interviewing, and hiring qualified candidates for job openings. If more people apply for jobs, there will be more opportunities to hire the appropriate person. This is also why diversity in the hiring process is essential. Job searchers, on the other hand, are looking for organizations that could be interested in hiring them. Recruitment is a connecting activity that links those who are working with others who are searching for work. In simple terms, recruiting is the act of discovering sources from which to choose potential employees (Nikolaou, 2021). Recruitment is a critical procedure that ensures the continuity of corporate operations. It assists businesses in evaluating their existing and future personnel needs, preventing disruption of business operations, and increasing hiring success rates. Recruitment also broadens the talent pool, attracts highly skilled candidates, and saves money.

Furthermore, a great recruiting procedure may boost an organization's reputation by reflecting its professionalism and honesty. Overall, a good recruiting process is critical for gaining fresh talent and developing a smarter workforce capable of meeting the organization's changing demands. Employees work hard to reach organizational goals, and the success of your company is dependent on your attitude and perspective on hiring. Companies that want to develop quickly but gradually recognize that they can only do so if they have the correct mix of people, which is why recruiting is so important. A scientific recruitment strategy offers advantages such as enhanced productivity, higher pay, improved morale, lower staff turnover, and greater credibility. It encourages individuals to apply for employment, which is beneficial (Kapse et al., 2012).

Based on Economic Times research, despite growing worries that artificial intelligence (AI) profiling could displace many jobs, people are eager to use AI to make their lives easier, both personally and professionally. Interestingly, according to Microsoft's Labor Trend Index Annual Report, they want AI to help them find the

DOI: 10.1201/9781003440901-4

right information and answers, and 70% of employees want to outsource as much labor as possible to work with AI to minimize the amount of work they do. AI technology has altered the hiring process, making it more efficient, objective, and accessible. Businesses have widely used several AI elements in recruitment to boost their talent acquisition approach. Candidate sourcing, evaluation, onboarding, and predictive analytics are examples of these components (Rana et al., 2021).

AI-powered systems may search job boards, social media networks, and résumé databases for the best candidates for specific job openings. Algorithms can evaluate applicants in ways that people cannot, such as pre-screening candidates by analyzing video interviews, voice samples, and written replies. AI technology may also be used to improve the onboarding process for new employees and forecast their likelihood of success in a certain function. Companies may entice customers by exploiting these AI components (Geetha & Bhanu, 2018; Oswal et al., 2020).

Advancement in new technology and AI have encouraged business practices to shift toward digitally managed services. The use of AI to accomplish commercial activities has attracted attention due to its tremendous effect on the business environment. Recruiting has always been a time-consuming and repetitive process. Many firms, for example, have used AI to quickly process massive amounts of data in order to improve efficiency, accuracy, and productivity. In its most basic form, AI is just automation that allows us to tackle large, complicated, repeating issues with high-quality outcomes. AI is there to help illuminate the way to success. AI in the hiring process, also known as AI in recruitment, helps talent acquisition teams to uncover passive prospects and leverage data-driven insights that influence decision-making and improve outcomes such as hiring quality. AI in recruiting tools can help match acceptable openings with appropriate prospects and appropriate candidates with appropriate recruiters.

According to recent data, 46% of organizations have difficulty finding and attracting competent people for open positions. AI programs may scan through online resumes and social media to find the best candidates for each vacancy based on predefined criteria. They can also deliver personalized messaging to prospective candidates on a wide scale, something that human recruiters cannot do. To overcome human biases in procurement and screening, AI is being educated. The goal is to train the program on data that seem to be gender-neutral while ignoring any.

An organized decision may end up with a much more diverse pool of candidates than if the HR team sourced them solely. According to a 2017 Deloitte study, 38% of respondents expected AI to be widely deployed in their organization within 3–5 years. In 2018, this proportion rose to 42%. It continues to rise. According to a LinkedIn study, 72% of CEOs believe AI will deliver significant business advantages in the next years, while 76% of recruitment managers believe AI will be at least somewhat critical in the future. AI is a very powerful tool as Figure 4.1. AI is a very successful approach for hiring top people who can thrive within a company because it uses massive amounts of data to generate predictions with greater accuracy than a person can achieve, said Eric Sydell, EVP of Innovation at Modern Hire (Trivella, 2023; Kot et al., 2021). Before the advent of AI in talent recruitment practices, there were several challenges that were faced by human resource personnel; some of them are listed below (Johansson & Herranen, 2019).

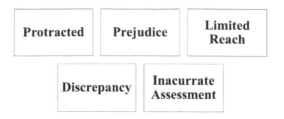

FIGURE 4.1　AI in talent recruitment.

- **Protracted**: Traditional recruiting processes were frequently time-consuming and labor-intensive, such as analyzing resumes, organizing interviews, and completing evaluations. This slowed and increased the cost of the procedure.
- **Prejudice**: Recruiters are susceptible to unintentional prejudices that might influence their decision-making; as a result, individual applicants face prejudice based on factors such as race, gender, age, and more.
- **Limited Reach**: Traditional means of recruiting, such as placing job advertisements on job boards, depending on employee recommendations, or visiting job fairs, have a limited reach and may not attract a wide pool of candidates. Job boards, for example, may only reach a subset of the job market, while depending only on employee recommendations may result in a lack of diversity in the candidate pool.
- **Discrepancy**: Traditional recruiting practices were also uneven, with various recruiters evaluating prospects using different criteria and processes. This made ensuring justice and impartiality in the selection process challenging.
- **Inaccurate Assessment**: Traditionally, resumes, cover letters, and interviews were used to assess a candidate's qualifications and suitability for the position. However, these approaches may not give a full picture of a candidate's talents, personality, work style, and other critical characteristics for job performance and satisfaction. As a result, businesses may choose applicants who are not a good match for the position, resulting in poor work performance, low job satisfaction, and, ultimately, greater turnover rates.

E-recruitment is one of the most popular non-traditional recruiting methods for discovering and attracting qualified job prospects. Organizations have turned to the internet to find possible job candidates since it is cheaper, quicker, and more efficient (Johansson & Herranen, 2019). One of the most notable developments among recruiting specialists was the use of AI in HRM. AI might be employed in the recruitment of new employees, particularly through data-gathering algorithms that can speed up the application screening and critical information extraction process (Sills, 2014).

For good reason, AI is taking over the recruitment business. Building relationships is critical in the industry. When you are able to develop genuine relationships with candidates, they get thrilled about your business's beliefs and ambitions, and your hires become genuinely committed to the firm, motivating them to stay longer and helping your organization to accomplish more. However, the truth is a little different.

Finding, connecting, and keeping people is a difficult process, and when a single bad hire may cost you $50,000 or more, you can't afford to recruit the wrong individual. The various benefits of AI in the recruitment process are listed below (Mitchel, 2021);

- AI may be used to increase candidate quality in a variety of ways. AI may assist in the creation of concise, comprehensive job descriptions that correctly express what you're looking for in candidates. AI may also be used to locate the best applicants online through programmatic job advertising, which targets individuals based on demographic and behavioral factors. This advertising attracts the correct people, increasing conversion rates.
- Dull manual duties are the last thing you want to deal with while you're busy organizing interviews, assessing prospects, and attempting to understand your team's ever-changing needs. When deployed properly, AI might serve as a personal assistant for recruiters. It may be used for automating many of the repetitive tasks that take up your time. By using software for scheduling or automatic resume reviewers, you may ensure that things get done without wasting time away from your demanding schedule.
- AI solutions in the recruitment field are intended to make the lives of job hopefuls and applicants simpler. Chatbots, for example, assist applicants more effectively through the interview process by removing barriers, roadblocks, or hassles that may dissuade them from completing an application or following up on a CV (curriculum vitae) or questions they've sent. When considering incorporating AI tools into your recruiting process, consider not just what it can do for you and your team, but also how it affects and enhances the applicant experience. Zappos, a US retail behemoth, used Harver in its recruiting process and prioritized candidate experience, and it paid off! After a year, 97% of candidates said they "agreed" or "agreed strongly" that they had a pleasant experience applying to Zappos.
- AI may assist in making your recruitment process more efficient and successful. It gathers and organizes data, enabling you to "grease the machine wherever it squeaks" in order to recruit the proper people. AI technologies collect data and organize them, allowing you to "grease the machine anywhere it squeaks."

An application tracking system (ATS) is a type of recruitment software that may assist store resumes, search the candidate database, and give insights into the recruiting process. Intelligent recruitment automation, on the other hand, has the potential to improve the recruiting process. AI may be utilized for resume screening, interview scheduling, and other tasks such as applicant sourcing, video interviewing, and recruitment marketing. Companies may also employ AI-powered recruiting automation software to take advantage of the majority of AI-enabled functions (Khanh et al., 2021).

Most AI-enabled features may be accessed using AI-powered recruiting automation software. Every firm values time, and the recruiting sector is no different. Many options for measuring candidates' talents are provided by AI. AI-based software just takes a few seconds to analyze large volumes of data and produce clear conclusions that decision-makers can evaluate. AI solutions not only save time but also money and resources.

4.2   LEVERAGING CANDIDATE SOURCING: EMBRACING THE POWER OF AI

The deliberate search for potential candidates to fill existing and future employment opportunities is known as candidate sourcing. The process of looking for, identifying, and contacting suitable applicants is referred to as sourcing. The HR procedures of screening, interviewing, and assessing applicants begin after sourcing. Sourcing is an important part of a company's entire recruitment strategy. Recruiters that engage in candidate sourcing benefit from greatly reduced time to fill openings. Recruiters may perform early assessments of their talents by proactively looking for and finding possible applicants, ensuring that only the most qualified individuals advance in the recruiting process.

Furthermore, sourcing enables recruiters to create talent pipelines, which are databases of qualified individuals. These pipelines serve as a great resource for future talent requirements as well as present employment demands. Recruiters may position themselves to accomplish long-term recruitment goals, such as attracting and employing a broad pool of applicants, by constantly developing connections with potential prospects. The deliberate application of AI in candidate sourcing may boost these efforts by providing recruiters with intelligent tools to more effectively and efficiently find, analyze, and interact with top talent (Hmoud & Laszlo, 2019; Ore & Sposato, 2022). Candidate sourcing software allows you to easily find, screen, and arrange interviews with applicants. Modern applicant sourcing software allows you to publish to several job boards with a single click, develop a personalized career page in minutes, easily filter prospects, and much more.

Implementing an Applicant Tracking System (ATS) improves the applicant sourcing process in a variety of ways. Recruiters may use an ATS to handle all of their sourcing channels, such as job boards, career sites, referrals, and social media, from a single platform. This single hub enables recruiters to publish job vacancies on major job sites such as Indeed, Adzuna, LinkedIn, and ZipRecruiter, making it easier to access a larger candidate pool. Applicants from multiple sourcing channels are automatically routed into the ATS, reducing the need for manual data entry and ensuring that all applicant information is centralized in a single system. This simplifies the application management process by allowing recruiters to assess, evaluate, and engage with candidates from a single spot.

All talks with applicants are logged and organized under a separate tab, allowing co-sources to easily reference and collaborate (Peicheva, 2022). An ATS is a comprehensive solution for candidate sourcing that streamlines sourcing channels, consolidates candidate information, encourages cooperation, enables talent pooling, simplifies social media sourcing, and provides vital statistics. Within the ATS platform, recruiters may interact with their team members, see applicant profiles, and write comments for each other. They may also save the profiles of exceptional individuals they meet during the sourcing process and invite them to apply for relevant opportunities. They can also submit jobs on social media networks and follow their status within the ATS, making it easier to find social talent. Finally, they may pull data such as time to fill, cost per hire, and pipeline throughput from the system to provide reports. Recruiters may optimize their efforts by utilizing the benefits of an ATS (Koteswari, 2021; Curiel et al., 2021).

AI plays a transformational and important role in talent sourcing. AI technologies have the potential to revolutionize the way recruiters locate and assess potential employees, and make the sourcing process far more effective and productive. AI-powered algorithms are capable of analyzing massive volumes of data from a variety of sources, including resumes, social media profiles, professional networks, and job boards. AI systems can rapidly and reliably extract essential information from various sources by utilizing natural language processing (NLP) and machine learning approaches, saving recruiters significant time and effort.

In candidate sourcing, AI can help to improve diversity and inclusion efforts using tools as shown in Figure 4.2. Traditional sourcing approaches are sometimes plagued by unconscious prejudice, resulting in homogenous candidate pools. AI may assist in finding talented applicants from under-represented groups when trained on diverse and inclusive data sets, creating a more diverse workforce (Uyarra, 2020).

- **Entelo**: Entelo is a candidate sourcing and engagement platform driven by AI. It analyzes large volumes of data from multiple sources, such as social media, professional networks, and public profiles, using machine learning algorithms. Entelo can find and attract top talent by offering recruiters' insights into applicant behavior, interests, and credentials by applying AI.
- **HiringSolved**: HiringSolved is a talent-sourcing platform driven by AI that uses NLP and machine learning algorithms. Analyzing and interpreting the content of job descriptions and applicant resumes assists recruiters in searching for and matching candidate profiles with job requirements. HiringSolved uses AI to assist recruiters in more successfully locating the best-fit candidates.
- **Textio**: Textio is anAI software that optimizes the wording used in job descriptions. It employs machine learning and predictive analytics to give

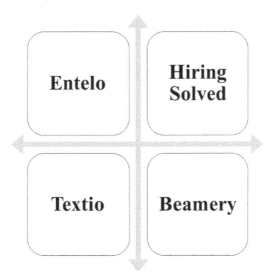

FIGURE 4.2 Some different AI-enabled software that are used for candidate sourcing are Entelo, HiringSolved, Textio, and Beamery.

real-time feedback and ideas for enhancing job posting efficacy. Textio's AI algorithms analyze language patterns to uncover possible biases or hurdles in job descriptions, assisting recruiters in creating more inclusive and attractive job adverts.

- **Beamery**: Beamery is a platform for AI-powered talent engagement and relationship management. To expedite the sourcing and engagement process, it integrates candidate sourcing, candidate relationship management (CRM), and analytics. The AI capabilities of Beamery enable recruiters to aggressively find, engage, and develop connections with applicants, resulting in a robust talent pipeline.

The application of AI to assist identify, screening, and engaging personnel is known as AI sourcing. AI sourcing technologies scour the internet for possible employees based on criteria like job title, skills, keywords, and geography to discover the best fit for your available openings. Once the pool of applicants has been narrowed down, AI technologies may often assist you in reaching out to possible prospects in a somewhat automated manner. Manually looking for potential prospects is time-consuming and sometimes leads to recruiters only being able to contact a limited number of people.

In contrast, AI sourcing allows you to swiftly and simply discover a wide pool of competent applicants, giving you a significantly greater chance of finding the appropriate individual for the position (Heymans, 2022). As the global talent pool gets more competitive and individuals are more easily connected to one another, AI sourcing is becoming more popular in the recruitment market. This has resulted in an increase in unqualified people seeking employment, causing job boards to become increasingly crowded and job postings to become less effective. AI-powered sourcing may assist by scouring the internet for data on possible candidates and providing it straight to recruiters, allowing recruiters to search more specifically and contact passive prospects that they might not have discovered otherwise (Khang & Misra et al., 2023).

AI pre-screening can be done more successfully, resulting in a smaller and more accurate pool of prospects for the recruiter to analyze. As the global talent pool gets more competitive and individuals become more easily linked to one another, recruiters must develop more effective ways to locate quality applicants. Many large-scale organizations, like TCS, Deloitte, IBM, Accenture, and Tech Mahindra, have already begun to integrate AI into their recruiting processes. The level of technology integration is just in its early stages. The majority of participants feel that technology will never be fully integrated. The primary areas where AI is presently being used in the recruiting process are applicant sourcing, screening, selection, and candidate experience (Mehrotra & Khanna, 2022). The field of recruiting is being transformed by AI (Khang & Hajimahmud et al., 2023).

As a result, it's no surprise that 43% of HR professionals are currently utilizing it in their hiring procedures. The reason for this quick acceptance is that organizations are beginning to recognize AI's potential and what it can do to boost their recruiting efforts. The worldwide AI market is predicted to exceed $1.8 billion by 2030, indicating that this shift is here to stay. AI-powered recruiting solutions can assist in mitigating this issue by automating the entire application process and making it effective and entertaining for prospects. You can drastically enhance the applicant experience and attract more quality candidates by employing AI to make the recruiting process more interesting and personalized (Agouridis, 2023).

4.3 SMART RECRUITMENT: ENHANCED PREDICTIVE ANALYTICS THROUGH AI

The use of data to forecast future trends and occurrences is known as predictive analytics. It forecasts future situations based on historical data, which may aid in strategic decision-making. Businesses are constantly seeking methods to optimize their recruiting process and swiftly and efficiently discover the appropriate candidate for a position in today's frantic and oversaturated labor market. One of the most exciting technologies to emerge in recent years is predictive analytics backed by AI. This technology has the potential to revolutionize the hiring process by supporting organizations in selecting individuals who are most inclined to flourish in a certain position.

Predictive analytics in recruitment comprises using AI algorithms to analyze massive volumes of data, such as job applications, resumes, and social media profiles, to identify trends that indicate a candidate's chance of success. These patterns may be seen in education, career experience, abilities, and even personality traits. By analyzing these data, predictive analytics may identify the most promising applicants and aid recruiters in making more informed hiring decisions (Thiyagarajan, 2021; Masputra et al., 2023). One of the most important benefits of predictive analytics in recruiting is the ability to save organizations time and resources. Instead of manually evaluating hundreds of resumes and applications, recruiters may swiftly discover the most potential prospects using AI algorithms. This can help organizations fill unfilled positions more rapidly, saving time and money spent on finding the perfect individual.

Predictive analytics can identify applicants who are most likely to succeed in a certain function by analyzing a wide variety of data points, including historical work performance and personality factors. This can assist organizations in avoiding costly recruiting errors and improving their entire recruitment process (Nijjer & Raj, 2020). Predictive analytics in recruitment is the utilization and analysis of historical data to provide future projections to drive future recruiting techniques, hiring decisions, and workforce planning.

By identifying previous trends in data, predictive analytics may provide recruitment and HR managers with insights into anticipated future occurrences. Though there is no such thing as a crystal ball, predictive analytics allows decision-makers to spot trends with high confidence, assisting them in optimizing their actions for better results. Talent analytics teams, in accordance with Gartner, are attempting to utilize them to effect business results rather than relying on capturing metrics from HR systems, preparing reports, and utilizing basic descriptive statistics to recognize trends and spot current talent issues and prospective problems (Davies, 2019).

Employee turnover may cost businesses up to 150% of an employee's annual salary due to the time and money spent on seeking a replacement. According to a study, 75% of the causes of turnover may be avoided. Organizations may utilize predictive analytics to anticipate which workers are most likely to leave or retire, as well as the financial implications. By tracking the workforce elements and circumstances that may cause each probable departure, predictive analytics may aid in targeted retention and/or hiring initiatives. Predictive algorithms attempt to screen applications before they reach the recruiter or hiring manager, proposing only the top candidates (Mishra & Lama, 2016).

Organizations may extract a lot more value from data by incorporating AI into predictive analytics. AI-powered predictive analytics, for example, may be utilized in settings identical to stand-alone predictive analytics (less AI, essentially), but the AI component is what will offer more nuanced, prescriptive outcomes. AI-powered predictive analytics is frequently required when real-time data from different data sources must be analyzed for quick decision-making.

In India, for example, a road-based pilot system is helping to reduce the frequency of automobile accidents by warning drivers of possible hazards in time for them to avoid crashes. To be completely successful, this pilot depends on the AI component. AI algorithms can analyze massive amounts of historical recruiting data, such as candidate profiles, resumes, evaluation findings, and hiring outcomes. AI can find patterns, correlations, and hidden insights in data that human recruiters may overlook by using machine learning techniques. This study aids in the development of prediction models for identifying top candidates and forecasting their prospective performance. By analyzing resumes, application forms, and other applicant data, AI-powered systems may automate the first screening process.

NLP algorithms can extract and match relevant information such as abilities, qualifications, and experience to job criteria. This automated screening method saves recruiters time and effort, allowing professionals to concentrate on higher-value duties (Deeba, 2020). AI-powered predictive analytics can aid in the reduction of unconscious bias throughout the hiring process. AI algorithms can eliminate human biases in applicant evaluation by relying on objective data-driven insights, resulting in fairer and more inclusive recruiting practices. However, it is critical to make sure that the training data employed by AI models are wide-ranging and representative in order to prevent propagating data biases (Beasley, 2021).

AI is anticipated to be a \$309 billion market by 2026, with 44% of CEOs reporting lower operating expenses as a direct outcome of AI implementation. AI technologies may assess applicants' abilities and competencies using a variety of ways, including online examinations, coding challenges, and simulations. By comparing individuals' skill sets to job requirements, AI algorithms may discover any gaps and offer focused training or development activities. AI technologies are transforming the traditional way of assessing candidates' abilities and competencies. These technologies use technology to deliver online examinations, coding challenges, or simulations, allowing recruiters to make objective and data-driven decisions. Online assessments examine particular knowledge and abilities, whereas coding challenges check technical ability. Candidates are put in immersive virtual worlds and assessed in simulations, which imitate real-world circumstances or job-related activities.

AI algorithms may detect any deficiencies and offer targeted training or development activities (Magomadov, 2020). NLP is a technique that allows AI to convert textual input into a computer-readable format. AI hiring solutions that use NLP can analyze job advertisements and CVs and match them based on the skills, expertise, and education necessary for a position. In a survey conducted by Glassdoor, 76% of hiring supervisors say one of their biggest issues is finding the perfect candidate. AI can assist you with this and automate the applicant-sourcing process by searching numerous databases and platforms for suitable individuals who meet the job criteria (Kucic, 2023; Maddumage et al., 2019). AI algorithms can foresee future employment market circumstances by analyzing market patterns, economic statistics, and industry projections. This data may help recruiters assess the availability of talent, identify possible skill shortages, and adjust their recruiting efforts as needed.

Data from previous recruiting procedures may be analyzed by AI to find high-potential individuals who were not chosen for a specific function but may be suited for future ones. Organizations may shorten time-to-hire for future opportunities and tap into a qualified applicant pool fast by keeping the talent pool and utilizing predictive analytics (Romanko & O'Mahony, 2022). AI algorithms are critical in analyzing employee data to detect trends related to attrition or turnover. These algorithms can estimate the possibility of a person leaving the organization by taking into account numerous aspects such as job satisfaction surveys, feedback from staff members, and demographic information. This predictive capability enables businesses to take proactive responses.

Organizations may design successful recruitment strategies, target particular areas of concern, and enhance overall employee happiness by utilizing AI information. For example, if the data reveals poor job satisfaction or repeated input about a specific issue, organizations can make focused efforts to address those characteristics and foster a more positive work environment. Furthermore, demographic data can identify tendencies that lead to attrition in certain groups, allowing organizations to customize retention programs to their specific requirements.

Organizations may manage possible turnover issues and cultivate a more engaged and loyal staff by using AI to analyze employee data and predict attrition (Baek et al., 2023). AI in predictive analytics transforms the CV-screening process by offering an efficient and independent assessment of candidate qualifications. AI can swiftly evaluate a huge number of CVs and extract important information, like abilities, certifications, and experience, using powerful algorithms and NLP techniques. AI reduces subjective biases and provides an equitable and uniform screening process by analyzing the content of CVs against established criteria for a given position. It may also match terms and phrases in CVs to job criteria, allowing recruiters to quickly discover individuals with the necessary qualifications.

Furthermore, AI algorithms evaluate individuals' work history, taking into account characteristics such as job titles and obligations, to establish their fit for a position. Because AI systems are adaptable, they may learn and improve over time depending on input and selections, continually improving the preciseness of CV screening. Overall, AI in predictive analytics improves the CV screening process, saving time, enhancing efficiency, and increasing the possibility of selecting applicants who best fit the job criteria (Lacroux & Martin-Lacroux, 2022).

4.4 AI IMPACT ON LEVERAGING CANDIDATE EXPERIENCE: IMPROVING THE RECRUITMENT PROCESS FOR SUCCESS

The term "candidate experience" refers to a candidate's view of an employer as formed by their interactions throughout the job hunt. A candidate's experience includes the entire recruiting process, from job search to application, interview, and onboarding. In order to further enhance the applicant background, every touchpoint throughout the recruitment process when an applicant comes into contact with a potential company, as well as how they feel about it, must be analyzed. This might be as basic as a phone conversation with a recruiter or as complicated as the user experience on the firm's recruitment page on its website.

Long before applying for a job, the candidate's experience begins. According to a recent IBM study, 48% of job seekers contacted the recruiting organization before applying. Friends or relatives working there, familiarity with the organization's track record or brand, being a customer, or having previously worked with them were all instances of interactions and experiences (Radka, 2021).

Candidate Experience is a live indicator of how your candidates perceive your hiring process and, by extension, your organization. Candidate Experience is important because actively engaging with your applicants and attentively considering their sentiments will help you stay under their radar and make you a more appealing company. In networked society principles, competitive advantage accumulates in organizations that are best equipped to use relevant networks in order to acquire excellent people. Recent advancements, such as the rising dominance of Facebook and LinkedIn, as well as the evolution of the web toward a more omnipresent, customized, linked, and highly personalized environment, have only heightened this opportunity. Good Candidate Experience may build trust and loyalty among your candidates, who will become ambassadors of your firm and eagerly raise your employer's reputation, even if they were denied an employment opportunity (Palenius, 2021; Allden & Harris, 2013).

Positive Candidate Experience is important since it leads to satisfied applicants promoting your firm in their social networks and guarantees that even rejected prospects stay as prospective clients. 77% of applicants who experience a favorable Candidate Experience will tell their networks about it. 65% of rejected applicants who had a good Candidate Experience will attract prospective talent for the organization. After having a favorable Candidate Experience, 38% of candidates are more inclined to accept a job offer.

Poor Candidate Experiences, on the other side, can impact your capacity to recruit and retain top people, as well as your income stream. It will also reduce the number of referrals that current applicants send your way. On average, 41% of applicants will not buy from a firm after a terrible Candidate Experience, and 72% will tell their friends and family about their unfavorable Candidate Experience. If they have a terrible Candidate Experience, 25% of candidates will "actively discourage" others from applying to a firm (Gupta & Mohan, 2019; Doverspike et al., 2019). A turn-down does not have to leave a bitter taste in the mouth of a job seeker. Indeed, how your organization handles rejections may help to improve the entire prospect experience. If job searchers are mistreated, they may be so outraged that they may never apply to your organization again.

Furthermore, because not all organizations reply to applications that do not make the initial cut, dismissing applicants who are not a good fit is required in and of itself. Those who do not receive an acknowledgment from your organization may not only ignore future vacancies, even if they are a good fit the next time, but they may also express their dissatisfaction to friends and family, harming your image. Chatbots and artificially intelligent assistants powered by AI may interact with candidates in real time, delivering personalized solutions to their questions and helping them through the recruiting process. This degree of responsiveness and targeted involvement improve the applicant experience overall (Pereyra, 2021).

To deliver personalized job suggestions, AI systems may analyze candidate profiles, resumes, and job criteria. This assists candidates in discovering relevant employment that matches their abilities and interests, boosting their entire experience by saving time and effort in looking for appropriate roles. Based on established criteria, AI algorithms can rapidly filter and shortlist resumes and applications. This expedites the initial review process, permitting recruiters to respond to prospects more quickly and reducing waiting periods. Candidates gain from faster feedback and a smoother selection procedure (Balasundaram et al., 2022). By offering virtual interview platforms and chatbots, AI has transformed the interview process. These technologies allow applicants to engage in remote interviews at their convenience, avoiding time-consuming scheduling issues.

AI-powered chatbots may interact with candidates in real time, responding to their questions and helping them through the application process. This personalized engagement improves the applicant experience by providing instant support and eliminating the customary waiting time associated with traditional recruitment procedures. AI has enhanced candidate screening and evaluation. AI algorithms can objectively assess a candidate's talents and credentials by analyzing numerous data points such as resumes, work samples, and assessment results. This computerized examination assures impartiality and removes potential human biases in the selection process.

Furthermore, AI-powered systems may provide applicants with feedback, providing vital insights into their talents and potential for growth, and assisting them in developing their professional abilities (Gusain et al., 2023). Global companies like Hirevue, Textio, Mya Systems, Entelo have AI used to find and identify possible candidates from a variety of internet platforms and databases. Its algorithms analyze data to forecast a candidate's chances of changing professions as well as their suitability for certain opportunities, assisting recruiters in identifying and engaging top talent.

For example, Mya, an AI-powered chatbot, connects with candidates during the recruiting process. Mya can answer applicant queries, arrange interviews, and offer updates, making the candidate experience more engaging and efficient (Ore & Sposato, 2022). AI is quickly becoming a standard component of the recruiting process, including initial discovery, sourcing, CV evaluation, and even job offer presentation. Employers may utilize AI to gain a deeper understanding of their existing applicant pools, including their diversity, engagement, and frequency of contact. This may also be used to reach out to new audiences, such as those who aren't actively looking for work but are interested in a certain opportunity. It can also be used to assist applicants.

According to a candidate experience survey, 73% of candidates say the job search process is one of their most stressful situations. Positive applicant experiences may help businesses recruit top talent. Overly difficult applications frequently result in applicants dropping out in the midst. This might also happen throughout the job application process owing to incorrect instructions from the company. As a result, candidates will give this a low rating. According to Zippia, if instructions are unclear, 92% of candidates will assess their application process poorly. This is where AI can make a significant difference by providing an AI-based application process to make the applicant experience interactive, better, and more favorable.

Understand their alternatives, such as the recruiting processes of specific businesses. Allowing applicants to track their applications and observe their own participation and interaction with the recruiting process would help eliminate emotions of doubt and misunderstanding, which frequently contribute to drop-off during the final stages of the hiring process. AI-powered recruiting chatbots use complex algorithms that may be examined to see if the continued conversation with prospects is beneficial. It enables businesses to have meaningful dialogues with potential applicants who are most likely to be hired in the future, hence improving the candidate experience (Black & Van Esch, 2020; Bartram, 2020).

For the company, the key advantage of employing a personalized strategy is to maximize conversions—turning more high-quality prospects into applications. Employers may use AI to put vacant positions in front of people with the correct talents and interests, allowing top prospects to source themselves for openings. The correct AI-driven technology may create smart suggestions for each individual candidate, highlighting possibilities that are a good fit for them based on their talents. Furthermore, the correct technologies enable applicants to inform companies about their content and style of communication. When recruiters are aware of a prospect's preferences, they may more readily deliver customized communications with tailored content and fresh employment possibilities that the candidate is interested in. This enables recruiters to develop and foster connections with applicants, distinguishing your organization from other companies they may be evaluating.

AI ensures that you can continue relevant and personalized contacts with applicants at all phases of the recruiting process by using AI-powered Candidate Portals. You can also keep them interested throughout the candidate process by displaying them additional opportunities that they would be a good fit for when they become available. And, with an understandable AI-driven solution, you can be confident that your staff will remain compliant and that your applicants will be able to govern and adjust their preferences over time (Palenius, 2021; Wilfred, 2018).

4.5 ROLE OF AI IN CANDIDATE ASSESSMENT: UNLOCKING THE POTENTIAL OF THE TECHNOLOGY

New technology is always being developed, and the talent acquisition industry is no exception. AI has been a hot issue for recruiters and hiring managers alike in recent years. AI may free up time for HR workers to focus on higher-level duties like relationship development and strategic goals. AI-based tests can assist an organization in totally automating the recruiting process; AI-driven assessments are frequently coupled with a proctoring module and technologies that automate candidate screening (Van den Broek et al., 2019; Huffcutt et al., 2001).

Candidate evaluation is critical in recruiting since it helps discover the best match for a certain post within an organization. Assessing applicants enables recruiters to completely examine their abilities, credentials, and experience to decide whether or not they fulfill the job criteria. Recruiters may guarantee that they pick individuals who have the essential competencies to perform effectively in the role and contribute to the organization's success by performing evaluations. Effective candidate evaluation includes a variety of procedures such as resume screening, interviews, skill testing, and cultural fit assessments. These assessments give vital information about a candidate's talents, expertise, and prospects for advancement. They let recruiters assess if candidates have the necessary technical skills, industry understanding, and relevant experience. Assessments also aid in the identification of individuals who possess required soft skills such as teamwork, communication, problem-solving, and leadership ability (Farashah et al., 2019; Adler, 2021).

Organizations can reduce the risks associated with bad hiring decisions by completing thorough applicant assessments. Hiring the incorrect applicant can result in diminished production, greater attrition, and increased recruiting expenditures. Assessments give a more complete picture of a candidate's credentials, ensuring that they have the required abilities to thrive in the job and positively contribute to the team and organizational goals (Bennett, 2021). With organizations under heightened performance pressure as a result of digital disruption induced by COVID-19, it has become critical to expedite talent acquisition and leadership succession. Talent evaluations are one of the most significant tools that more and more businesses are using to properly manage this predicament.

Talent evaluation is a screening procedure that assesses your candidates' abilities, aptitude, personality, and viewpoints in relation to the job requirements. It is used to forecast future performance and make sound recruiting decisions. While the notion of talent evaluation as a part of the organizational hiring process is not new, it is quickly gaining traction. This is mostly because talent evaluations offer a clear indicator of your prospects' skills to do the position for which they are being hired. Talent evaluations may also help you forecast whether or not a candidate will fit in with your organization's team and culture (Galli, 2023). AI plays an important role in candidate assessment in recruiting. The following are some of the important implications of AI in this broader context in the below-mentioned in Figure 4.3.

- **Time and Effort Savings**: AI can automate several elements of applicant evaluation, including resume screening and early candidate shortlisting. Recruiters can handle a huge volume of applications quickly and efficiently by employing AI algorithms, saving substantial time and effort compared to human screening.
- **Bias-Free Evaluation**: In the recruitment process, human prejudice is an inherent obstacle. By objectively evaluating individuals based on preset criteria and qualifications, AI-based evaluation systems can help eliminate prejudice. This decreases the possibility of unjust judgments based on criteria such as gender, color, or other irrelevant characteristics.

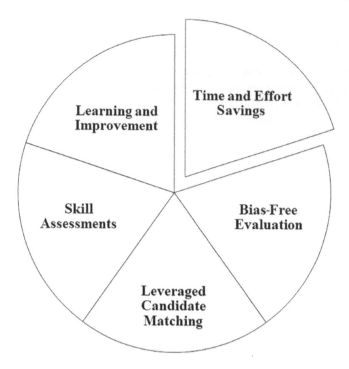

FIGURE 4.3   Importance of AI in candidate assessment.

- **Leveraged Candidate Matching**: Machine learning algorithms are used by AI-powered recruiting platforms to analyze job specifications and match them with candidate profiles. AI can deliver more accurate and personalized applicant suggestions by taking into account aspects like skills, knowledge, schooling, and other appropriate information, resulting in better matches among candidates and job opportunities.
- **Skill Assessments**: Through automated examinations, simulations, or coding challenges, AI may be used to assess candidates' technical or job-specific skills. This gives a standardized and scalable method of assessing candidates' talents, assisting recruiters in identifying the most qualified candidates for the post.
- **Learning and Improvements**: AI systems can improve their efficacy and precision over time by learning from feedback and human interactions. AI algorithms may enhance their evaluation techniques by analyzing the results of previous assessments and adding comments from recruiters, resulting in more effective applicant assessments in the future.

AI-powered applicant evaluation improves the recruiting process's objectivity and uniformity. AI can objectively assess candidates based on their credentials, talents, and experience by using established criteria and algorithms. This lowers the possibility of subjective bias and assures a fair and uniform evaluation for all candidates. It contributes to the development of a standardized evaluation procedure that is used uniformly by all candidates, independent of personal preferences or unconscious prejudices (Tambe et al., 2019; Raghavan et al., 2020).

AI systems can extract important insights from unstructured data using machine learning and NLP techniques. Text from resumes, internet profiles, and social media is analyzed to find relevant abilities, experiences, and certifications. AI algorithms can also detect trends and correlations in applicant data, allowing recruiters to make data-driven judgments and find individuals with specific characteristics that match the needs of the organization. AI-powered evaluations can give useful predictive insights to aid decision-making.

AI systems can discover patterns and trends in successful candidates by analyzing past recruiting data. 50% of organizations want to invest more in chatbot assistants rather than mobile apps. Because chatbots are more like intelligent software programs, they may be evaluated from multiple perspectives such as usability and human variables, and they can focus on particular measurements such as customer satisfaction, job completion, time management, and so on. It is merely an attempt to provide a more human-like contact, after which the discussion is analyzed by AI (Thakuria, 2022).

4.6 GLOBAL COMPANIES THAT ARE LEVERAGING AI IN THEIR RECRUITMENT PROCESS

In today's labor market, using AI technologies in the hiring process is important for firms to remain competitive. AI might help attract top talent by speeding and personalizing the application process for candidates, increasing the employee experience through skill matching and job alerts, freeing up recruiters' time to focus on other important tasks, and more. The following are the top global businesses that use AI in their recruitment process (Blehar, 2023):

- **Magellan Health**: Magellan Health, a healthcare organization with over 10,000 workers, used an AI recruitment platform, which resulted in substantial increases in applicant engagement and conversion rates. Personalized job suggestions, easier navigation on the career site, a conversational chatbot, and a Customer relationship management (CRM) that links recruiters to excellent prospects more effectively are all benefits of AI technology in recruiting. AI has expedited the job application process and provided prospects with a more personalized experience. It has made it possible for applicants to rapidly grasp how their talents and expertise fit with the organization, increasing their engagement and interest. Magellan Health has boosted its brand and prevented possible damage caused by a poor applicant experience by providing an extraordinary candidate experience. At Magellan Health, AI technologies have proven beneficial in recruitment, resulting in more apply clicks and completed applications. A conversational chatbot was extremely effective, generating 14,500 click-to-apply leads in 6 months. Furthermore, the average site visit duration has more than doubled, and 81% of applicants have given the site 4 or 5 stars, suggesting excellent satisfaction with the candidate encounter.

- **Kuehne+Nagel**: Kuehne+Nagel, a worldwide logistics giant, transformed their internal recruiting approach by using an intelligent talent marketplace powered by AI. They were able to use this platform to promote available opportunities to workers, personalize job suggestions, and offer recruiters with a smart tool for identifying great talent. Within 2.5 months, they increased the conversion rate for internal applicants by 22% and cut the time it took to complete internal requisitions by 20%. Furthermore, with the better experience offered by AI, they earned a high employee satisfaction percentage of 74%. This indicates that using AI technology in recruiting not only helps the organization by optimizing internal talent management, but also improves the employee experience by making employees feel appreciated and empowered in their career growth.
- **Brother International Corporation**: Brother International Corporation, a renowned manufacturer of home office and business equipment, used AI in their recruiting approach to boost their employer brand and attract top talent. To do this, they developed a career site that showed their real employer brand while also providing prospects with a personalized experience. They also put in place a CRM system to identify passive prospects and gather quality leads. They saw a 140% increase in completed applications, a 45% rise in total page visits, a 40% increase in job seekers, and a 15% increase in returning job searchers within 3 weeks. Furthermore, the time it takes to fill vacancies has been reduced by 25%. This technique not only attracted more individuals but also improved the likelihood that they would be interested in future possibilities.
- **Stanford Health Care**: Stanford Health Care's use of an artificially intelligent chatbot has transformed the applicant experience and proved the value of AI in recruiting. By leading candidates through a series of questions and recommending opportunities that fit their talents and expertise, the chatbot generates suitable employment matches. It also makes the application process easier by allowing candidates to do it at their leisure. The chatbot's interaction with the CRM system streamlines communication, allowing for efficient and personalized communication. With a quarter-million conversations, 35,000 new visits to the website, and over 11,000 applicant leads and 12,000 apply clicks, the impact of the chatbot adoption has been tremendous. The data demonstrate AI's success in engaging prospects and attracting competent professionals.
- **Mercy:** Mercy, which operates 40 acute care, managed care, and specialty hospitals, recognized the critical need for skilled nurses in the midst of a nursing shortage. Mercy included an AI recruiting platform and other AI-powered technologies in their plan to solve these difficulties. An intelligent employment portal featuring a conversational chatbot for lead collection, automatic personalized job searches, interview scheduling, and FAQs (frequently asked questions) were among the options available. They also used an AI-powered CRM to discover and interact with active and passive prospects, as well as events and university recruitment tools to secure early talent, and Short Message Service (SMS) campaigns and quick-apply features to ensure efficient yet personalized communication. Mercy saw a 10% boost in nurse

hiring and a 14% rise in overall hires after implementing AI-powered recruitment technologies. Furthermore, recruiter productivity improved, as did the number of candidates, but the time required to apply and organize interviews was reduced. Mercy's achievement highlights the importance of AI in overcoming recruiting hurdles and optimizing talent acquisition tactics.

4.7   CONCLUSION

The introduction of AI has transformed the talent recruiting process, allowing organizations to build a smarter workforce. This study report examined four major components that show the important impact of AI in applicant sourcing, candidate assessment, and candidate experience, all of which contribute to improving the recruiting process. AI is sweeping over the recruiting sector, and this sophisticated intelligence is rapidly expanding as it assists businesses in doing previously laborious, time-consuming, and costly tasks (Khang & Shah et al., 2023).

AI-powered applicant sourcing solutions have transformed the way businesses find and recruit qualified individuals. AI-powered assessment systems may analyze and evaluate applicant resumes, application answers, and video interviews, revealing important information about a prospect's abilities, credentials, and cultural fit. AI has also enhanced the candidate experience throughout the recruiting process, offering real-time replies to applicant inquiries, personalized communication, and coaching at various phases of the trip. Candidates get a favorable perception of the employer brand as a result of this. This study report highlighted numerous multinational organizations that have effectively utilized AI in their recruiting procedures (Khang & Kali et al., 2023).

Companies like Mercy, Stanford Healthcare, and Magellan Health have used AI technology to improve their talent acquisition activities, resulting in increased efficiency, higher hiring quality, and overall productivity. These success stories demonstrate AI's transformational impact on recruiting. By using the abilities it has in candidate sourcing, evaluation, and candidate experience, AI has brought about a new era of talent recruiting, producing a smarter workforce. As technology advances, organizations must appropriately adopt AI-driven solutions, using their capacity to complement human talents and make well-informed choices that promote organizational success (Khang & Muthmainnah et al., 2023).

4.8   ACKNOWLEDGMENT

I thank Atmika Patnaik, Senior Professor Ipseeta Satpathy, Professor Dr. Alex Khang[PH], and Professor B.C.M. Patnaik for their valuable contribution to this study.

REFERENCES

Abdalla Hamza, P., Jabbar Othman, B., Gardi, B., Sorguli, S., Mahmood Aziz, H., Ali Ahmed, S. & Anwar, G. (2021). Recruitment and selection: the relationship between recruitment and selection with organizational performance. *International Journal of Engineering, Business And Management (IJEBM)*, 5, 1–13. https://papers.ssrn.com/sol3/papers.cfm?abstract_id=3851335

Adler, L. (2021). *Hire with Your Head: Using Performance-based Hiring to Build Outstanding Diverse Teams*. John Wiley & Sons, Hoboken, NJ. https://www.google.com/books?hl=en&lr=&id=StYNEAAAQBAJ&oi=fnd&pg=PA1&dq=Using+Performance-based+Hiring+to+Build+Outstanding+Diverse+Teams&ots=cfK2IE8dQc&sig=QOw7i8FexCY5N8K2ZAPsdWTQ-uw

Agouridis, A. (2023). *How AI is Transforming the World of Recruitment*. Jobylon, Stockholm. https://www.jobylon.com/blog/how-ai-is-transforming-the-world-of-recruitment

Allden, N. & Harris, L. (2013). Building a positive candidate experience: Towards a networked model of e-recruitment. *Journal of Business Strategy*, 34(5), 36–47. https://www.emerald.com/insight/content/doi/10.1108/JBS-11-2012-0072/full/html

Baek, S., Marques, S., Casey, K., Testerman, M., McGill, F., & Emberson, L. (2023). Attrition rate in infant fNIRS research: A meta-analysis. *Infancy*, 28(3), 507–531. https://onlinelibrary.wiley.com/doi/abs/10.1111/infa.12521

Balasundaram, S., Venkatagiri, S., & Sathiyaseelan, A. Using AI to enhance candidate experience in high volume hiring: a conceptual review and case study. https://papers.ssrn.com/sol3/papers.cfm?abstract_id=3521915

Bartram, D. (2000). Internet recruitment and selection: kissing frogs to find princes. *International journal of selection and assessment*, 8(4), 261–274. https://onlinelibrary.wiley.com/doi/abs/10.1111/1468-2389.00155

Beasley, K. (2021). Council post: unlocking the power of predictive analytics with AI. Forbes. https://www.forbes.com/sites/forbestechcouncil/2021/08/11/unlocking-the-power-of-predictive-analytics-with-ai/?sh=462c9aa76b2a

Bennett, M. M. (2021). The impact of providing realistic job previews for teacher candidates during the hiring process: a qualitative multiple case study (Doctoral dissertation, Northcentral University). https://search.proquest.com/openview/e1d74e4349807df116f19e50d1e1fad8/1?pq-origsite=gscholar&cbl=18750&diss=y

Black, J. S. & Van Esch, P. (2020). AI-enabled recruiting: What is it and how should a manager use it?. *Business Horizons*, 63(2), 215–226. https://www.sciencedirect.com/science/article/pii/S0007681319301612

Blehar, M. (2023). 6 companies successfully using AI in their recruiting strategies. 6 Examples of Companies Successfully Using an AI Recruiting Platform. https://www.phenom.com/blog/examples-companies-using-ai-recruiting-platform

Curiel, A., Gutiérrez-Soto, C., & Rojano-Cáceres, J. R. (2021). An online multi-source summarization algorithm for text readability in topic-based search. *Computer Speech & Language*, 66, 101143. https://www.sciencedirect.com/science/article/pii/S0885230820300760

Davies, R. (2019). How predictive analytics can improve hiring. Top Business Software Resources for Buyers -2022 | Software Advice. https://www.softwareadvice.com/resources/predictive-analytics-recruitment-hiring/

Deeba, F. (2020). The role of data analytics in talent acquisition and retention with special reference to SMEs in India: a conceptual study. *IUP Journal of Knowledge Management*, 18(1), 7–23. https://search.ebscohost.com/login.aspx?direct=true&profile=ehost&scope=site&authtype=crawler&jrnl=09729216&AN=142016763&h=rJlz8%2BjCPKTUXuiff1N3pn%2FjpSzKo3gQquSgWwzdrEBeHvfmAjzYgWVexlAVTAF2wH1WvOu7s1hZy1eZqq6wCg%3D%3D&crl=c

Doverspike, D., Flores, C., & VanderLeest, J. (2019). Lifespan perspectives on personnel selection and recruitment. *Work Across the Lifespan*. (pp. 343–368). Academic Press, Cambridge, MA. https://www.sciencedirect.com/science/article/pii/B9780128127568000141

Farashah, A. D., Thomas, J., & Blomquist, T. (2019). Exploring the value of project management certification in selection and recruiting. *International Journal of Project Management*, 37(1), 14–26. https://www.sciencedirect.com/science/article/pii/S0263786318301327

Galli, E. (2023). The relevance of talent assessment in recruiting. Starred. https://www. starred.com/blog/talent-assessment-relevance#:~:text=Testing%20various%20 job%2Drelevant%20skills,which%20helps%20your%20Candidate%20Experience

Geetha, R. & Bhanu, S. R. D. (2018). Recruitment through artificial intelligence: a conceptual study. *International Journal of Mechanical Engineering and Technology*, 9(7), 63–70.

Gupta, J. & Mohan, D. (2019). Candidate experience in recruitment cycle facilitating employer brand: a case study of idea cellular limited in the Delhi and NCR circle. *International Journal of Human Resources Development and Management*, 19(1), 37–46.

Gusain, A., Singh, T., Pandey, S., Pachourui, V., Singh, R., & Kumar, A. (2023). E-recruitment using artificial intelligence as preventive measures. *2023 International Conference on Sustainable Computing and Data Communication Systems (ICSCDS)*, pp. 516–522. IEEE, New York.

Heymans, Y. (2022). AI sourcing: the ultimate data driven recruitment tools. RSS. https://www.hero-hunt.ai/blog/ai-sourcing-data-driven-recruitment-tools#:~:text=AI%2Dpowered%20 sourcing%20can%20help,may%20not%20have%20otherwise%20found.

Hmoud, B. & Laszlo, V. (2019). Will artificial intelligence take over human resources recruitment and selection. *Network Intelligence Studies*, 7(13), 21–30.

Huffcutt, A. I., Conway, J. M., Roth, P. L., & Stone, N. J. (2001). Identification and meta-analytic assessment of psychological constructs measured in employment interviews. *Journal of applied psychology*, 86(5), 897.

Johansson, J. & Herranen, S. (2019). The application of artificial intelligence (AI) in human resource management: current state of AI and its impact on the traditional recruitment process. Bachelor Thesis.

Kapse, A. S., Patil, V. S., & Patil, N. V. (2012). E-recruitment. *International Journal of Engineering and Advanced Technology*, 1(4), 82–86.

Khang, A., Hajimahmud, V. A., Gupta, S. K., Babasaheb, J., & Morris, G. (2023). *AI-Centric Modelling and Analytics: Concepts, Designs, Technologies, and Applications* (1 ed.). CRC Press, Boca Raton, FL. https://doi.org/10.1201/9781003400110

Khang, A., Kali, C. R., Suresh Kumar, S., Amaresh, K., Sudhansu Ranjan, D., & Manas Ranjan, P. (2023). Enabling the future of manufacturing: integration of robotics and IoT to smart factory infrastructure in industry 4.0. *AI-Based Technologies and Applications in the Era of the Metaverse* (1st ed., pp. 25–50). IGI Global Press, Hershey, PA. https://doi.org/10.4018/978-1-6684-8851-5.ch002

Khang, A., Misra, A., Gupta, S. K., & Shah, V. (2023). *AI-aided IoT Technologies and Applications in the Smart Business and Production*. CRC Press, Boca Raton, FL. https://doi.org/10.1201/9781003392224

Khang, A., Muthmainnah, M., Seraj, P. M. I., Yakin, A. A., Obaid, A. J., & Panda, M. R. (2023). AI-aided teaching model for the education 5.0 ecosystem. *AI-Based Technologies and Applications in the Era of the Metaverse* (1 ed., pp. 83–104). IGI Global Press, Hershey, PA. https://doi.org/10.4018/978-1-6684-8851-5.ch004

Khang, A., Shah, V., & Rani, S. (2023). *AI-Based Technologies and Applications in the Era of the Metaverse* (1 ed.). IGI Global Press, Hershey, PA. https://doi.org/10.4018/978-1-6684-8851-5

Khanh, H. H. & Khang, A. (2021). The role of artificial intelligence in blockchain applications. Reinventing Manufacturing and Business Processes through Artificial Intelligence, 2, 20-40. https://doi.org/10.1201/9781003145011-2

Kot, S., Hussain, H. I., Bilan, S., Haseeb, M., & Mihardjo, L. W. (2021). The role of artificial intelligence recruitment and quality to explain the phenomenon of employer reputation. *Journal of Business Economics and Management*, 22(4), 867–883.

Koteswari, B. (2021). Use of Applicant Tracking System (ATS) in Talent Acquisition Padmaja P. https://jems.net.in/wp-content/uploads/2021/08/Use-of-Applicant-Tracking-System-ATS-in-Talent-Acquisition.pdf

Kucic, L. (2023). The role of Artificial Intelligence in the hiring process. HROS Community. https://community.hros.io/ai-recruitement-maximising-artifical-intelligence-in-hiring/

Lacroux, A. & Martin-Lacroux, C. (2022). Should I trust the artificial intelligence to recruit? Recruiters' perceptions and behavior when faced with algorithm-based recommendation systems during resume screening. *Frontiers in Psychology*, 13, 895997. https://www.frontiersin.org/articles/10.3389/fpsyg.2022.895997/full

Maddumage, C., Senevirathne, D., Gayashan, I., Shehan, T., & Sumathipala, S. (2019). Intelligent recruitment system. *2019 IEEE 5th International Conference for Convergence in Technology (I2CT)*, pp. 1–6. IEEE, New York.

Magomadov, V. S. (2020). The application of artificial intelligence and big data analytics in personalized learning. *Journal of Physics: Conference Series* (vol. 1691, no. 1, pp. 012169). IOP Publishing, Bristol. https://iopscience.iop.org/article/10.1088/1742-6596/1691/1/012169/meta

Masputra, H., Nilasari, B. M., & Nisfiannoor, M. (2023). The role of big data predictive analytics as a mediator of the influence of recruitment and selection, remuneration and rewards, training, and development on employee retention. *Return: Study of Management, Economic and Bussines*, 2(04), 353–366. https://journals.sagepub.com/doi/abs/10.1177/09726225211066220

Mehrotra, S. & Khanna, A. (2022). Recruitment through AI in selected Indian companies. *Metamorphosis*, 21(1), 31–39. https://journals.sagepub.com/doi/abs/10.1177/09726225211066220

Mishra, S. N. & Lama, D. R. (2016). A decision making model for human resource management in organizations using data mining and predictive analytics. *International Journal of Computer Science and Information Security (IJCSIS)*, 14(5), 217–221. https://www.academia.edu/download/46568117/28_Paper_30041662_IJCSIS_Camera_Ready_Paper_pp._217-221.pdf

Mitchel. (2021). 7 Top benefits of AI in recruiting. Harver. https://harver.com/blog/benefits-ai-in-recruiting/

Nijjer, S. & Raj, S. (2020). Predictive Analytics in Human Resource Management: A Hands-On Approach. Taylor & Francis, New York

Nikolaou, I. (2021). What is the role of technology in recruitment and selection?. *The Spanish journal of psychology*, 24, e2. https://www.cambridge.org/core/journals/spanish-journal-of-psychology/article/what-is-the-role-of-technology-in-recruitment-and-selection/451DF5C763B110A845EEAD50F5BAF851

Ore, O. & Sposato, M. (2022). Opportunities and risks of artificial intelligence in recruitment and selection. *International Journal of Organizational Analysis*, 30(6), 1771–1782. https://www.emerald.com/insight/content/doi/10.1108/IJOA-07-2020-2291/full/html

Oswal, N., Khaleeli, M., & Alarmoti, A. (2020). Recruitment in the era of industry 4.0: use of artificial intelligence in recruitment and its impact. *PalArch's Journal of Archaeology of Egypt/Egyptology*, 17(8), 39–47. https://mail.palarch.nl/index.php/jae/article/download/1113/1165

Palenius, L. (2021). The Importance of Candidate Experience as a Part of the Recruitment Process.

Peicheva, M. (2022). Data analysis from the applicant tracking system. *HR and Technologies, Creative Space Association*, 2, 6–15. https://journal.cspace-ngo.com/arb/Issues/2022/2/1-Peicheva.pdf

Pereyra, C. (2021). 6 reasons candidate experience matters. CareerArc social recruiting. https://www.careerarc.com/blog/why-candidate-experience-matters/

Radka, O. (2021). Automatická lokalizace hdEEG elektrod (Bachelor's thesis, České vysoké učení technické v Praze. Vypočetní a informační centrum.). https://dspace.cvut.cz/handle/10467/94537

Raghavan, M., Barocas, S., Kleinberg, J., & Levy, K. (2020). Mitigating bias in algorithmic hiring: evaluating claims and practices. *Proceedings of the 2020 Conference on Fairness, Accountability, and Transparency*, pp. 469–481. Association for Computing Machinery, New York. https://dl.acm.org/doi/abs/10.1145/3351095.3372828

Rana, G., Khang, A., Sharma, R., Goel, A. K., & Dubey, A. K. (2021). Reinventing manufacturing and business processes through artificial intelligence. CRC Press, Boca Raton, FL. https://doi.org/10.1201/9781003145011

Romanko, O. & O'Mahony, M. (2022). The use of online job sites for measuring skills and labour market trends: a review. Economic Statistics Centre of Excellence (ESCoE) Technical Reports (ESCOE-TR-19). https://escoe-website.s3.amazonaws.com/wp-content/uploads/2022/05/30133155/TR-19.pdf

Sills, M. (2014). E-recruitment: a comparison with traditional recruitment and the influences of social media: a qualitative and quantitative review.

Tambe, P., Cappelli, P., & Yakubovich, V. (2019). Artificial intelligence in human resources management: challenges and a path forward. *California Management Review*, 61(4), 15–42. https://journals.sagepub.com/doi/abs/10.1177/0008125619867910

Thakuria, D. (2022). Importance of AI-based chatbot assessment when hiring: Turbohire. Resources. https://turbohire.co/resources/blog/importance-of-chatbot-assessment-when-hiring/

Thiyagarajan, R. (2021). The role of recruitment analytics and metrics in targeted recruitment post pandemic. *Proceedings of the First International Conference on Combinatorial and Optimization, ICCAP 2021*, Chennai, India. https://eudl.eu/doi/10.4108/eai.7-12-2021.2314764

Trivella, C. (2023). The role of artificial intelligence in the hiring process. Peoplehum. https://www.peoplehum.com/blog/the-role-of-artificial-intelligence-in-the-hiring-process#:~:text=Once%20your%20AI%20program%20sources,feedback%20and%20suggest%20next%20steps

Uyarra, E., Zabala-Iturriagagoitia, J. M., Flanagan, K., & Magro, E. (2020). Public procurement, innovation and industrial policy: Rationales, roles, capabilities and implementation. *Research Policy*, 49(1), 103844. https://www.sciencedirect.com/science/article/pii/S0048733319301635

Van den Broek, E., Sergeeva, A., & Huysman, M. (2019). Hiring algorithms: an ethnography of fairness in practice. https://core.ac.uk/download/pdf/301385085.pdf

Wilfred, D. (2018). AI in recruitment. *NHRD Network Journal*, 11(2), 15–18. https://journals.sagepub.com/doi/pdf/10.1177/0974173920180204

5 Challenges of Communication with Gen-Z in the Era of Artificial Intelligence-Interceded Digital Economy

Zakia Tasmin Rahman, Ruhi Lal, and Ravinder Rena

5.1 INTRODUCTION

When it comes to climate change, Gen-Z is the most vocal and proactive generation yet (Ariestya et al., 2022). They are often described as the "climate change generation" because they are coming of age at a time when the effects of climate change are already being felt by communities around the world (Antoniades, 2023). In addition to speaking out about the need for action, Gen-Z is also taking concrete steps to reduce its carbon footprint and promote sustainable living (BANGKOK & S. C. I., 2020). For example, many members of Gen-Z are choosing to live car-free life-styles and are advocates for renewable energy sources (Bogueva & Marinova, 2022). Mental health is another important issue for Gen-Z (Pichler et al., 2021). Due to the high demands of modern life, this generation is more prone to anxiety and stress than any other before it (Kannan & Kumar, 2022). In response, members of Gen-Z are working to destigmatize mental health issues and break down barriers to seeking help (Schroth, 2019). They are also using technology to their advantage by seeking out online resources and support groups (Rue, 2018).

Economic concerns are also top of mind for many Gen-Zers. This generation is coming of age in a time of great economic uncertainty, with rising college costs and limited job prospects (Schroth, 2019). In response, members of Gen-Z are advocating for affordable education and workplace policies that promote work-life balance (Della Volpe, 2022). They are also using their entrepreneurial spirit to start their own businesses and create their own job opportunities (Mahapatra et al., 2022). As the newest entrants to the workforce, Gen-Z is in the spotlight—and the news isn't all good (Brower, 2023). There is an emphasis on their challenges, their stress, their

DOI: 10.1201/9781003440901-5

worry, and their lack of engagement (Baldonado, 2018). But despite the sobering statistics, there are bright spots for Gen-Z and so much to be hopeful about for this youngest generation (Maloni et al., 2019). Employers, family, friends, and Gen-Zs themselves can influence their experiences and create a positive future (Karr, 2023).

5.2 LITERATURE REVIEW

The focus of the research study is on Gen-Z's requirements in the workplace which will help in the growth and development of the organizations and the Gen-Z employees as well. The process of Communication with Gen-Z should be the way that will help them to understand better and accordingly contribute. The various traits of Gen-Z in the workplace are discussed as given by various available literature (Khang & Shah et al., 2023).

5.2.1 PREFERENCE FOR TRADITIONAL COMMUNICATION

Even though Generation Z grew up with texting and instant messages, studies show that they prefer to speak face-to-face in the workplace (Bredbenner, 2020). This could be because they find the nuances of written communication difficult to interpret and would rather have the reassurance that comes with personal interaction (Raslie, 2021). It is commonly known that 98% of Generation Z owns a Smartphone and that they spend 10 hours per day online on average. Instant messaging is glorified among this generation but when it comes to the work environment, statistics show that 72% of Gen-Z prefers in-person communication with their boss and colleagues. In fact, 40% of Gen-Z employees expect daily feedback from their boss on their performance (Nguyen, 2023).

Lack of constant interaction with the higher-ups might make them feel something is wrong, which will negatively impact the quality of their work (Pichler et al., 2021). Another reason why Gen-Zers prefer in-person communication in the workplace is that they want to be taken seriously by Baby Boomers and Millennials alike (Goldring & Azab, 2021). Instant messaging has made Gen-Z communication extremely informal with abbreviations like lol, brb, ttyl, and their frequent use of emojis to convey emotions (Rue, 2018). They fear that if they use such with Baby Boomers, it might be seen as childish (Dabija & Lung, 2019).

5.2.2 DESIRE TO WORK INDIVIDUALLY

Team environments are not a problem for Gen-Z, but many young employees prefer to work on individual projects as much as possible. By working independently, Gen-Zers can showcase their skills and abilities to prove themselves to employers (Patel, 2017). Gen-Zers' independence ties into their competitiveness, but they generally like to work alone (Szymkowiak et al., 2021). Many of them prefer to have office space to themselves, rather than an open, collaborative workspace (Mahapatra et al., 2022). Many also want to manage their own projects so that their skills and abilities can shine through (Kim et al., 2022). They do not want to depend on other people to get their work done (Mărginean, 2021).

5.2.3 MOBILE-FIRST HABITS

Generation Z is used to smartphones and relies heavily on productivity apps in the workplace (Freer, 2019). Additionally, the advent of voice command technology has transformed the smartphone into a necessary work tool for Generation Z. Employers who are aware of this should use apps that work best on mobile devices (Witt & Baird, 2018).

Most Gen-Z Smartphone users (64%) say they are constantly connected online with 57% admitting that they feel insecure without their mobile phone (Hansen, 2023). According to a recent Snapchat analysis that analyses user behaviors and highlights important distinctions between Gen-Z and Millennials, this is the case. Gen-Z spends an average of 4 hours and 15 minutes per day on the mobile with 95% of them owning a smartphone (Karr, 2023). 78% of them consider their mobile devices their most important device to go online compared to 74% of Millennials (Brower, 2023). However, multi-device usage tends to be common among all generations and marketers and brands should adopt strategies that consider various screens equally (Fromm & Read, 2018).

5.2.4 MOTIVATED BY STABILITY

Gen-Z is less risk-averse than Millennials because they were raised during a period of severe economic hardship. Thus, they value the stability that comes from having a predictable job with a clearly defined compensation package (Critical, 2016). Gen-Z does have it harder than previous generations—even if not on par with a violent civil war (Schrager, 2022). Their college and high school lives were rocked by the COVID pandemic and now, just as they're trying to get their careers going, they're entering an uncertain economy where a recession is likely (Kurt, 2022). Both experiences can have a lasting impact. Graduating during a recession can mean lower earnings for decades (Bewicke, 2023). A COVID-dominated college term robbed them of important socialization and the chance to learn how to manage their time and interact with adults (Schlott, 2023).

5.2.5 NATURALLY COMPETITIVE

Generation Z is used to competition and enjoys the challenge of putting themselves to the test against someone else (Scion Advisory Services, 2023). If you can encourage a healthy sense of competition in your workplace, particularly during the training stage, you can keep young employees motivated and help them do their best work (Henderson, 2023). Gen-Z Brits have been revealed as the most money-focused generation, according to a new survey about the British public's competitive habits and behaviors (Belmonte, 2018). The research, commissioned by Paddy Power Games, asked respondents which scenarios would bring out their competitive side with friends, with over half of Gen-Z's (totally 52%) having money-hungry motives as shown in Table 5.1 (Shaw, 2022).

The above table states the differences between Millennials and Gen-Z specifically in the workplace. It shows that comparatively Gen-Z is more aggressive than the Millennials. The thinking process of both generational cohorts is different.

TABLE 5.1
Gen-Z and Millennials in the Workplace

S/No.	Millennials	Gen-Z
1.	Desire to work toward a goal	Driven by financial gain and job security
2.	More interested in ongoing discussions than yearly evaluations	Require regular performance feedback
3.	Interested in working in teams and collaboration	Driven by competition and personal performance
4.	Consider their career to be one of their most essential priorities	Give a good work-life balance a top priority

Source: https://www.betterteam.com/5-traits-of-gen-z-in-the-workplace.

5.3   OBJECTIVES OF THE RESEARCH STUDY

To learn the challenges of Gen-Z
To study the career cluster of Gen-Z
To know the pathway models for Gen-Z Skills
To find out solutions for the challenges of Gen-Z
To find the relationship of Gen-Z with Gen-X and Millennials

5.4   RESEARCH METHODOLOGY

- The research study is qualitative and explorative.
- Primary data and Secondary data used.
- Sample Units: Youngsters of Gen-Z, Academicians, Recruiters, Corporate and Government Employees
- Sample Size: 100
- Gen-Z -50
- Aca-10
- Sample Area: Delhi/NCR
- Secondary data: Newspapers, magazines, journals, websites, research papers, government reports, records, surveys
- Primary data: Youngsters of Gen-Z, Academicians, Recruiters, Corporate and Government Employees
- Questionnaire, Interview method (focus group interview) (5 were conducted)
- After data collection, coding and tabulation methods were used to analyze the findings.

5.5   FINDINGS AND ANALYSIS

5.5.1   THE CHALLENGES OF GEN-Z

Today's youth are prepared to quit their jobs if they do not get the work-life balance, fair pay, and value alignment they desire as shown in Table 5.2.

TABLE 5.2

Challenges of Gen-Z

S/No.	Challenges of Gen-Z
1.	Younger people today are prepared to quit their professions if they do not get all they desire, including work-life balance, fair pay, and value alignment
2.	The top priority for this cohort of workers is higher pay
3.	Inclusive company culture
4.	Collaboration and autonomy
5.	Short appreciation process
6.	Millennials entering the workplace valued career progression and personal development
7.	Gen-Z faces particularly acute stressors, especially as rising inflation outpaces salary growth
8.	Willing to sacrifice corporate social responsibility for companies they admired as consumers
9.	Gen-Zers spend less time in a role than millennials
10.	Reflecting and Reevaluating
11.	Fostering Well-being
12.	Expanding Engagement
13.	Connecting with Colleagues
14.	Developing Financial Acumen
15.	Regular reassessment and reevaluation

Source: Authors' own compilation.

According to a Cigna survey, many members of Generation Z are concerned about the future. In fact, 65% of workers say they've spent more time than they did 2 years ago reviewing their priorities in life, while 71% of Gen-Z members say the same. The research on stress and health is conflicting. On the one hand, according to the Cigna study, 91% of Gen-Zers stated they felt stressed out and 98% claimed they were burned out. According to Stress in America research, 62% of women and 51% of males between the ages of 18 and 34 reported feeling completely stressed out. On the other side, a Gympass study found that 59% of Gen-Zers felt better about their well-being in 2022. In addition, Gen-Zers are giving greater importance to their job experiences, with 78% of them admitting that their compensation is just second to their workplace well-being.

People are now working more remotely and in hybrid arrangements, which has reduced engagement. Only 32% of adults claim to be actively engaged, and 17% actively disengage, according to Gallup. Additionally, the Cigna study found that Gen-Z workers claim to be present but not fully engaged for 25% of the time they are at work. Having a sense of community is essential to well-being, regardless of age or one's tendency toward introversion or extroversion. Unfortunately, a lot of people experience loneliness and a lack of friends; this is especially true for younger people. It's depressing to see how professional friendships are currently faring. Money is a problem for Gen-Zers in addition to other things. According to the Cigna study, 39% of respondents cited financial concerns as a major source of stress (Khang & Rani et al., 2023).

5.5.2   THE CAREER CLUSTER OF GEN-Z

Even if Gen-Z could have had technology when they started their careers that older generations did not, understanding generational disparities is helpful for guiding your personnel strategy and workforce planning. Knowing what they want in a work is a key characteristic that sets Gen-Z apart as Table 5.3.

Most of the Gen-Zers are interested in the same conventional occupations as generations before them, notwithstanding the popularity of social media and the creative economy. Young people nowadays are more inclined to move professions, but they are also pursuing stability by choosing careers as CEOs, doctors, and engineers, according to a recent Axios/Generation Lab survey. Businesspersons, doctors, and engineers were named as the top three job ambitions of Gen-Z respondents. In addition, Gen-Zers tend to favor larger corporations; 58% of them said they would choose a job with a large or mid-sized company over one with a startup or the government. Only 14% of people wanted to work in government.

Personal fulfillment (49%) is the primary driver of job advancement for young people, followed by income (25%) and success. Only 9% of respondents claimed societal effect is their top motivator for working, in contrast to research that suggests Gen-Z cares more about their companies' opinions on social issues than prior generations. The "quiet quitting" trend is also being driven by Gen-Z, with 82% of them believing it's attractive to do the bare minimum to maintain employment. For them, achieving personal fulfillment and happiness is best accomplished in this manner. Instead of seeing their career to an end, they view it as a means.

A LinkedIn survey found that Gen-Z wants to climb the corporate ladder, and smart businesses are giving them the rungs to do so. In 2021, reskilling and upskilling will be more crucial than ever, according to 59% of learning & development experts. And Generation Z is quite content with this, feeling right at home (often literally) in their efforts to pick up new abilities. Despite being recent entrants into the workforce—the oldest Gen-Zer alive today is about 24. To develop in their careers,

---

TABLE 5.3

Career Clusters of Gen-Z

S/No.	Career Cluster of Gen-Z
1.	Businessperson
2.	Doctor
3.	Engineer
4.	Artist
5.	Coder/IT Tech
6.	Financer/Banker
7.	Nurse
8.	Lawyer
9.	Author
10.	Others

Source: Authors' own compilation.

Gen-Z is highly motivated, and 76% of them believe that learning is the key. This need for information is giving people a big competitive advantage in a world that is changing more quickly than ever.

5.5.3 THE PATHWAY MODELS FOR GEN-Z SKILLS

To communicate with Gen-Z it is necessary to prepare certain strategies which will help in better understanding.

5.5.3.1 Gen-Z's Advantage

The Gen-Z cohort stands out as one that is enthusiastically embracing continuous learning at work among the report's varied findings. Future employees are those who develop new skills. Gen-Z is prepared to grow quickly and advance both their skills and careers by keeping one eye on career advancement and the other on their online course. The concept of lifelong learning was instilled in this generation of digital natives at a young age, and it is still the best way to win their hearts, minds, and cooperation at work.

The generation of persons born between 1995 and 2012 is referred to as "Generation Z." They are already enrolled in schools and colleges and range in age from 10 to 27. Five observations regarding Gen-Z were made, including what they are interested in studying, how they study, what degrees they hope to earn, what worries them about higher education, and how they prefer to communicate. Even though it's crucial to comprehend each of these factors, using the proper communication techniques is of utmost importance as Table 5.4.

Specialized, personalized communication that takes the individual into account is the secret to marketing to and connecting with Generation Z. Despite being the most digitally savvy and connected generation to date, Gen-Z values personalized methods even more. The key finding is that this generation is just now entering the world of higher education. It is necessary to make sure that the strategies of marketing communication should be agile, customizable, and ready for the momentum of Generation Z. In the above table, the respondents agreed to the various communication strategies that can be applied to communicate with Gen-Z for education, recruitment, brand promotion, and various other purposes.

5.5.3.2 Solutions to the Challenges of Gen-Z

There is arguably a more influential change: the rise of a new generation that plans to make its mark in the workplace. By 2025, Gen-Z, or youngsters born between 1997 and 2012, will account for 27% of the labor force and 30% of the global population. Because of this, the nature of work as we know it might shift even further as firms compete to find and keep the best personnel as Table 5.5.

In the above table, the respondents highly agree with the solutions provided for the challenges of Gen-Z. Gen-Zs are already having a positive impact on the workforce and business. Gen-Z employees are important because they are the future of companies. But it is necessary to understand them and meet their expectations, which differ from other generations. And provide them with purposeful employee experiences, development, and future careers. By doing these, employers can attract some amazing talent.

TABLE 5.4
Pathway Models for Gen-Z Skills

S/No.	Various Strategies for Communication with Gen-Z	Responses in Percent
1.	Gen-Z is used not only to the availability of information but also for the immediacy of information	98%
2.	Gen-Z communicates with images	96%
3.	Gen-Z uses numerous screens to multitask	97%
4.	The decreasing length of Gen-Z's attention spans also explains why they prefer visual material to written content	95%
5.	They are the ideal audience for snack media	92%
6.	Communication needs to be in bite sizes	95%
7.	Longer sections or wordy headlines do not have the same impact as shorter, more concise language	90%
8.	More time is involved in digital interaction than in offline interactions	85%
9.	Smartphones truly becoming an extension of self and the use of platforms and apps in it	98%
10.	One-on-one communication is the most effective way to reach Gen-Z	87%
11.	The digital method of communication is preferred to be the best	80%
12.	More comprehensive expression of self through selfies is the order of the day	90%
13.	Influencer marketing is highly influential and preferable	95%
14.	The multiculturalism of Gen-Z requires consideration	88%
15.	Gen-Z feels the significance of Gen-X and Gen-Y's support	79%
16.	Gen-Z presents themselves as aspirational	98%
17.	Generation Z shares genuine memories on Snapchat	97%
18.	Gen-Z learns the news on Twitter	98%
19.	Facebook is where Gen-Z gathers information	98%

Source: Authors' own compilation.

5.5.4 CAREER EXPECTATIONS OF GEN-Z

For Gen-Z, diversity and inclusion are important. Additionally, they expect their employer to be ethically and transparently strict and to care about the environment Instead of greenwashing, they want sincere dedication to reversing climate change and safeguarding the environment. For employment and shopping, they seek out businesses that are sustainable. This could drive a shift in companies' mindsets and strategies.

Each generation has its own unique style, needs, goals, and traits. So, it's vital to create an environment in which all employees feel engaged, inspired, and energized and can thrive, work together, and learn from each other. That is why some organizations offer first-work experiences programs such as apprenticeships and graduate

TABLE 5.5

The Solutions to the Challenges of Gen-Z

S/No.	Solutions to the Challenges of Gen-Z	Responses in Percent
1.	Providing purposeful employee experiences, development, and future careers	100%
2.	Diversity and inclusion in the workplace	100%
3.	Ethics and transparency of employers	
4.	Real commitment from employers	85%
5.	Sustainable companies could drive a shift in companies' mindsets and strategies	90%
6.	All employees should feel engaged, inspired, and energized	87%
7.	Organizations to offer first work experience programs	93%
8.	Changes in the work environment which will help to give the best by the employees	80%
9.	Taking care of the mental health and wellness of the employees will help in the growth and development of the organizations by Gen-Z	95%
10.	Brands becoming more purpose- and value-driven due to Gen-Z	97%
11.	A future that has equal opportunities for all	85%
12.	Companies should directly get in touch with Gen-Z for their feedback	91%

Source: Authors' own compilation.

opportunities. In return, the Gen-Zs bring new skills, innovative ideas, and diverse perspectives.

From the inclusion of women in the workforce during World War II to the way Millennials raised awareness of topics like mental health, each generation affects the workplace. But as Generation Z reaches adulthood, the human species is facing an unprecedented threat from the climatic catastrophe and the aftermath of a historic epidemic. It is likely that what they want and won't tolerate at work will have an impact for a long time.

In the current labor market, it's worthwhile to explore what the next generation of employees wants from their employers (Francis, 2022). Otherwise, employers risk losing out on the competitive talent market (Brower, 2023). Gen-Z employees stay at their jobs for 2 years on average, and more than a third intend to stay at their jobs for more than 4 years. So, if a company gets it right, the youngest employees will show up and contribute (Banov, 2022). It is necessary to take the insights above into account when assessing current and aspirational workplace culture and values, as successful recruiting and retention will depend on it (Perna, 2021). Yet, it won't be enough to cultivate an attractive culture that appeals to Gen-Z (Luttrell & McGrath, 2021). Tools that enable the youngest generation to perform at their best, such as employee engagement software and mentoring software, will increase a company's chances of recruiting and retention success (Patel, 2017).

5.5.5 THE RELATIONSHIP OF GEN-Z WITH GEN-X AND MILLENNIALS

When asked Gen-Z whether they need support from Gen-X and Millennials to Succeed in Life, 89% of the respondents agreed that they needed help and support from these two generational cohorts as Table 5.6.

However, 11% of the respondents feel that their success was dependent on their own hard work and they do not have to depend on anyone as in Table 5.7.

5.5.5.1 Gen-Z Differs from Previous Generations

While everyone seeks a rewarding job, respectable salary, and favorable working conditions, there are three essential traits that differentiate Gen-Z from their fore-bears: Baby Boomers, Generation X, and Millennials. Being born in the digital era, they are the first generation to have no memory of existence prior to the internet. A paper map, a floppy disc, a CD player, an analog camera, or analog film have never been used by them.

TABLE 5.6
Whether Gen-Z Seeks Support from Gen-X and Millennials to Succeed in Life

S/No.	Responses of Gen-Z	Responses in Percent
1.	Yes	89%
2.	No	11%

TABLE 5.7
Major Details of Gen-X, Gen-Y, and Gen-Z

S/No.	Generations	Time Period	Technological Developments	Values of Various Generations
1.	Gen-X	Generation of Americans born from 1965 to 1980	Use of e-mail and telephone to communicate. Use of Smartphones Access to Apps Social Media Internet	Work-life balance, technology products, and experiential travel
2.	Millennials or Gen-Y	Born from 1980 to 1994	Grown up with social media and Smartphones. Broadcast technology usage	Personalization, authenticity, socially conscious branding, and digital experiences.
3.	Gen-Z, iGen or Centennials	Born from 1995 to 2012	Use Handheld communication accessories and devices. On messaging Apps	Convenience, accessibility, and technology

Source: Authors' own compilation.

Gen-Z has been dubbed the "first global generation" in a culture where one-click purchases of goods from any country are commonplace and where access to world-wide content and information is generally more readily available. They might use this to pioneer mobile jobs and create fresh online revenue streams. One might assume that Gen-Z has evolved into a practical, risk-averse, non-entrepreneurial group driven by pay and job security given their upbringing in the wake of the recession brought on by the global financial crisis in 2008, as well as the economic fallout from the COVID-19 pandemic and Russia's war on Ukraine. Not always the case, though.

Even though pay is the most important factor to consider when choosing a vocation, Gen-Z values it less than any other generation. When given the option to choose between a job that paid more but was dull and one that paid less but was more fascinating, Gen-Z was divided equally. A survey of American undergraduates found that between the ages of 18 and 34, Gen-Z may switch occupations up to ten times. Something much more ad hoc and adaptable may take the place of the antiquated idea of a career ladder ascending from the mailroom to the executive suite. While looking for work, 42% of Gen-Z employees emphasize work-life balance, remote work, and flexible leave, according to a recent U.S. survey.

Additionally, this generation gives jobs that let them advance their abilities, talents, and experience a higher priority. Therefore, employers will need to change how they locate, nurture, and retain talent while promoting personal development. And that might help any future generation. In fact, their exposure to two significant global crises in a decade may have strengthened their resilience and prepared them for future difficulties like climate change.

5.5.5.2   Gen-Z Seeks Support from Gen-X and Gen-Y

Gen-Z is diverse, therefore institutions in the US and abroad should be ready to welcome and assist incoming first-generation students who might be less familiar than their peers with the demands of higher education if they want to draw in US students. Support is an essential kind of communication for Generation Z in general. Institutions can benefit from the fact that Gen-Z students in the United States have less life experience than their Millennial siblings and Gen-X parents when they showcase campus resources that provide support and assistance.

Although most companies are now quite familiar with Millennials, a sizable chunk of the workforce will soon be made up of Generation Z. Employers may build a culture that plays to this generation's strengths and makes their business an appealing alternative for job seekers by understanding this generation's characteristics and preferences.

5.5.5.3   Conceptual Framework

The Gen-Z generation, usually real digital natives, those born between 1995 and 2010, are people who have had access to the internet, social media, and mobile devices since they were young. A hypercognitive generation has been formed by this environment that excels in cross-referencing information from various sources and fusing it with offline and online experiences. As the world becomes more linked, generational shifts may begin to influence behavior more so than socioeconomic distinctions as Figure 5.1.

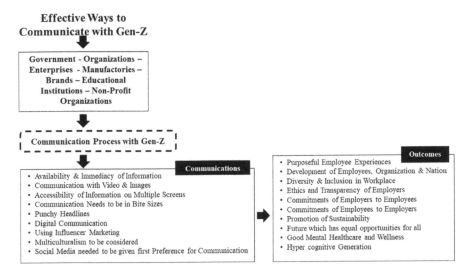

FIGURE 5.1 Effective ways of communication with Gen-Z leading to fruitful outcomes (Khang, 2021).

Generation Z is pro-individual expression and against categorization. For many reasons, they organize among themselves. They firmly believe that conversation may help resolve disputes and advance society. Finally, people consider their choices and interact with institutions in sensible and useful ways. This generation accepts several identities. Its quest for authenticity encourages greater expression freedom and a stronger desire to understand all types of individuals. As an extremely diverse generation, Gen-Z, does not distinguish between buddies they make online and real-life friends. They make use of the extensive mobility that technology provides to move between organizations that support their objectives. Online networks are treasured by Gen-Zers because they enable people from all economic backgrounds to interact and come together around issues and interests.

Young people have historically been the best representatives of a society's zeitgeist since they have a big influence on both trends and behaviors. The first generation of real digital natives, Gen-Z, is now having an external influence, with the pursuit of truth driving much of its distinctive behavior and consumption patterns. Thanks to technology, the connections between young people and the rest of society are stronger than ever today. As a result, demographic shifts are becoming more prominent and technology trends are speeding. Businesses will face challenges as well as profitable opportunities as a result of this transition. And remember that the first step in seizing chances is to be open to them.

If the recruiters, employers, Gen-X, and Gen-Y try to fulfill the needs of Gen-Z, Gen-Z's contribution to the respective organizations and nations will help to flourish which can lead to enhanced growth and development. Theoretical Applicability of the Law of Demand and Supply of Economics in the Job Market where Recruiters and Gen-Z Job seekers are involved as in Figure 5.2.

Demand and Supply of Gen-Z Employees with Various Skills

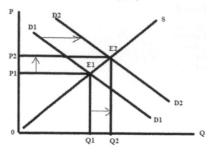

There are many
interesting jobs
and new hires in
local and global
labor market

Recruiters and
Employers
providing
Facilities and
Higher Salary

Gen-Z with Various Kinds of Skills

FIGURE 5.2   Supply and demand curve of job market for Gen-Z.

The above figure describes the law of demand and law of supply in terms of Gen-Z employees with various kinds of skills like digital skills, time management, leadership skills, multitasking, giving extra time to the job, creative skills, presentation skills, communication skills, and crisis management. D1 is the demand curve of Gen-Z job seekers by the recruiters. S is the supply curve of the Gen-Z job seekers. The point of intersection of D1 and S is E1 which is the point of equilibrium where remuneration, perks, allowances, incentives, etc. of employees are decided by the recruiters. P1 is the price, that is, remuneration and incentives that are given in the first instance. Q1 is the quantity, that is, employment of Gen-Z. When there is an increase in demand for employees with skill development the supply also increases with demand. P2 is the increase in remuneration with the increase in Gen-Z employees with skill development. The increase in remuneration in the above figure is stated by P1 and P2. The increase in employment of Gen-Z employees with skill development is from Q1 to Q2.

The new equilibrium point is the point of intersection of D2 and S which is E2 at the point higher than E1 the first point of equilibrium where D1 intersects S. From the above figure, it can be stated that the law of demand and supply of economics applies to the job market where the recruiters seek for Gen-Z employees with skill development and Gen-Z employees, on the other hand, try to qualify themselves in various kind of skills development as per the requirement in the job market (Snehal et al., 2023).

5.6   CONCLUSION

Gen-Zs are wise to reflect on what matters most to them (Raslie, 2021). Happiness is significantly correlated with feeling a sense of purpose and meaning, so when Gen-Zs are encouraged to think about what they love to do, what they want to learn, and the difference they want to make, it can be tremendously helpful for their sense of meaning and identity (Luttrell & McGrath, 2021). Regular reassessment and reevaluation are important so Gen-Zs can take action to create the conditions for a great experience—spending time on what means the most to them in work and life (Mahesh et al., 2021).

Stress is a significant experience that can have broad impacts—so it's a good idea for Gen-Zs to be aware of their experience and get help—and others can support Gen-Zs in doing this as well (Schroth, 2019). At the same time, it's healthy for Gen-Zs to remind themselves some stress is natural (Casey, 2021). Happiness is not defined by constant contentment, rather there are ebbs and flows in positive feelings and some stress can be constructive (Online.sbu.edu, 2022).

Far from the narrative that all work is bad and doing as little as possible is the ideal work is an important source of esteem and meaning (Shaw, 2022). It's the place where people express their talents, contribute to the community, and learn new things (Mărginean, 2021). And all of these are correlated with happiness and joy (Nguyen, 2023). Gen-Zs can seek work that aligns with what they like to do and remind themselves about how their work contributes to their colleagues, their team, and their organization (Dabija & Lung, 2019).

In turn, employers can support Gen-Zs in creating the conditions for meaningful work by ensuring access to inspirational leaders, giving meaningful feedback, and providing growth opportunities (Bredbenner, 2020). All of these encourage greater engagement and a more positive experience of work (Goldring & Azab, 2021). Gen-Z can prioritize friendship and invest time in inviting people for coffee, meeting up, and hanging out with colleagues and people outside of work as well (Pichler et al., 2021). Others can support Gen-Z by reaching out and creating relationships with them, offering a listening ear, advice, and companionship (Mahapatra et al., 2022).

With the reduction of superficial interactions (think ordering coffee on an app or getting products delivered rather than shopping for them in person), work takes on an increasingly important role in relating with others (Karr, 2023). It is the context for making friends over time, seeing people in both task-focused and relationship-focused settings, and in getting to know people through good days and down days (Banov, 2022). Employers can create cultures where team members value each other, where people can collaborate on projects, and in which affinity groups can thrive—all of these contribute to Gen-Z's positive relationships and connections (Hughes, 2022).

Gen-Zs can actively seek to develop their financial acumen—by taking classes or learning from mentors about investment strategies and lessons learned about managing money (Henderson, 2023). Employers can support Gen-Zs by providing fair wages and meaningful benefits which are thoughtfully designed with nudges that encourage smart saving and investing for the future (Cigna 360 Global Well-Being Survey, 2022). Far from automatic, smart money management is more often learned than inborn and the knowledge younger workers need is frequently something they need from employers because it wasn't a topic of study in school (Harter, 2022).

The "law of Demand and Supply in Economics" applies to the current context. The law of demand and supply can be applied in the Gen-Z's job market very appropriately in terms of various skills development. For instance, the demand for skilled, competent, and qualified employees is always there in the job market. The Gen-Z job seekers are highly qualified but they lack soft skills, interpersonal communication, are not empathetic, lacking communication skills, cannot express thoughts clearly, and lack other requisite skills, etc (Babasaheb et al., 2023).

In such a situation, there will be an increase in the rate of educated unemployment. Educated people will have to work for lower remuneration, with no perks, allowances, qualifications, or designations will not match, and no increments, or promotions despite adequate qualifications and serving for several years in the same organization. On the other hand, if all the employees have all the qualities in that case also, supply will be more than demand, and the recruiters will offer low remuneration with less or no perks, allowances, incentives, etc. Henceforth, there should be a proper balance between the supply of Gen-Z job seekers and the demand of recruiters for Gen-Z. The point of equilibrium of both curves is the point of mutual consent of the Gen-Z seekers and the recruiters where the remuneration, perks, allowances, incentives, etc. is decided (Babasaheb et al., 2023).

5.7   A BRIGHT FUTURE

There are good reasons for optimism in considering the future of Gen-Zs. They've survived a global pandemic and arguably some of the biggest historical struggles, but they can learn, develop their resilience, and bond with others through hard times (Cigna 360 Global Well-Being Survey, 2022). Life hasn't been easy for them, and there will be challenges ahead, but with their own spirit and support from others, they can create a bright tomorrow (United Way NCA, 2022).

5.8   LIMITATIONS

The limitations are written based on the responses of the respondents and also on the basis of personal observations.

- At times it becomes difficult for the employers or the recruiters of organizations and enterprises to fulfill the demands of the Gen-Zers.
- With the fluctuations of the economic situation of a country it is not possible to fulfill all the expectations of the Gen-Zers.
- It has been observed that for petty reasons the employees or students of the Gen-Z cohort quit their organizations and later when they find out the flaws of organizations it becomes too late for them to make a comeback to their previous organizations.
- The Gen-Zers at times become highly self-centric and do not think about the organization's or nations' well-being and development.

Many Gen-Z people like to make independent decisions in organizational matters without consulting their seniors and feel proud of themselves (Khang & Muthmainnah et al., 2023).

The following recommendations are given based on the responses of the respondents and personal observations and experiences.

- It is necessary for Gen-X and Gen-Y to understand the needs of Gen-Z.
- As seniors to Gen-Z, both Gen-X and Gen-Y should provide transparency, and better opportunities in the workplace to make the Gen-Z people happy which will help the Gen-Z showcase their talents and help in developing the organizations in every aspect.

- Higher salaries and an acceptable work environment will help to retain the Gen-Z workforce in the organizations.
- Sustainable development of the organization and creating a positive environment in the workplace will highly motivate the Gen-Z employees to in remain the same organization for a longer time and become highly productive.

Gen-Z workforce can bring higher profit and can highly contribute to the wholesome development of an organization. Hence, it is necessary for the senior generations as recruiters and seniors to understand Gen-Z and their needs better and provide them opportunities for their self-growth and development which will motivate them to work for the future growth and development of the respective organizations the Gen-Z employees work with (Khang & Kali et al., 2023).

REFERENCES

2022 EY US generation survey reveals impact company culture plays in employee retention (Press Release). (2022). Retrieved from: https://www.ey.com/en_us/news/2022/10/ ey-generation-survey-reveals-impact-company-culture-plays-in-employee-retention

Antoniades, N. (2023). Packaging global warming products: the CIS strategy as a driver of Gen Z's satisfaction and loyalty. *Review of Marketing Science*, 2023, 187–194. https://www. degruyter.com/document/doi/10.1515/roms-2022-0084/html

Ariestya, A., Paramitha, G., & Elmada, M. A. G. (2022). Climate change awareness of Gen Z: the influence of frame and jargon on online news. *Jurnal Studi Komunikasi*, 6(3), 753–770.

Babasaheb, J., Sphurti, B., & Khang, A. (2023). Industry revolution 4.0: workforce competency models and designs. *Designing Workforce Management Systems for Industry 4.0: Data-Centric and AI-Enabled Approaches* (1st ed., pp. 14–31). CRC Press, Boca Raton, FL. https://doi.org/10.1201/9781003357070-2

Babasaheb, J., Sphurti, B., & Khang, A. (2023). Design of competency models in the human capital management system. *Designing Workforce Management Systems for Industry 4.0: Data-Centric and AI-Enabled Approaches* (1st ed., pp. 32–50). CRC Press, Boca Raton, FL. https://doi.org/10.1201/9781003357070-3

Baldonado, A. M. (2018). Leadership and Gen Z: motivating Gen Z workers and their impact to the future. *International Journal of Managerial Studies and Research (IJMSR)*, 6(1), 56–60. https://doi.org/10.1201/9781003440901

BANGKOK, S. C. I. (2020). Independent Study: The Gen Z's Perspective towards Sustainable Clothing in Bangkok (Doctoral dissertation, Thammasat University). https://ethesisarchive.library.tu.ac.th/thesis/2020/TU_2020_6216120151_13592_16610.pdf

Banov, N. (2022). How Gen Z Is Shaping the Future of Business. Retrieved from: https:// moonshotpirates.com/blog/how-gen-z-is-shaping-the-future-of-business

Belmonte, A. (2018). Gen Z Author: We're 'Much More Competitive' than Millennials. Retrieved from: https://finance.yahoo.com/news/gen-z-author-much-competitive-millennials-232220111.html?

Bewicke, H. (2023). Gen Z Consumer Behavior: What You Need to Know. Retrieved from: https://www.talon.one/blog/gen-z-consumer-behavior-what-you-need-to-know

Bogueva, D. & Marinova, D. (2022). Australian generation Z and the Nexus between climate change and alternative proteins. *Animals*, 12(19), 2512. https://www.mdpi.com/ 2076-2615/12/19/2512

Bredbenner, J. (2020). Generation Z: A Study of Its Workplace Communication Behaviors and Future Preferences (Doctoral dissertation, Wichita State University).

Brower, T. (2023). Gen Z Is Struggling: 5 Things They Need For A Bright Future. Retrieved from: https://www.forbes.com/sites/tracybrower/2023/01/15/gen-z-is-struggling-5-thin gs-they-need-for-a-bright-future/?sh=1c570de43410

Casey, A. E. (2021). Generation Z and Mental Health. Retrieved from: https://www.aecf.org/blog/generation-z-and-mental-health

Cigna 360 Global Well-Being Survey. (2022). Exhausted by Work – The Employer opportunity. Retrieved from: https://www.cigna.com.hk/iwov-resources/docs/Cigna-360-Global-Well-being-Survey.PDF

Critical, V. (2016). The Everything Guide to Generation Z. Retrieved from https://www.vision-critical.com/resources/the-everything-guide-to-gen-z.

Dabija, D. C. & Lung, L. (2019). Millennials versus Gen Z: online shopping behavior in an emerging market. *Applied Ethics for Entrepreneurial Success: Recommendations for the Developing World: 2018 Griffiths School of Management Annual Conference (GSMAC) on Business, Entrepreneurship and Ethics 9*, pp. 1–18. Springer International Publishing, New York. https://link.springer.com/chapter/10.1007/978-3-030-17215-2_1

Della Volpe, J. (2022). *Fight: How Gen Z is Channeling Their Fear and Passion to Save America.* St. Martin's Press, New York https://www.google.com/books?hl=en&lr=&id=OggQEAAAQBAJ&oi=fnd&pg=PT138&dq=Fight:+How+Gen+Z+is+channeling+th eir+fear+and+passion+to+save+America.+St.+Martin%27s+Press.&ots=V-0J6wLLb S&sig=3dqcG10yXKGt53cHZzzcnAfaha0

Francis, A. (2022). Gen Z: The Workers Who Want It All. Retrieved from: https://www.bbc.com/worklife/article/20220613-gen-z-the-workers-who-want-it-all

Freer, A. (2019). A look at Gen Z Mobile Behaviors - 64% of Mobile Users are Always Connected. Retrieved from: https://www.businessofapps.com/news/a-look-at-gen-z-mobile-behaviours-64-of-mobile-users-are-always-connected/

Fromm, J. & Read, A. (2018). Marketing to Gen Z: The Rules for Reaching This Vast--and Very Different--Generation of Influencers. Amacom. https://www.google.com/books?hl=en&lr=&id=kVBLDwAAQBAJ&oi=fnd&pg=PP1&dq=Marketing+to+Gen+Z:+The +rules+for+reaching+this+vast--and+very+different--generation+of+influencers.+A macom&ots=fbZCuP1eKk&sig=bPq9aIZpOvzidUGZOF-Z4pDvmuo

Goldring, D. & Azab, C. (2021). New rules of social media shopping: personality differences of US Gen Z versus Gen X market mavens. *Journal of Consumer Behavior*, 20(4), 884–897. https://onlinelibrary.wiley.com/doi/abs/10.1002/cb.1893

Hansen, L. (2023). Gen Z Values and How Companies Can Prepare for the Workplace of the Future. Retrieved from: https://technologyadvice.com/blog/human-resources/generation-z-in-the-workplace.

Harter, J. (2022). U.S. Employee Engagement Slump Continues. Retrieved from: https://www.gallup.com/workplace/391922/employee-engagement-slump-continues.aspx

Henderson, A. (2023). 7 characteristics of Gen Z in 2023. Retrieved from: https://blog.gwi.com/marketing/generation-z-characteristics.

Hughes, J. (2022). Communicating with Generation Z: Everything You Need to Know. Retrieved from: https://www.keg.com/news/communicating-with-generation-z-every thing-you-need-to-know.

Kannan, L. & Kumar, T. P. (2022). Social media-the emotional and mental roller-coaster of Gen Z: an empirical study. *Managing Disruptions in Business: Causes, Conflicts, and Control*, pp. 81–102. Springer International Publishing, New York. https://link.springer.com/chapter/10.1007/978-3-030-79709-6_4

Karr, D. (2023). Generational Marketing: How Each Generation Has Adapted To and Utilizes Technology. Retrieved from: https://martech.zone/generation-technology.

Khang A. (2021). "Material4Studies," Material of Computer Science, Artificial Intelligence, Data Science, IoT, Blockchain, Cloud, Metaverse, Cybersecurity for Studies. Retrieved from https://www.researchgate.net/publication/370156102_Material4Studies

Khang, A., Rani, S., Gujrati, R., Uygun, H., & Gupta, S. K. (2023). *Designing Workforce Management Systems for Industry 4.0: Data-Centric and AI-Enabled Approaches*. CRC Press, Boca Raton, FL. https://doi.org/10.1201/9781003357070

Khang, A., Shah, V., & Rani, S. (2023). *AI-Based Technologies and Applications in the Era of the Metaverse* (1 ed.). IGI Global Press, Hershey, PA. https://doi.org/ 10.4018/978-1-6684-8851-5

Khang, A., Kali, C. R., Suresh Kumar, S., Amaresh, K., Sudhansu Ranjan, D., & Manas Ranjan, P. (2023). Enabling the future of manufacturing: integration of robotics and IoT to smart factory infrastructure in industry 4.0. *AI-Based Technologies and Applications in the Era of the Metaverse* (1st ed., pp. 25–50). IGI Global Press, Hershey, PA. https:// doi.org/10.4018/978-1-6684-8851-5.ch002

Khang, A., Muthmainnah, M., Seraj, P. M. I., Yakin, A. A., Obaid, A. J., & Panda, M. R. (2023). AI-aided teaching model for the education 5.0 ecosystem. *AI-Based Technologies and Applications in the Era of the Metaverse* (1st ed., pp. 83–104). IGI Global Press, Hershey, PA. https://doi.org/10.4018/978-1-6684-8851-5.ch004

Kim, S., Jang, S., Choi, W., Youn, C., & Lee, Y. (2022). Contactless service encounters among Millennials and Generation Z: the effects of Millennials and Gen Z characteristics on technology self-ef ficacy and preference for contactless service. *Journal of Research in Interactive Marketing*, 16(1), 82–100. https://www.emerald.com/insight/content/ doi/10.1108/JRIM-01-2021-0020/full/html

Kurt, D. (2022). How the Financial Crisis Affected Millennials. Retrieved from: https://www. investopedia.com/insights/how-financial-crisis-affected-millennials/

Luttrell, R. & McGrath, K. (2021). *Gen Z: The Superhero Generation*. Rowman & Littlefield Publishers, Lanham, MD. https://www.google.com/books?hl=en&lr=&id=E-cZEAA AQBAJ&oi=fnd&pg=PR7&dq=Gen+Z:+The+superhero+generation.+Rowman+%2 6+Littlefield+Publishers&ots=9TbbkBiuRR&sig=4Zi57nlw65P6f0DH0yllvcE-ONw

Mahapatra, G. P., Bhullar, N., & Gupta, P. (2022). Gen Z: an emerging phenomenon. *NHRD Network Journal*, 15(2), 246–256. https://journals.sagepub.com/doi/abs/ 10.1177/26314541221077137

Mahesh, J., Bhat, A. K., & Suresh, R. (2021). Are Gen Z values the new disruptor for future educational institutions?. *Journal of Higher Education Theory and Practice*, 21(12), 102–123. https://search.proquest.com/openview/fd70e1d34fad5cad8367e8d552de3375/ 1?pq-origsite=gscholar&cbl=766331

Maloni, M., Hiatt, M. S., & Campbell, S. (2019). Understanding the work values of Gen Z business students. *The International Journal of Management Education*, 17(3), 100320. https://www.sciencedirect.com/science/article/pii/S1472811719300126

Mărginean, A. E. (2021). Gen Z perceptions and expectations upon entering the work-force. *European Review of Applied Sociology*, 14(22), 20–30. https://sciendo.com/ article/10.1515/eras-2021-0003

Nguyen, T. V. (2023). 5 Best Ways to Communicate with Gen Z in Workspace. Retrieved from: https://blog.grovehr.com/generation-z-communication-in-the-workplace

Online.sbu.edu. (2022). How Gen Z is Changing the Conversation on Mental Health. Retrieved from: https://online.sbu.edu/news/gen-z-mental-health.

Patel, D. (2017). 8 Ways Generation Z Will Differ From Millennials In The Workplace. Retrieved from: https://www.forbes.com/sites/deeppatel/2017/09/21/8-ways-generation-z-will-differ-from-millennials-in-the-workplace/?sh=37b2624a76e5

Perna, M. C. (2021). Why Skill And Career Advancement Are The Way To Gen-Z's Heart. Retrieved from: https://www.forbes.com/sites/markcperna/2021/03/02/why-skill-and-career-advancement-are-the-way-to-gen-zs-heart/?sh=5ab5705e22b

Pichler, S., Kohli, C., & Granitz, N. (2021). DITTO for Gen Z: a framework for leveraging the uniqueness of the new generation. *Business Horizons*, 64(5), 599–610. https://www. sciencedirect.com/science/article/pii/S0007681321000239

Raslie, H. (2021). Gen Y and gen Z communication style. *Studies of Applied Economics*, 39(1), 22–28. https://ojs.ual.es/ojs/index.php/eea/article/view/4268

Rue, P. (2018). Make way, millennials, here comes Gen Z. *About Campus*, 23(3), 5–12. https://journals.sagepub.com/doi/abs/10.1177/1086482218804251

Schlott, R. (2023). Employers reveal why Gen Z is the hardest generation to work with. Retrieved from: https://nypost.com/2023/04/25/employers-reveal-why-gen-z-is-hardest-generation-to-work-with/

Schrager, A. (2022). Gen Z Really Does Have It Harder Than Millennials. Retrieved from: https://www.washingtonpost.com/business/gen-z-really-does-have-it-harder-than-millennials/2022/11/14/fc833ae8-6413-11ed-b08c-3ce222607059_story.html

Schroth, H. (2019). Are you ready for Gen Z in the workplace?. *California Management Review*, 61(3), 5–18. https://journals.sagepub.com/doi/abs/10.1177/0008125619841006

Scion Advisory Services. (2023). Competitive or Collaborative? Understanding Generation Z Characteristics That May Impact Student Housing. Retrieved from: https://scionadvisory.com/competitive-or-collaborative-understanding-generation-z-characteristics-that-may-impact-student-housing

Snehal, M., Babasaheb, J., & Khang A. (2023). Workforce management system: concepts, definitions, principles, and implementation. *Designing Workforce Management Systems for Industry 4.0: Data-Centric and AI-Enabled Approaches* (1st ed., pp. 1–13). CRC Press, Boca Raton, FL. https://doi.org/10.1201/9781003357070-1

Shaw, N. (2022). Gen Z are Most Competitive Generation - And Most Money-Focused. Retrieved from: https://www.walesonline.co.uk/whats-on/whats-on-news/gen-z-most-competitive-generation-24488988

Szymkowiak, A., Melović, B., Dabić, M., Jeganathan, K., & Kundi, G. S. (2021). Information technology and Gen Z: the role of teachers, the internet, and technology in the education of young people. *Technology in Society*, 65, 101565. https://www.sciencedirect.com/science/article/pii/S0160791X21000403

United Way NCA. (2022). Generation Z Social Issues & Their Impact on Society. Retrieved from: https://unitedwaynca.org/blog/gen-z-social-issues

Witt, G. L. & Baird, D. E. (2018). *The Gen Z frequency: how brands tune in and build credibility*. Kogan Page Publishers, London. https://www.google.com/books?hl=en&lr=&id=cwVrDwAAQBAJ&oi=fnd&pg=PP1&dq=The+Gen+Z+frequency:+How+brands+tune+in+and+build+credibility.+Kogan+Page+Publishers&ots=1k6DZd9KFT&sig=E_6js8VyH-fSKLd6sy-yta-Rjks

6 Modern Work Ecosystem and Career Prospects with Artificial Intelligence Integration

Sugandha Chebolu and Sakthi Srinivasan K.

6.1 INTRODUCTION

Organizations throughout the globe strive for excellence in the field of their respective business domains. Organizational effectiveness is a key to the success of any organization. Organization effectiveness is not only reflected in the financial statements and balance sheets but also in employee job satisfaction and employee effectiveness. Organizations can achieve success by effectively using their employees and ensuring they are satisfied in the job roles offered by the company. Effective employee usage can ensure companies generate greater output in less time. Companies can also have a competitive edge over their competitors one such classic example of employee effectiveness and organizational success is given below.

Apple developed iOS 10, with 600 engineers who developed and deployed the new programming in 2 years. While Microsoft developed and executed the programming on Vista with almost 6,000 engineers, for more than 6 years. The study conducted by the leadership consulting firm Bain & Company revealed that companies like Apple, Google, and Netflix are 40% more productive than the average company. Some may think that this is the result of the hiring talent pool; big companies attract more talent. However, this is not the case, as Google and Apple have found a way to answer the most fundamental dilemma in management: how do you balance productivity and employee satisfaction and commitment? Companies like Google have the same percentage of 'talent pool' as other companies, but instead of spreading out the talent, they group dynamically to achieve more throughout the day. The grouping is focused on the grouping of talented people in the most business-critical roles and is the key to the success of the overall company (OpenStax, 2019). This makes the company's overall productivity go higher. This proves the fact that top-level companies are concerned about the management of their talent pool, as their success depends on them.

In the real-world scenario, most companies are fighting for the talent pool in the markets, and it is difficult to keep the turnover rate low in the fierce competition as competitors come up with higher compensation packages and better incentives. Moreover, Modern-day managers are witnessing changes in technologies, markets, competition, workforce demographics, employee expectations, and ethical standards.

DOI: 10.1201/9781003440901-6

Therefore, it is becoming increasingly difficult to retain talented people in the organization. Most of the organizations are working toward ensuring job satisfaction in the workplace and turning to new methods to retain employees (Khang & Gupta et al., 2023).

6.2 MODERN-DAY WORK ECOSYSTEM

Before understanding the employees' aspirations and attitude toward the work. We must understand how the work itself is changing and evolving, as shown in Figure 6.1. The future of the work depends on the changes in the three deeply interconnected areas of an organization, namely, the work, the workforce, and the workplace (GenZ, 2023).

The work systems are integrated across the world more than ever. The work nowadays requires employees to collaborate with peers throughout the world. It is normal for an employee working in India to collaborate with an employee in the US working in the same organization for the same project. The collaboration tools are increasingly adapted by the organization to make the work process more effective. Cloud computing has also helped increase the mobility of work. Workers can access the required documents from anywhere in the world. This helps to keep the workers engaged and increases the overall productivity of the organization. Moreover, work is increasingly becoming automated with robotic process automation (RPA). The workers must be trained in automation systems to adapt to new work processes. Work hours also have undergone a drastic change; earlier, the work hours were restricted to 32–40 hours a day and confined to 9:00 AM–5:00 PM on working days. Post-pandemic work from home has changed the work hours.

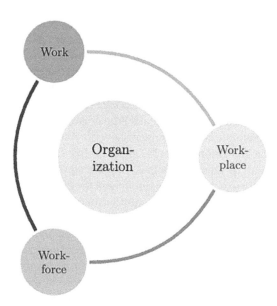

FIGURE 6.1 Relationship between interconnected areas of an organization, namely, the work, the workforce, and the workplace.

Earlier, the workforce of the organizations had different traits and perceptions about the organization. The main traits include loyalty, hard work, etc. There is also a change in attitude toward the organization. These changes may be attributed to three main elements (Bryson & Forth, 2011), namely, the

- Compositional Change
- Economic Cycle
- Underlying Time Trend

The compositional change includes a change in the composition of the workforce. The new workforce in most organizations consists of more Gen Z employees rather than millennial employees (Lee & Wei, 2018), whose outlook toward work is entirely different. The method of working and the expectation of work benefits of Gen Z are entirely different. Gen Z needs flexible working hours and high-growth jobs rather than fixed work hours and high salaries. The loyalty of Gen Z is more influenced by the growth offered by the job rather than the higher packages.

The economic lifecycle also influences the attitude of the employees (workforce). If the economy is facing a recession, the employees are not motivated to work as the organization is not able to incentivize their work, affects productivity. For example, when there are multiple employees laid off in the major tech companies, employees are less motivated to deliver their optimum performance due to several factors.

Third, there is an underlying time trend that can explain the reasons, which are not dealt with in the first two factors, such as societal trends. For example, the activity of businesses was production earlier, but now the contribution of service sectors is more to the economy. Earlier, a job as a product manager was perceived of a higher status in society; now, the tech jobs are considered a status symbol of employees by society. The organization must carefully evaluate these diversities in the work force before drafting any employee policies (Luke et al., 2023).

The workplace has undergone changes from time to time, earlier employees used to wear formals and ties to work, then casual was adopted in most tech offices. The workplace has undergone a drastic change since the pandemic. After the pandemic, the world has experienced many unprecedented situations, like lockdown, etc. Organizations also faced difficulties such as supply chain bottlenecks, a standstill of economic activity, etc. With the advent of digitalization and cloud computing, workplaces are more integrated than ever; employees can work from anywhere and collaborate easily (Subhashini & Khang, 2023).

With the changing composition of the determinants of the forces of work, the organizations must constantly evolve their strategies to stay relevant. It is obvious that most organizations want to retain Gen Z employees. This will not only help in the success of the organization but also help the branding of the organization in the eyes of the employees.

6.3 ARTIFICIAL INTELLIGENCE

Artificial intelligence (AI) is the simulation of human brain activity into machines, enabling them to work on their own without having to preprogram everything. With AI,

inputs

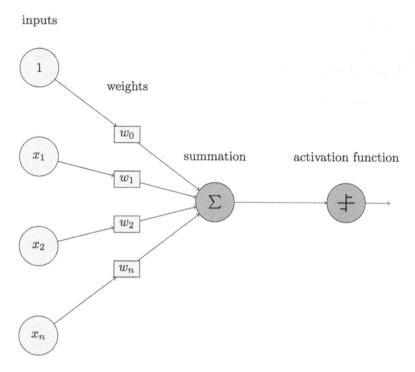

FIGURE 6.2 Diagram of the simulation of human brain activity into machines.

machines become more intelligent and work on their own to solve a set of problems and perform a set of activities. This will help human evolution to new heights with more computational abilities to solve existing complex problems. AI roots can be traced back to the philosophers who theorized human thinking as the mechanical manipulation of symbols. Few scientists have experimented with the thought and produced a digital programmable computer machine based on the abstract idea of mathematical reasoning. This piqued the interest of many researchers. They wanted to develop a machine that was capable of thinking like the human brain. AI was taking shape at that point in time, as shown in Figure 6.2.

The perceptrons were connected in a densely connected fashion to form dense layers, which form neural networks. Neural networks are further applied in the tasks such as classification tasks, prediction tasks, etc. This research was further carried out and spearheaded in various directions different domains of computing consider the application of AI in solving complex problems. There are several classifications of AI. For our understanding, we will consider the following AI classification: AI is divided into 6 sub branches based on the application, as shown in Figure 6.3 below.

- Neural Networks
- Robotics
- Natural Language Processing

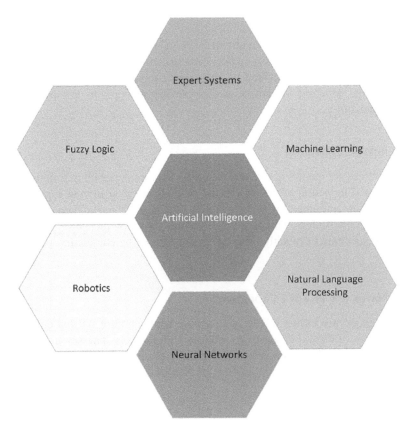

FIGURE 6.3   Six classifications of AI.

- Fuzzy Logic
- Machine Learning
- Expert Systems

Let us look at the branches in detail.

6.3.1   EXPERT SYSTEMS

Information systems are important for the existence of the organizations. There are several kinds of information systems, such as electronic data processing, decision support systems, etc. Expert systems are a special type of decision support system, a complex computer program that is designed to extract information from a knowledge base to support the decision-making process of executives. Expert systems are designed to substitute human experts and were a successful application of AI in the 1970s. Expert systems can read facts from the database and prepare information without any bias and heuristics. This makes decision-making efficient.

6.3.2   MACHINE LEARNING

Machine learning is a branch of artificial learning which deals with human imitation of problem solving without being explicitly programmed. The machine learning algorithms ingest the data provided by the user to solve problems such as classification problems. Machine learning is used in speech recognition, computer vision, email filtering, etc.

6.3.3   NEURAL NETWORKS

Neural networks are models of machine learning that are designed to imitate human brain activity. The neural networks consist of three main layers: an input layer, a hidden layer, and an output layer. Neural networks train from the data given at the input layer and give output. This helps the machines work on their own without having to be programmed explicitly. The layers are formed by connecting perceptrons in series.

6.3.4   NATURAL LANGUAGE PROCESSING

Natural language processing (NLP) is a specific application of AI that employs AI in the synthesis of natural language to interact with humans. The key areas include voice assistants, etc. The main concept of NLP is semantic understanding and performing repetitive tasks. The applications of NLP include chatbots, email filtering, questioning and answering, etc.

6.3.5   ROBOTICS

Robotics deals with the design, construction, and implementation of robots in various stages of human labor. Robots help in reducing human intervention in complex and risky tasks. Robotics help in developing autonomous unmanned vehicles to reduce the risk in human life. Robotics is applied in agriculture, healthcare, manufacturing applications, etc.

6.3.6   FUZZY LOGIC

Fuzzy logic tries to implement human reasoning abilities in machines. Classic logic has true or false, 0 and 1, to deduce logic output. Whereas fuzzy logic tries to model the intermediary values between YES or NO, such as certainly yes and little no. It tries to model the degree of truth into the computing systems. This helps in dealing with uncertainty and in solving engineering problems.

AI and machine learning have revolutionized many fields in recent times. AI is disrupting the lifestyle of an individual, the way we interact with machines, the way we work, the way we receive information, etc. Most organizations are increasing investments and recruiting new people to adapt AI into their work culture. The recent popularity of AI chatbots is revolutionizing the way individuals retrieve information. Companies are increasingly trying to integrate AI into their products (both manufacturing and service). NLP will enable machines to consider the spoken words as an input to the system rather than typing the input, which will make the user communicate with the system naturally, i.e., as if communicating with any other human. Additionally, AI can detect complex patterns in the datasets, behaviors, etc. Artificial neural networks can help in classification problems and pattern recognition problems.

6.4   CAREER DEVELOPMENT ECOSYSTEM

Most organizations are under pressure from both external factors, namely, economic, political, environmental, and legal factors, and internal factors such as employee training, chain of command, etc. Research by Chris Zook and James Allen of Bain shows that 94% of the issues faced by organizations are internal. They interviewed CEOs and top management to understand the issues faced by the organizations, such as attrition, a lack of skilled employees, etc. Organizational behavior is the science that deals with the study of individual interactions in groups that affect organizational effectiveness and productivity. It studies the individual, group, and organizational factors that affect the overall productivity of an organization (Shah et al., 2024).

According to organizational behavior, the individual factors include perception, motivation, learning, personality, etc. These factors are also crucial in determining group and organizational factors. There are several models that try to characterize personality types, such as the Big Five model and the Myers-Briggs model. Psychometric tests can be used to evaluate the personality of the individual. These tests are increasingly being adapted by the organizations to evaluate the personality type to determine personality job fit.

The second factor affecting organizational performance is that most individuals join organizations with personal goals such as job security, salary, social status, etc. Each organization has its own goals, such as sustainability, cost reduction, and profitability. Most Gen Z employees have a high growth orientation when it comes to work; they need more flexible work hours and challenging work. They also prioritize financial security. However, the nature of the job should be enterprising and secure.

Another issue most organizations are facing is the skill gap of employees. The skill gap is the difference between employer expectations and employee skill levels. One of the solutions for reducing the skill gap is to acquire talent pools. The competition for acquiring talent gap is extremely high, companies are offering high packages, perks and benefits, and additional benefits to attract the talent pool. Even though the first step of acquiring a talent pool is difficult, the second challenging task is to retain the talented employees acquired. Organizations spend a lot of time and resources training employees to fulfill the requirements of the job. Therefore, it is important to retain employees for an organization, as employee turnover can cause a loss to the organization in terms of money and time.

Therefore, companies want to retain their talent pool, and human resource department managers throughout the world in every company are under pressure to keep the attrition rate under control. Most companies in the world draft policies keeping in mind the welfare of their employees. One famous Walmart company founder

Sam Walton said, "If you'll take care of your employees, your employees will take care of your customers, and your business will take care of itself". This reflects his perception toward the importance of Walmart associates in running the Walmart. Sam Walton believed that most of the success of Walmart depends on the associates (Walmart calls its employees associates); they are a part of the core competitive advantage of Walmart.

Given the scenario, the organization must focus on internal issues to improve its efficiency. There are different methods to address different internal issues. One of the methods to improve employee retention is to maintain a career development that motivates employees looking for career growth and advancement.

Career development theories are a set of frameworks that help individuals understand how people choose and transform their work lives. They provide tools, strategies, and insights to help professionals achieve their career goals. A career development ecosystem (CDE) is a system of interrelated components that interact with each other to form a large CDE. The components of the system vary across different resources. However, we will only consider the common ones, as shown in Figure 6.4.

- Career Coaching
- Career Portals
- Assessments
- Employee Training
- Governance
- People Training Leader

Let us look at each one of these in detail

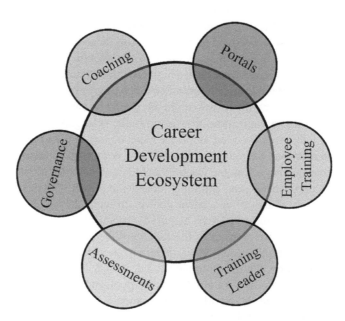

FIGURE 6.4 Components of the CDE.

6.4.1 CAREER COACHING

The goal of career coaching is to identify the traits and characteristics of an individual and the job fit of that individual. We have seen organizational behavior; it consists of several models that explain the personality job fit of a person, such as Trait and Factor Theory, Hackman Oldham Job Characteristics, etc. A career coach is aware of these models and tools to help individuals choose their job roles.

A career coach can be an internal employee who is trained in these models to identify another employee's strengths and weaknesses and help him identify his inherent interests and passion. Using this information, he can guide him to choose the job role out of all the roles in the company. He will mentor and guide him through the job role, making him more effective in his job. Career coaches can also be external human resource individuals hired by the organization to help the employees (Prashasti et al., 2023).

6.4.2 CAREER PORTALS

Portals are synonymous with gateways to access the World Wide Web. In general, portals are the access points available to access information. Portals are websites that are designed to access information in a required format. There are various kinds of portals, which are accessed across organizations, such as human resource portals, supply management portals, etc. The portals serve as gateways to organizational resources such as workforce, capital, financial information, etc. One such specific portal, which is of importance to the CDE is the career portal. The career portals have information regarding job roles in the company, the skill set expected by the company, and future career prospects.

Employees can access these portals with the credentials they need to access the information and look for career growth opportunities. These portals are maintained in the intranet of the company to keep the information safe and secure.

6.4.3 ASSESSMENTS

Assessments are a means to analyze the performance of an individual. However, the scope of assessments is not limited to the knowledge of the employees. There are several other kinds of tests available that will help reveal the personality, attitude, and behavior of the test. The test in this context refers to the tests related to both skill tests and personality tests. These tests reveal the skills and aspirations of the employee.

Organizations can conduct these tests on a regular basis and record the results. Based on the results of the assessments, the company can design the roles of employees. They can design the training programs. This will help employees improve their skill set and reduce skill gaps. This will also help companies use the workforce in appropriate job roles. This process can help in the overall improvement of productivity.

6.4.4 PEOPLE LEADER TRAINING

As discussed in career coaching, most of the career coaches are internal managers who volunteer in an organization to help people progress in their career and transition, but they may fall short of the required skills, which will be counterproductive for the organization as this may lead to undesirable results.

To avoid such consequences, organizations can provide employee people leader training to train employee volunteers with the necessary skills to guide other employees. The people leader training may involve organizing workshops with human resource experts, brief period training courses with the experts, etc. People leaders can mentor employees to choose the right paths, help them deal with psychological issues, provide guidance, etc.

6.4.5 EMPLOYEE TRAINING

Most employees carry their own perspective of their careers and want to progress. But they lack the knowledge to navigate through their career and advance in it. Organizations can help their employees navigate through their careers by organizing employee training programs that suit various internal roles.

Employee training can be both in terms of improving the skillset of the employees and psychological help of employees. They include training for job roles, training for working in teams, and training to improve communication skills. Organizations across the US spend around 101.6 billion USD on employee training, according to Training Magazine on the U.S training industry. The employees of the organization also spent 62.4 hours on training. These statements put stress on the fact that employment training is important for organizations.

6.4.6 GOVERNANCE

Governance includes support teams and people monitoring the CDE closely. They help employees address grievances regarding career development in the organization and address these issues. They also monitor the managers and employees to create policies and processes to reduce employee turnover. For example, what practices are prompting people to leave their jobs? Are the managers capable of handling employee career growth?

Organizations can retain their talent pool by maintaining a progressive CDE. Most Gen Z have peculiar choices regarding their career choices, when given the option between a high-paying salary and a high-growth career, the results split evenly (Gen Z, 2023). AI can help organizations improve their CDEs in many ways.

6.5 AI ROLE IN CAREER DEVELOPMENT ECOSYSTEM

Career development is not only restricted to organizations but also has a wider scope, it can also be useful for educational systems and government organizations. Several researchers have proposed real-time engines with AI that can be used by these entities. The engines will forecast future opportunities using machine learning, which will forecast job opportunities based on current job availability. Simply put, these engines will estimate and forecast the demand for jobs in the future based on the current data.

The engines can be used by educational institutes to prepare their students for future job roles that will arise that are not currently in the market. This will help educational institutes estimate the demand in the job market and design courses according to the requirements of the jobs. This can also be used by governments to design policies and take initiatives to reduce unemployment in the economy. The engines can be used to reduce the skill gap of the youth in the country. This will attract investments from foreign countries to help the country progress.

There are arguments that it is possible to evaluate the social media site usage of students and employees and assign a social score to the students. Here, we can use machine learning models to evaluate the social media score. This will help in understanding the social behavior of the students. There are other metrics that can be used to evaluate the overall performance of students. Metrics like social scores and other metrics can be used to improve the behavior of students and give guidance to them (Muthmainnah et al., 2023).

As seen, AI is very capable of revolutionizing many fields, and the capabilities of AI are very vast. Organizations are increasing investments in AI to improve the working efficiency of their employee. However, the applicability of AI is not limited to the work aspect of the organization but also to many other areas of the organization, one such area is the CDE. As we have seen, the CDE consists of six elements that are important to an organization. AI can improve all the elements of this CDE. Implementing AI systems into the CDE of an organization, as shown in Table 6.1.

TABLE 6.1

The Applicability of AI in Various Components of the CDE along with the Benefits

Component of CDE	Role of AI	Benefits
Career Coaching	Virtual coaching assistant	Easy to access Available all the time Can be trained as per the requirement. Reduced bias in suggestions
Career Portals	NLP and expert systems can be incorporated.	Improved communication abilities. Improved analytics.
Assessments	Machine learning can be used.	Improved test effectiveness. Improves evaluation criterion. Decreased inefficiencies of the testing methods. Improved specificity of test.
People Training Leader	AI-based Skill mapping.	Improved efficiency of the leader to map skills Better training is provided to employees.
Employee Training	AI-based virtual Learning.	Improved effectiveness of training. Immediate evaluation in the form of test.
Governance	AI-based expert systems	Governance will be improved as the expert systems will generate better reports.

AI can help in improving the career coaching of employees, NLP can help in designing virtual career coaches who can understand and respond to employees as a substitute for human coaches. There are certain benefits of employing AI career coaches because they are easy to deploy in computing systems to interact with employees. Employees can use the AI career coaches to seek help regarding their growth. AI career coaches have a neutral approach, they give advances without biases and heuristics. AI career coaches are available 24 hours a day throughout the year. However, they must be improved by including more data for addressing specific requirements (Khang & Muthmainnah et al., 2023).

As discussed above, the skillset of an employee is an asset to the organization. AI can help in mapping the skillset of an employee and the needs of organizational roles. This is explained a skills and career mapping model using AI. The model talks about the possibilities of mapping individual skillsets to company job role. AI chatbots can be integrated into social media networks and help students and working professionals understand their personality types with the help of psychometric tests such as Big Five and Myer-Briggs test to choose their career options and progress wisely (Silva et al., 2020).

The paper (Jain et al., 2023) proposes a career support model that uses Random Forest classifier with an underlying collaborative filtering model to suggest career prospects based on interests, passions, and the Myers-Briggs Test (MBIT). Additionally, this model also uses web scraping and NLP to improve user recommendations. AI can also be used in employee training, in which robot instructors can map the learning level of the students and prepare new methods to teach them based on their understanding level (Jia et al., 2018). This will generate reports on the effect of employee training using data analytics for the organization.

Knowledge management is a key area which focuses on documenting the knowledge of employees of the company for creation, access, and reuse (O'Leary et al., 1998). The tacit knowledge of an employee is a key component to the organization, which is lost when an employee leaves the organization. Documenting tacit knowledge is a key challenge faced by knowledge managers. AI can help in documenting the tacit knowledge of employees. AI applications can help in knowledge management at various levels, such as creation, storing and retrieving, sharing applications, etc (Jarrahi et al., 2023).

Apart from these, AI can also be used in the automation of the recruitment process, one such example is discussed in (Ruparel et al., 2023) which talks about professional social media platforms (PSMP). PSMPs have revolutionized the process in which the employers look for employees. The platforms enable individuals to share their professional achievements, awards, and activities on the platform. This will help recruiters identify potential talent and call for interviews. In these many ways, AI can transform the CDE (Pooja et al., 2023).

6.6   CONCLUSION

Every organization wants to stay relevant and successful in their relevant domain of business. It can be achieved by hiring and retaining the top talent pool from the job markets in the organization. Even though organizations are finding new ways to retain their talent pool of the company. There are inefficiencies in the process. AI is one such phenomenon which is revolutionized many fields at a rapid pace in recent times. It is changing the way in which an individual interacts with machines. It can help organizations to retain employees by improving the CDE of the organization. The improved CDE will ensure growth opportunities are communicated across all employees of the organization. This will help in the career growth and career transition of the employees, which will help lead to their satisfaction. As most of the Gen Z are focused on career growth, this will attract them to consider the organization. This CDE will keep the employees engaged in the organization, leading to a low attrition rate (Khang & Rani et al., 2023).

6.7   FUTURE SCOPE OF WORK

Even though AI has developed in many dimensions, it still has a long road ahead. NLP, machine learning, and artificial neural networks have shortcomings of their own. The chatbots also must be trained to deal with the specificity of the topics because most of the results are generic in nature. Organizations must implement AI in full scale to get the benefit. The moral and ethical concerns of AI are still in discussion among the research community. However, there is a strong positive sentiment about AI implementation across various sectors. AI is not a substitute for human intelligence; it is a tool to amplify human creativity and ingenuity (Khang & Shah et al., 2023).

REFERENCES

Bryson, A. & Forth, J. (2011). The Evolution of the Modern Worker: Attitudes to Work. https://www.torrossa.com/gs/resourceProxy?an=4912596&publisher=FZ7200#page=126

Gen Z Understanding Generation Z in the workplace. (2023). Retrieved on 2023 from https://www2.deloitte.com/us/en/pages/consumer-business/articles/understanding-generation-z-in-the-workplace.html

Jain, A., Durairaj, S., Paduri, A. R., Krishnan, P., Chalaiah, P., Chanda, J., & Darapaneni, N. (2023). Career support platform for older adults powered by AI. *2023 IEEE 13th Annual Computing and Communication Workshop and Conference (CCWC), Computing and Communication Workshop and Conference (CCWC), 2023 IEEE 13th Annual*, pp. 47–53. IEEE, New York. https://ieeexplore.ieee.org/abstract/document/10099112/

Jia, Q., Guo, Y., Li, R., Li, Y., & Chen, Y. (2018). A conceptual artificial intelligence application framework in human resource management. *ICEB 2018 Proceedings*. Guilin, China. https://aisel.aisnet.org/iceb2018/91/

Jarrahi, M. H., Askay, D., Eshraghi, A., & Smith, P. (2023). Artificial intelligence and knowledge management: a partnership between human and AI. *Business Horizons*, 66(1), 87–99. https://www.sciencedirect.com/science/article/pii/S0007681322000222

Khang, A., Gupta, S. K., Rani, S., & Karras, D. A. (2023). *Smart Cities: IoT Technologies, Big Data Solutions, Cloud Platforms, and Cybersecurity Techniques*. CRC Press, Boca Raton, FL. https://doi.org/10.1201/9781003376064

Khang, A., Muthmainnah, M., Seraj, P. M. I., Yakin, A. A., Obaid, A. J., & Panda, M. R. (2023). AI-aided teaching model for the education 5.0 ecosystem. *AI-Based Technologies and Applications in the Era of the Metaverse* (1st ed., pp. 83–104). IGI Global Press, Hershey, PA. https://doi.org/10.4018/978-1-6684-8851-5.ch004

Khang, A., Rani, S., Gujrati, R., Uygun, H., & Gupta, S. K. (2023). *Designing Workforce Management Systems for Industry 4.0: Data-Centric and AI-Enabled Approaches*. CRC Press, Boca Raton, FL. https://doi.org/10.1201/9781003357070

Khang, A., Shah, V., & Rani, S. (2023). *AI-Based Technologies and Applications in the Era of the Metaverse* (1 ed.). IGI Global Press, Hershey, PA. https://doi. org/10.4018/978-1-6684-8851-5

Lee, J. M. & Wei, L. (2018). Bloomberg News. Retrieved from "Gen Z is Set to Outnumber Millennials Within a year," https://www. bloomberg.com/news/articles/2018-08-20/ gen-z-to-outnumber millennials-within-a-year-demographic-trends"

Luke, J., Khang, A., Vadivelraju, C., Pravin, A. R., & Sriram, K. (2023). Smart city concepts, models, technologies and applications. *Smart Cities: IoT Technologies, Big Data Solutions, Cloud Platforms, and Cybersecurity Techniques*. CRC Press, Boca Raton, FL. https://doi.org/10.1201/9781003376064-1

Muthmainnah, M., Khang, A., Seraj, P. M. I., Yakin, A. A., Oteir, I., & Alotaibi, A. N. (2023). An innovative teaching model - the potential of metaverse for English learning. *AI-Based Technologies and Applications in the Era of the Metaverse* (1st ed., pp. 105–126). IGI Global Press, Hershey, PA. https://doi.org/10.4018/978-1-6684-8851-5.ch005

OpenStax | Free Textbooks Online with No Catch. (2019). Retrieved from https://openstax.org/ details/books/organizational-behavior?Book%20details

O'Leary, D. E. (1998). Using AI in knowledge management: knowledge bases and ontologies. *IEEE Intelligent Systems and their Applications*, 13(3), 34–39. https://doi.org/ 10.1109/5254.683180

Pooja, K., Babasaheb, J., Ashish, K., Khang, A., & Sagar, K. (2023). The role of blockchain technology in metaverse ecosystem. *AI-Based Technologies and Applications in the Era of the Metaverse* (1st ed., pp. 228–236). IGI Global Press, Hershey, PA. https://doi. org/10.4018/978-1-6684-8851-5.ch011

Prashasti, P., Ipseeta, S., Patnaik, B. C. M., Patnaik, A., & Khang, A. (2023). Role of the internet of things (IoT) in enhancing the effectiveness of the self-help groups (SHG) in smart city. *Smart Cities: IoT Technologies, Big Data Solutions, Cloud Platforms, and Cybersecurity Techniques*. CRC Press, Boca Raton, FL. https://doi.org/10.1201/9781003376064-14

Ruparel, N., Bhardwaj, S., Seth, H., & Choubisa, R. (2023). Systematic literature review of professional social media platforms: development of a behavior adoption career development framework. *Journal of Business Research*, 156, 113482. https://www.science direct.com/science/article/pii/S014829632200947X

Silva, D., Jani, G., Jadhav, M., Bhoir, V., & Amin, P. (2020). Career counselling chatbot using cognitive science and artificial intelligence. *Advanced Computing Technologies and Applications: Proceedings of 2nd International Conference on Advanced Computing Technologies and Applications-ICACTA 2020*, pp. 1–9. Springer, Singapore. https://link. springer.com/book/10.1007/978-981-15-3242-9

Shah, V., Suketu, J., & Khang, A. (2023). Automotive IoT: accelerating the automobile industry's long-term sustainability in smart city development strategy. *Smart Cities: IoT Technologies, Big Data Solutions, Cloud Platforms, and Cybersecurity Techniques*. CRC Press, Boca Raton, FL. https://doi.org/10.1201/9781003376064-9

Subhashini, R. & Khang, A. (2023). The role of internet of things (IoT) in smart city framework. *Smart Cities: IoT Technologies, Big Data Solutions, Cloud Platforms, and Cybersecurity Techniques*. CRC Press, Boca Raton, FL. https://doi.org/10.1201/9781003376064-3

7 Artificial Intelligence Competency Framework

Navigating the Shift to an Artificial Intelligence-Infused World

Bharathithasan S. and Sakthi Srinivasan K.

7.1 INTRODUCTION

The rapid advancement of Artificial Intelligence (AI) has transformed industries, societies, and economies, ushering in an era of AI-infused technologies. Organizations across the globe are embracing AI to enhance efficiency, productivity, and decision-making. However, the successful integration of AI into various domains necessitates a skilled workforce proficient in AI competencies (Bozdag, 2023). This chapter presents an AI Competency Framework designed to guide individuals, educational institutions, and businesses in navigating the shift to an AI-infused world. By identifying the key skills, knowledge, and attributes required for AI proficiency, this framework aims to empower stakeholders to adapt, innovate, and thrive in an increasingly AI-driven landscape (Khang & Misra et al., 2023).

7.2 UNDERSTANDING THE AI LANDSCAPE

The proliferation of AI technologies has accelerated in recent years, impacting diverse sectors such as healthcare, finance, transportation, and entertainment. AI-driven innovations, such as autonomous vehicles, natural language processing, and smart assistants, have reshaped industries and redefined job roles. Despite the opportunities presented by AI, many organizations face challenges in adopting these technologies effectively. Limited AI competencies among the workforce, concerns about job displacement, and ethical considerations are some of the barriers hindering widespread AI integration (Zahlan et al., 2023).

As organizations strive to leverage AI to stay competitive and innovative, the importance of a skilled workforce becomes evident. An AI-infused world demands individuals capable of understanding, designing, and deploying AI solutions responsibly. This section emphasizes the urgency of developing a comprehensive AI Competency Framework to address the skills gap and foster a culture of continuous learning and adaptation (Collins et al., 2023).

DOI: 10.1201/9781003440901-7

7.3 DEVELOPING THE AI COMPETENCY FRAMEWORK

The development of the AI Competency Framework involves a rigorous and iterative process to ensure its accuracy, relevance, and adaptability. The methodology combines literature review, consultations with AI experts, surveys, and case studies. Existing AI-related competency models and frameworks from reputable sources are analyzed to identify commonalities and gaps. Interviews and focus group discussions with AI researchers, industry leaders, and educators provide valuable insights into the skills and attributes they consider crucial for AI proficiency (Cannas et al., 2023).

The iterative nature of the framework's development allows for feedback and refinement at various stages. Pilot studies involving educational institutions and businesses validate the framework's effectiveness in addressing their specific needs. This section highlights the collaborative approach employed to create a well-rounded, future-proof AI Competency Framework (Sarraipa et al., 2023).

7.4 KEY COMPONENTS OF THE AI COMPETENCY FRAMEWORK

The AI Competency Framework encompasses a wide range of technical and non-technical skills required for AI proficiency. The framework acknowledges the multidisciplinary nature of AI, emphasizing the significance of both domain-specific knowledge and cross-cutting abilities (Msweli et al., 2023).

7.4.1 TECHNICAL SKILLS

- **Programming Languages and Tools**: Proficiency in programming languages like Python, R, and Java, along with familiarity with AI libraries and frameworks.
- **Machine Learning and AI Algorithms**: Understanding of various machine learning algorithms, such as regression, classification, clustering, and neural networks, and their applications.
- **Data Analytics and Data Science**: Ability to collect, process, and analyze large datasets to extract meaningful insights and support AI-driven decision-making.
- **Big Data and Cloud Computing**: Knowledge of big data technologies and cloud computing platforms to handle large-scale AI applications efficiently.
- **Computer Vision and Natural Language Processing**: Familiarity with computer vision techniques and natural language processing for image and text analysis.
- **AI Model Deployment and Optimization**: Skills in deploying AI models in real-world environments and optimizing their performance.

7.4.2 NON-TECHNICAL SKILLS

- **Critical Thinking and Problem-Solving**: The ability to think critically, analyze complex problems, and design effective AI solutions.
- **Ethics and Responsible AI Practices**: Understanding the ethical considerations associated with AI development and ensuring responsible AI deployment.

- **Communication and Collaboration**: Effective communication skills to convey technical concepts to non-technical stakeholders and collaborate in multidisciplinary teams.
- **Continuous Learning and Adaptation**: A growth mindset and a commitment to continuous learning in the dynamic field of AI.
- **Business and Domain Knowledge**: Understanding of specific industries and domains to develop AI solutions that align with business objectives.

By incorporating both technical and non-technical competencies, the AI Competency Framework ensures a holistic approach to AI proficiency, fostering well-rounded professionals capable of tackling complex AI challenges.

7.5 APPLICATIONS OF THE AI COMPETENCY FRAMEWORK

The AI Competency Framework finds applications in various contexts, supporting individuals, educational institutions, and businesses in different ways (Brauner, 2023).

7.5.1 THEME OF THE FRAMEWORK

In general, this flow diagram presents a hierarchical depiction of the proficiencies needed to navigate the transition toward a world infused with AI as Figure 7.1. It can be utilized as an initial framework for the subsequent enhancement and elaboration of the involved competencies and sub-competencies (Qin et al., 2023). Each node within the hierarchy is linked to the subsequent level below it through a connecting line, symbolizing the sub-competencies that are encompassed within each overarching competency.

Likewise, within the realm of machine learning fundamentals, there may exist subordinate proficiencies encompassing comprehension of supervised and unsupervised learning algorithms, acquaintance with prevalent machine learning libraries and frameworks, and a rudimentary grasp of model evaluation metrics and techniques. The lowermost node symbolizes the ongoing process of acquiring knowledge and enhancing one's professional skills, which is a crucial element in effectively adapting to the transition toward a world influenced by AI. This is due to the perpetual advancements and enhancements observed in AI technologies (Khang & Rani et al., 2022).

7.5.2 FOR INDIVIDUALS

- **Career Development and Upskilling**: The framework enables individuals to assess their current AI competencies and identify areas for improvement. They can use it as a guide to develop personalized learning paths and enhance their employability in AI-related roles.
- **Career Transitions and Reskilling**: Professionals from non-AI backgrounds can leverage the framework to navigate career transitions into

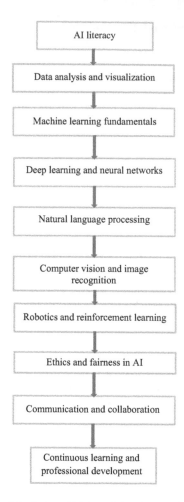

FIGURE 7.1 A hierarchical depiction of the proficiencies needed to navigate the transition toward a world infused with artificial intelligence.

AI-focused roles. The framework provides a structured roadmap to acquire the necessary skills and knowledge.

- **Continuous Professional Development**: AI practitioners can utilize the framework to stay updated with the latest trends and advancements in the AI domain, ensuring their skills remain relevant and cutting-edge.

7.5.3 FOR EDUCATIONAL INSTITUTIONS

- **Curriculum Design and Development**: Educational institutions can use the framework as a foundation for designing AI-focused curricula that align with industry demands. The framework assists in identifying the essential topics and learning outcomes to equip students with AI competencies.

- **Pedagogical Approaches**: Educators can adopt the framework to develop effective teaching methodologies that foster critical thinking, problem-solving, and hands-on experience in AI projects.
- **Assessment and Certification**: The framework provides a structured basis for evaluating students' AI competencies, enabling institutions to issue certifications that hold industry-wide recognition.

7.5.4 FOR BUSINESSES AND ORGANIZATIONS:

- **Talent Acquisition and Recruitment**: Businesses can refer to the framework to define clear job descriptions and skill requirements for AI-related positions. This helps attract talent with the right expertise to meet organizational needs.
- **Employee Training and Development**: The framework facilitates the identification of skill gaps among the existing workforce, allowing businesses to invest in targeted training programs for upskilling and reskilling.
- **AI Strategy and Implementation**: Organizations can align their AI strategy with the framework, ensuring that AI initiatives incorporate ethical considerations and responsible practices.

The applications of the AI Competency Framework demonstrate its versatility and broad utility across diverse stakeholders, fostering a workforce capable of navigating the AI-infused world effectively.

7.6 ROLE OF AI IN THE LABOR MARKET TODAY

The advent of AI has swiftly emerged as a catalyst for profound change within the labor market, fundamentally altering industries, job functions, and the necessary skill sets. The incorporation of this technology into diverse industries is causing significant upheaval and advancement, exerting an impact on the characteristics of employment and the proficiencies sought after by employers. AI technologies have a broad scope of applications, including automated data analysis, decision-making, natural language processing, and robotics. These capabilities are significantly influencing business operations and ways in which employees participate in their work (Khang & Hajimahmud et al., 2023).

In recent years, there has been a growing integration of AI within recruitment and talent acquisition procedures. Organizations are utilizing AI-based algorithms to analyze resumes, evaluate the appropriateness of candidates, and even facilitate preliminary interview stages. The primary objective of AI-driven systems is to optimize and accelerate the hiring process, thereby minimizing the time and resources needed to identify prospective candidates. For example, AI-driven platforms such as HireVue employ facial expression analysis, vocal tone analysis, and word choice analysis during video interviews to evaluate candidates' soft skills and personality characteristics (Joamets & Chochia, 2020).

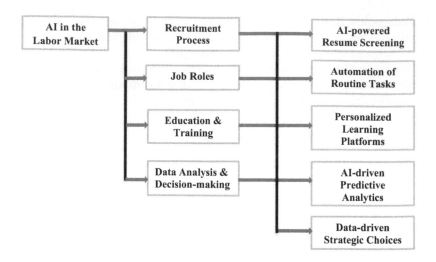

FIGURE 7.2 Robo-advisors in the financial industry utilize AI algorithms to deliver customized investment suggestions.

In addition, the implementation of AI-driven automation is having a substantial impact on job responsibilities within various sectors. The automation of routine and repetitive tasks is becoming more prevalent, enabling human workers to allocate their efforts toward strategic and creative endeavors. The manufacturing industry has experienced the implementation of robotics and AI-enabled machines, which are capable of efficiently and accurately performing repetitive tasks as Figure 7.2.

Nevertheless, there have been discussions regarding the future of work in response to apprehensions surrounding job displacement resulting from the automation of AI. As automation continues to streamline routine tasks, there is a growing need for individuals with expertise in the design, development, maintenance, and oversight of AI. The current transition requires a workforce that possesses the necessary skills to effectively collaborate with AI systems. In order to stay pertinent in the ever-changing job market, it has become imperative for employees to engage in ongoing upskilling and reskilling efforts.

The education and training sector is utilizing AI to optimize and improve learning experiences. Personalized learning platforms employ AI algorithms to customize educational content based on the unique pace, learning style, and knowledge deficiencies of each student. The capacity to adapt facilitates enhanced learning outcomes and provides students with pertinent skills for the labor market. Furthermore, the utilization of AI-driven virtual reality (VR) and augmented reality (AR) technologies is being observed in the simulation of authentic situations, thereby providing immersive training opportunities in various sectors such as healthcare and aviation (Khang & Gupta et al., 2023).

The impact of AI is not solely confined to the realm of automation, but rather encompasses a broader scope that includes data analysis and decision-making processes. Organizations are employing AI algorithms to efficiently analyze large quantities of data, extract valuable insights, and make informed strategic decisions. For example, the utilization of AI-driven predictive analytics enables businesses to proactively anticipate market trends, optimize their supply chains, and make well-informed financial decisions. The utilization of data-driven methodologies enhances operational efficiency and competitiveness, thereby emphasizing the pivotal role that AI assumes in contemporary business operations.

The multifaceted nature of AI's role in the contemporary labor market is evident. The phenomenon under consideration exerts an impact on the procedures of personnel acquisition, the delineation of occupational responsibilities, and the qualifications demanded, all the while offering prospects for advancement as well as posing obstacles. The ongoing progress of AI calls for a proactive stance toward skill enhancement and lifelong learning in order to effectively incorporate AI into the labor force. By acknowledging and embracing the potential of AI and adequately preparing for its implications, both individuals and organizations can effectively navigate the continuously evolving labor market and effectively harness the transformative power of AI to achieve sustainable growth.

7.7 IMPACT OF CHATGPT ON BUSINESS

AI has garnered considerable momentum in diverse sectors, leading to the restructuring of conventional business frameworks and augmenting operational efficacy. An exemplary demonstration of the impact of AI is the incorporation of sophisticated language models such as ChatGPT into various operational aspects of businesses. The language model known as ChatGPT, which was created by OpenAI, represents a cutting-edge advancement in the field. It possesses the ability to participate in conversations that closely resemble those between humans, rendering it an invaluable asset for enterprises aiming to enhance their customer service, optimize their operational processes, and foster innovation (Zarifhonarvar, 2023).

7.7.1 IMPROVING CUSTOMER SUPPORT AND SERVICE

The impact of ChatGPT on business is notably significant, particularly in its capacity to transform customer support and service. By incorporating ChatGPT into chatbots or virtual assistants, organizations can offer immediate and continuous customer support. AI-powered conversational systems have the capability to effectively handle a diverse array of customer inquiries, deliver troubleshooting support, and provide valuable product recommendations. For example, corporations such as Shopify and Capital One have employed chatbots powered by AI to improve their customer engagements, resulting in decreased response durations and heightened levels of customer contentment (Shah & Khang, 2023).

7.7.2 EFFICIENT INFORMATION RETRIEVAL: A STUDY IN OPTIMAL SEARCH TECHNIQUES

The natural language processing capabilities of ChatGPT facilitate the efficient retrieval and organization of information for businesses. Organizations have the capacity to employ ChatGPT for the purpose of developing intelligent knowledge bases or internal databases, which can effectively address employee inquiries in a timely manner, thereby diminishing the duration spent on information retrieval. This enhances the efficiency of employees and guarantees the precise and uniform distribution of information throughout the organization (Pooja et al., 2023).

ChatGPT can also play a significant role in facilitating innovation and fostering product development. Businesses have the potential to utilize their creative text generation capabilities in order to facilitate the generation of new product concepts, produce marketing content, and even compose preliminary draughts of reports or proposals. The integration of ChatGPT into ideation processes enables companies to accelerate the generation of ideas and prototypes, thereby facilitating more rapid cycles of innovation. The concept of personalized marketing and communication refers to the practice of tailoring marketing strategies and communication efforts to meet the specific needs and preferences of individual consumers.

AI-driven language models such as ChatGPT facilitate the ability of businesses to tailor their marketing and communication endeavors to individual customers. Through the examination of customer data and interaction history, organizations have the ability to customize their messages in accordance with individual preferences and behaviors. The implementation of this degree of personalization significantly amplifies customer engagement and augments the efficacy of marketing campaigns. The influence of ChatGPT on enterprises is significant, yet it also presents noteworthy challenges and ethical considerations. The diligent oversight of AI-generated content is imperative in order to uphold accuracy and mitigate the propagation of misinformation. In addition, it is imperative for businesses to acknowledge and tackle apprehensions pertaining to the safeguarding of data privacy and security while employing AI systems for customer interactions (Khang & Shah et al., 2023).

The incorporation of ChatGPT into organizational workflows holds the capacity to fundamentally transform multiple facets of business operations, encompassing customer support and innovation. The versatility of this tool lies in its capacity to engage in conversations resembling those between humans, offer information, and aid in decision-making, thereby augmenting efficiency and enhancing the customer experience. Nevertheless, it is imperative for businesses to effectively manage ethical considerations and guarantee the responsible utilization of AI-generated content as Figure 7.3.

7.7.3 RESEARCH IMPLICATIONS OR FINDINGS

The research presented in this paper has several key implications and findings:

- The AI Competency Framework provides a comprehensive and structured approach to identify, develop, and assess AI-related skills in individuals and organizations. Its integration into education and professional development can significantly enhance the readiness of the workforce to embrace AI technologies.
- The framework emphasizes the importance of a multidisciplinary approach to AI competence, highlighting technical and non-technical skills that are equally vital in an AI-infused world. This recognition promotes the development of well-rounded AI professionals capable of addressing real-world challenges.

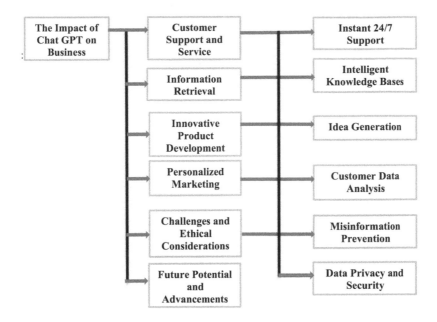

FIGURE 7.3   The flowchart of the responsible utilization of AI-generated content.

- By bridging the AI skills gap, the framework addresses the challenges of AI adoption in various sectors and fosters a workforce capable of leveraging AI technologies effectively. As organizations become more AI-driven, the framework can help them build competitive advantages through skilled personnel.

Ethics and responsible AI practices are integral components of the AI Competency Framework, underscoring the importance of AI being developed and deployed with societal and ethical considerations in mind. This emphasis promotes the adoption of AI technologies that prioritize transparency, fairness, and inclusivity.

7.8 CONCLUSION

The AI Competency Framework serves as a valuable roadmap for all stakeholders navigating the transition to an AI-infused world. As AI continues to reshape industries and societies, it is crucial to invest in building a skilled and adaptable workforce that can harness the full potential of AI. The framework's comprehensive approach, encompassing technical expertise, critical thinking, and ethical considerations, ensures that AI technologies are deployed responsibly and for the collective benefit of humanity.

The framework's adoption holds promising prospects for individuals seeking career growth, educational institutions striving to equip their students with relevant skills, and businesses aiming to drive innovation through AI adoption. By fostering a culture of continuous learning and adaptability, the AI Competency Framework becomes an invaluable asset in preparing the global workforce for the challenges and opportunities of an AI-infused world (Shyam & Khang, 2023).

7.9 FUTURE SCOPE OF WORK

While the AI Competency Framework represents a significant step forward, there are several avenues for future research and improvement:

- **Continuous Refinement**: The framework should be periodically updated to keep pace with the evolving AI landscape, incorporating emerging technologies and new skill requirements. Stakeholders' feedback and real-world experiences should drive this iterative process.
- **Global Applicability**: Future work should focus on adapting the framework to different cultural contexts and economies, ensuring its relevance and effectiveness on a global scale. Consideration of regional nuances and industry-specific demands will enhance its applicability.
- **Impact Assessment**: Longitudinal studies should be conducted to measure the impact of the framework on organizations' AI adoption, workforce performance, and overall competitiveness. Long-term assessment can provide valuable insights into the framework's effectiveness and areas for improvement.
- **Inclusivity and Diversity**: Further research is needed to explore how the AI Competency Framework can address issues of diversity and inclusivity in AI-related fields, promoting equal opportunities for all. It should encourage diversity in AI development teams and consider the societal implications of AI systems.

The AI Competency Framework is a valuable tool in the journey toward an AI-infused world, fostering a skilled workforce that can navigate and lead in this transformative era of technology. By promoting responsible AI practices and cultivating diverse talent, we can collectively shape a future where AI serves as a force for positive change. The successful implementation of the framework will empower individuals, organizations, and societies to thrive in the AI era and contribute to a sustainable and equitable future (Khang & Kali et al., 2023).

REFERENCES

Azadnia, A. H., Movahhedinia, N., & Cao, G. (2021). Industry 4.0 technologies and their impact on supply chain sustainability. *Journal of Cleaner Production*, 307, 127229. https://www.sciencedirect.com/science/article/pii/S095965262100473X

Bozdag, A. A. (2023). AIsmosis and the pas de deux of human-AI interaction: exploring the communicative dance between society and artificial intelligence. *Online Journal of Communication and Media Technologies*, 13(4), e202340. https://doi.org/10.30935/ojcmt/13414

Brauner, S., Murawski, M., & Bick, M. (2023). The development of a competence framework for artificial intelligence professionals using probabilistic topic modelling. *Journal of Enterprise Information Management*, 13(4), e202340 https://doi.org/10.1108/jeim-09-2022-0341

Cannas, V. G., Ciano, M. P., Saltalamacchia, M., & Secchi, R. (2023). Artificial intelligence in supply chain and operations management: a multiple case study research. *International Journal of Production Research*, 2023, 1–28. https://doi.org/10.1080/00207543.2023.2232050

Collins, C., Kauppaymuthoo, K. J., Chowdhury, A., & Steen, O. (2023). Artificial Intelligence Adoption in Software Development. https://lup.lub.lu.se/luur/download?func=downloadFile&recordOId=9123274&fileOId=9123420

Joamets, K. & Chochia, A. (2020). Artificial intelligence and its impact on labour relations in Estonia. *Slovak Journal of Political Sciences*, 20(2), 255–277. ttps://sjps.fsvucm.sk/index.php/sjps/article/view/136 (Accessed: 18August2023).

Khang, A., Gupta, S. K., Rani, S., & Karras, D. A. (2023). *Smart Cities: IoT Technologies, Big Data Solutions, Cloud Platforms, and Cybersecurity Techniques*. CRC Press, Boca Raton, FL. https://doi.org/10.1201/9781003376064

Khang, A., Hajimahmud, V. A., Gupta, S. K., Babasaheb, J., & Morris, G. (2023). *AI-Centric Modelling and Analytics: Concepts, Designs, Technologies, and Applications* (1 ed.). CRC Press, Boca Raton, FL. https://doi.org/10.1201/9781003400110

Khang, A., Kali, C. R., Suresh Kumar, S., Amaresh, K., Sudhansu Ranjan, D., & Manas Ranjan, P. (2023). Enabling the future of manufacturing: integration of robotics and IoT to smart factory infrastructure in industry 4.0. *AI-Based Technologies and Applications in the Era of the Metaverse* (1st ed., pp. 25–50). IGI Global Press, Hershey, PA. https://doi.org/10.4018/978-1-6684-8851-5.ch002

Khang, A., Misra, A., Gupta, S. K., & Shah, V. (2023). *AI-aided IoT Technologies and Applications in the Smart Business and Production*. CRC Press, Boca Raton, FL. https://doi.org/10.1201/9781003392224

Khang, A., Rani, S., & Sivaraman, A. K. (2022). *AI-Centric Smart City Ecosystems: Technologies, Design and Implementation* (1st ed.). CRC Press, Boca Raton, FL. https://doi.org/10.1201/9781003252542

Khang, A., Shah, V., & Rani, S. (2023). *AI-Based Technologies and Applications in the Era of the Metaverse* (1 ed.). IGI Global Press, Hershey, PA. https://doi.org/10.4018/978-1-6684-8851-5

Msweli, T. N., Mawela, T., & Twinomurinzi, H. (2023). Data science education - a scoping review. *Journal of Information Technology Education Research*, 22, 263–294. https://doi.org/10.28945/5173

Pooja, K., Babasaheb, J., Ashish, K., Khang, A., & Sagar, K. (2023). The role of blockchain technology in metaverse ecosystem. *AI-Based Technologies and Applications in the Era of the Metaverse* (1st ed., pp. 228–236). IGI Global Press, Hershey, PA. https://doi.org/10.4018/978-1-6684-8851-5.ch011

Qin, Y., Hu, S., Lin, Y., Chen, W., Ding, N., Cui, G., Zeng, Z., Huang, Y., Xiao, C., Han, C., Fung, Y. R., Su, Y., Wang, H., Qian, C., Tian, R., Zhu, K., Liang, S., Shen, X., Xu, B., & Sun, M. (2023). Tool Learning with Foundation Models. https://doi.org/10.48550/ARXIV.2304.08354

Sarraipa, J., Zamiri, M., Marcelino-Jesus, E., Artifice, A., Jardim-Goncalves, R., & Moalla, N. (2023). A learning framework for supporting digital innovation hubs. *Computers*, 12(6), 122. https://doi.org/10.3390/computers12060122

Saxena, A. C., Ojha, A., Sobti, D., & Khang, A. (2023). Artificial intelligence (AI) centric model in metaverse ecosystem. *AI-Based Technologies and Applications in the Era of the Metaverse* (1 ed., pp. 1–24). IGI Global Press, Hershey, PA. https://doi.org/10.4018/978-1-6684-8851-5.ch001

Shah, V. & Khang, A. (2023). Metaverse-enabling IoT technology for a futuristic healthcare system. *AI-Based Technologies and Applications in the Era of the Metaverse* (1st ed., pp. 165–173). IGI Global Press, Hershey, PA. https://doi.org/10.4018/978-1-6684-8851-5.ch008

Shyam, R. S. & Khang, A. (2023). Effects of quantum technology on metaverse. *AI-Based Technologies and Applications in the Era of the Metaverse* (1st ed., pp. 104–203). IGI Global Press, Hershey, PA. https://doi.org/10.4018/978-1-6684-8851-5.ch009

Zahlan, A., Ranjan, R. P., & Hayes, D. (2023). Artificial intelligence innovation in healthcare: literature review, exploratory analysis, and future research. *Technology in Society*, 74(102321), 102321. https://doi.org/10.1016/j.techsoc.2023.102321

Zarifhonarvar, A. (2023). Economics of ChatGPT: a labor market view on the occupational impact of artificial intelligence. Journal of Electronic Business & Digital Economics. https://ssrn.com/abstract=4350925 or http://dx.doi.org/10.2139/ssrn.4350925

8 Building the Perfect Match Using Artificial Intelligence in Career Development

Arpita Nayak, Alex Khang, Ipseeta Satpathy,
Sasmita Samanta, and B. C. M. Patnaik

8.1 INTRODUCTION

The process of obtaining and refining skills, information, and experiences that have contributed to an individual's professional success and personal improvement is referred to as career development. It entails a lifelong journey of investigating, planning for, and adjusting to diverse work-related possibilities and obstacles. Career development refers to a lifetime of behaviors that lead to the discovery, formation, achievement, and fulfillment of one's vocation.

Career development includes not only the learning of occupational skills and abilities, but also the development of self-awareness, values, interests, and objectives that correspond to one's chosen career path. It includes making educated choices concerning school, training, employment options, and career changes. Furthermore, internal and external elements such as human qualities, social and cultural settings, organizational environments, and technology breakthroughs all have an impact on career progression.

Finally, career development aims to assist individuals in achieving personal fulfillment, professional success, and a sense of accomplishment. By the end of the 20th century, the National Career Development Association's definition of career development had broadened to include a wide variety of criteria such as psychological, social, educational, physical, economic, and chance components. These elements impact an individual's career behavior during the course of their life (Lent & Brown, 2013; Herr, 2001). In the latest 14-month research project carried out by Integral Talent Systems, 1,200 Millennials throughout the United States ranked achieving career aspirations as well as possibilities for professional development as their number one and two factors, accordingly, that would attract them to an organization and encourage them to stay more time (Khang & Rani et al., 2023).

Career development is critical to an individual's professional development and overall workplace happiness. It includes a variety of characteristics such as employment options, skill development, work satisfaction, and career growth. Based on a study from Statista, 84% of businesses feel that investing in AI will result in stronger

DOI: 10.1201/9781003440901-8

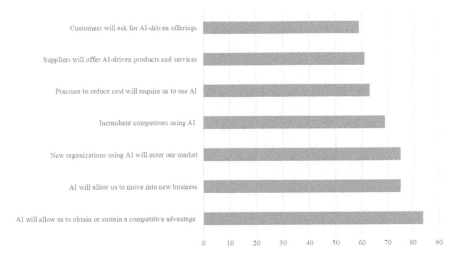

FIGURE 8.1 Graphical representation of Businesses investing in AI.

Source: Adopted from Statista.

competitive advantages. 75% say AI will create new enterprises while also offering rivals new opportunities to enter their markets. 63% feel that cost-cutting pressures will necessitate the deployment of AI shown in the graph below (Columbus, 2018).

AI technology has been integrated into the field of career development in recent years. AI has the ability to transform career development by developing creative tools and solutions that improve the efficacy and efficiency of numerous career-related procedures. Individuals may acquire important insights, receive customized assistance, and make educated career decisions by employing AI-driven examinations, personalized skill development platforms, and performance feedback systems (Westman et al., 2021; Gati & Kulcsar, 2021). The corporate climate has deteriorated, resulting in fewer hierarchical jobs. Organizations, on the other hand, must increase production and adapt to changing technologies. They often promote current staff, necessitating careful succession planning and ongoing development for potential top-level jobs.

Organizational career development helps both people and businesses, but it is important to recognize that unanticipated changes and repercussions might occur, affecting the whole landscape. Employees and employers must both be adaptive and respond well to changing conditions. Personnel must constantly improve their skills and abilities to meet changing expectations, while organizations must have capable personnel who can manage obstacles successfully and minimize vulnerability to changing conditions. As a result, both sides must recognize the importance of professional development and be willing to embrace it as a method of thriving in the face of shifting (Vondracek et al., 2019).

According to the McKinsey Global Institute, "between 400 million and 800 million jobs could be eliminated by automation and robotics by 2030." However, the same article states that "automation could boost growth in productivity and the economy, as well as generate better and novel employment opportunities for workers."

This means that, while some professions may become obsolete, new possibilities will emerge for people with the necessary abilities. Individuals must therefore stay current with technology breakthroughs and continually expand their abilities in order to remain employed.

According to the World Economic Forum, "by 2022, no less than 54% of all employees are going to need considerable re- and upskilling." Individuals must be proactive in expanding their abilities and learning new technology. Amazon is one example of a firm that actively promotes lifelong learning. Amazon Career Choice pays 95% of tuition and fees for programs in high-demand professions including healthcare, technology for computers, and transportation (Brunette, 2020). This program encourages workers to learn new skills that will benefit them as well as the firm in order to keep ahead of other businesses (Snehal et al., 2023).

With the growth of AI and automation, some talents, such as data analysis, coding, and problem-solving are becoming increasingly valued. Professionals should investigate the talents in demand in their sector and take action to acquire such skills. To remain competitive, individuals will need to upskill or reskill as AI and automation become increasingly integrated into the workplace. This entails learning new skills or honing old ones in order to adapt to changing employment demands. The various ways in which AI impacts career development are mentioned in Figure 8.2 below (Chan et al., 2022; Saxena et al., 2023).

- **Career Guidance**: AI-powered tools can help people explore various job choices based on their talents, interests, and aptitudes. AI systems can analyze massive volumes of data to deliver personalized career advice and job prospects.
- **Skill Assessment**: By recognizing skill gaps and proposing suitable training programs or online courses, AI may help with personalized skill development. Virtual instructors or chatbots driven by AI can provide personalized learning experiences while delivering real-time feedback and advice.
- **Job Opportunities**: By connecting people with relevant job opportunities, AI algorithms can improve the productivity of job search procedures. AI-powered systems may analyze resumes, descriptions of jobs, and

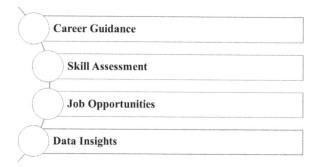

FIGURE 8.2   AI impacts on career development.

Source: Author's own framework.

candidate profiles to discover the greatest fit, increasing the likelihood of landing the appropriate job.

- **Data Insights**: AI may help employees grow in their careers by analyzing employee data and suggesting possible candidates for promotions or leadership positions. It can assist organizations in identifying high-potential individuals and developing customized career development strategies.

Employers have a social duty to prepare the people they hire for a world that is still in its infancy in a labor market that is constantly evolving. Because the typical shelf life of skills is less than 5 years, assisting individuals in learning and upskilling is critical to both their continuing performance and the financial line of a firm (Bayireddi, 2020). AI may help people shift between jobs by finding transferable talents, examining alternative career routes, and offering appropriate training or upskilling opportunities. This can assist people in navigating professional shifts and pursuing new possibilities. AI analytics can analyze massive amounts of data relevant to careers, such as job market trends, demands from industries, and necessary skills. This information can help individuals make educated professional decisions, such as picking in-demand occupations or learning in-demand skills (Omae et al., 2023; Brooks et al., 1995).

AI tools are growing increasingly vital for job advancement in a variety of sectors. Professionals skilled in the use of AI tools will have a competitive advantage in the labor market as AI expands across sectors. Furthermore, many businesses are implementing AI technologies to improve their operations and services, opening up new possibilities for individuals who can create and apply these technologies. Professionals who are knowledgeable about AI technologies and techniques will be better positioned to capitalize on these possibilities and enhance their careers. AI tools can be beneficial for job advancement and development. Here are some examples of how AI technologies may assist (Anand, 2023):

- **Resume Screening**: AI-powered technologies can assess your resume and make suggestions to help you stand out from the crowd. They can also assist recruiters in identifying relevant keywords and credentials rapidly.
- **Job Search**: AI-powered job search engines can assist you in locating relevant job listings that match your abilities, experience, and interests. They can also suggest occupations that you may not have considered before.
- **Personalized Learning**: AI-powered systems may give personalized learning experiences based on your requirements and objectives. These systems can assess your strengths and shortcomings and make specific recommendations to help you improve your abilities.

Employees benefit greatly from a career development system since it offers them useful guidance and assistance when making career decisions. Employees acquire insights into their own aims, goals, and interests as a result of this method, allowing them to properly plan their professional routes. Individuals who use this technique can create more feasible and doable objectives for themselves throughout their careers.

Furthermore, it promotes successful engagement at all levels of the organization by fostering improved interaction between employees and supervisors. One significant advantage is the ability for employees to get opinions about their progress, allowing them to improve their work style and learn new skills. This procedure eventually results in job enrichment and greater job satisfaction (Wood & Lane, 2023).

8.2 THEORETICAL FRAMEWORK

Arnold and Evangelia presented the JD-R Model in 2006 as a supplement to current employee well-being frameworks. The JD-R Model (Job Demands-Resources) is a well-known theoretical paradigm that describes the link between job features, human resources, and work results. This comprehensive model is a renowned model in the field of career development because it gives a complete view of the subtle interplay between job demands, job resources, individual well-being, and performance (Babasaheb & Sphurti et al., 2023a, b).

The JD-R Model may also be described as a heuristic model that investigates two essential characteristics of the workplace: job demands that employees confront and job resources that they can change. Work demands and job resources, according to the JD-R Model, are both major drivers of employee outcomes. Job demands include the physical, psychological, social, and organizational components of work that need long-term effort and may cause strain or stress. Workplace resources, on the other hand, are the supporting features of work that assist goal accomplishment, alleviate workplace pressures, and generate happy work experiences (Lee, 2018).

Researchers as well as practitioners in the field of career development can get significant insights into the elements that impact individuals' career trajectories, well-being, and performance by studying the complete framework of the JD-R Model. The model serves as a guide for comprehending the intricate interplay of job features and individual resources, eventually assisting in the creation of effective career development initiatives and approaches. Key components of JD-R Model are;

- **Job Demand**: Job demands include a variety of issues that people encounter in the workplace. These demands might include a heavy workload, tight deadlines, mental stress, physical exertion, and difficult working circumstances. When a person's employment expectations surpass his or her resources, it may contribute to elevated stress, burnout, and negative health results.
- **Job Resources/Job Positives**: Job resources are the components of work that enable employees to efficiently satisfy job expectations. Social support from colleagues and supervisors, opportunities for skill development and improvement, autonomy in decision-making, performance feedback, and access to essential tools and equipment are all examples of resources. Job resources improve motivation, engagement, and overall well-being.
- **Personal Resources**: Individual traits and characteristics that individuals bring to their profession are referred to as personal resources. These resources include self-efficacy, resilience, optimism, and a proactive personality. Personal resources operate as buffers against the negative

consequences of occupational expectations, assisting individuals in dealing with difficulties, adapting to changes, and maintaining well-being in the face of pressure.

- **Work Outcomes**: Work consequences are the outcomes of the interaction of employment demands, job resources, and personal capabilities. These outcomes include job satisfaction, performance, engagement, work-life balance, and general well-being. Individuals with enough job and personal resources are more likely to have excellent work results.

8.3 JOB DEMANDS AND JOB POSITIVES IN CAREER DEVELOPMENT

8.3.1 JOB DEMANDS IN CAREER DEVELOPMENT

The JD-R Model (employment Demands-Resources) is a theoretical framework that studies the relationship between work features, personal resources, and employment outcomes. It assesses the influence of occupational demands on individuals' well-being and performance, such as workload, time constraints, and stress. Job resources including social support, autonomy, and growth opportunities can help motivate and engage employees. The method emphasizes the need to manage job demands and develop job resources with the goal of fostering an atmosphere that promotes employee well-being and successful performance. Organizations and people may build methods to boost career development, support work-life balance, and create a good work environment by understanding the interplay between job needs and resources.

The JD-R Model is well-known for its comprehensive view of the complicated interaction of job characteristics (Demerouti & Bakker, 2011). Job demands in career development relate to the different pressures and obstacles that people face at work that might affect their career paths. Workload, time limits, job stress, and physical or mental demands are examples of such demands (Yang & Shiu, 2023). Organizations compete in more competitive marketplaces, which raises productivity, efficiency, and innovation requirements. This competitive pressure increases the strain on staff to perform at a high level.

Work overload and role conflict are two common instances of employment expectations. Coping with job demands can deplete one's assets and exhaust one's energy, a mechanism similar to the renowned stressor-strain relationship (Zekavat, 2022).On one hand, there are job demands—that is, "items that have to be done"—that have an impact on employees' mental health in the form of emotional or cognitive stress, and on the contrary, there are employment resources available inside the workplace, such as social contact. Job expectations have a huge impact on people's employment experiences and outcomes. One of the most convincing reasons individuals leave their employment is a lack of professional advancement.

Employees who work without a clear path to advancement are more likely to experience burnout, according to experts. When this occurs, it is usual for their job motivation and service quality to suffer workloads and time constraints can impair an individual's ability to manage their responsibilities successfully and maintain work-life balance, affecting their job satisfaction and general well-being. Job stress caused by stressful work environments can lead to burnout and impede professional advancement. Job demands can have an impact on career choices and decisions. Individuals should examine the number of employment expectations while analyzing prospective career choices because the demands imposed on employees vary by vocation and industry (Babasaheb & Khang et al., 2023a, b).

Job demands can also influence people's desire to take on new tasks or advance in their professions (Parker et al., 2023). Career Development is a method of enhancing one's individual job skills in order to embark on a desired career. Professionals in fast-paced sectors, such as marketing, where short turnaround times and multiple client projects produce high workload demands, frequently experience difficulties juggling multiple activities at the same time. The stress of meeting deadlines and dealing with a high workload can lead to greater stress and possibly burnout. This can harm job satisfaction, motivation, and general well-being, eventually impeding professional advancement.

A heavy workload can disturb the work-life balance, leaving people with little time and energy for their personal lives and self-care activities. This mismatch might have an influence on job satisfaction and career choices. Individuals may seek job possibilities that provide a better work-life balance, which may impact their career choices and growth path (Hayati et al., 2022). According to a Gartner poll of over 3,300 employees performed in March 2022, less than one-third of employees are aware of how to develop their careers in the next 5 years. Only half of employees feel their managers adapt comments to their preferred career routes, indicating a lack of trust in their managers' abilities to give direction. As a result, a rising number of employees are considering quitting their existing employers.

According to the head of the Gartner HR practice, the poll found that employees are leaving their positions for greater professional development chances (45%) at a similar proportion to those departing for higher pay (48%). Employees have fewer chances to engage with colleagues in other professions as work becomes more decentralized, making it difficult to explore new career options. Employees desire the opportunity to explore several employment opportunities, allowing them to envisage themselves in various jobs. Employees are actively looking out for those

progressive organizations that are investing in technologies that give more personal insight into diverse professions and career trajectories, such as internal networking opportunities and career management assistance (job shadowing, formal and informal talks around career objectives and ambitions). These investments can increase workers' trust in their present organizations by 31% (Baker, 2022). Time constraints are an important part of job demands in career growth. Professionals in event management frequently confront tight timetables and demanding deadlines in order to properly organize and execute events. Event management necessitates precise coordination, rigorous preparation, and effective execution within strict time constraints. The existence of time constraints produces a sense of intensity and necessitates persons working quickly and effectively to fulfill deadlines.

Event organizers must manage numerous activities at the same time, such as venue selection, vendor collaboration, controlling the budget, and logistical planning, all within a constrained time frame. Excessive time constraints can cause stress and have an influence on job satisfaction and work-life balance. Long hours of work, increased stress, and eventual burnout may arise from the relentless struggle against time. This can have a negative influence on one's general well-being, output, and work satisfaction deviating from career advancement (Santos, 2016; Briggs et al., 2012). Mental demands are especially important in jobs that involve cognitive exertion, complicated problem-solving, and analytical thinking. Data analysis, handling finances, and strategic consulting are all fields with strong mental demands.

Excessive mental demands might also be difficult. Under time restrictions, the pressure to solve complicated issues and make vital judgments can lead to stress and mental exhaustion. This may have an impact on job satisfaction, work-life balance, and general well-being (Dispenza, 2021). Managing job demands is critical for professional advancement. This includes adopting effective coping methods, requesting help from bosses and coworkers, and applying work-life balance measures. Organizations can also help to create a supportive work environment by addressing excessive job expectations through workload management, flexible schedules, and stress management services.

8.3.2   JOB POSITIVES IN CAREER DEVELOPMENT

The good features or advantages of a job that are beneficial to an individual's professional growth, work happiness, and overall career progression are referred to as job positives. These advantages provide an atmosphere in which individuals may grow, develop their abilities, and accomplish their professional objectives. When career development programs are linked with an organization's continuing training and development initiatives, their efficacy is maximized. This necessitates the deployment of a well-planned career development strategy that is tailored to the organization's specific aims and demands (Lannin & Townsend, 2020).

A well-designed automated career management system not only helps the organization but also improves communication between employees and managers or supervisors. This integration provides distinct benefits to all stakeholders, helping to the creation of an organizational culture that fosters such practices inside the firm. When an organization has a thorough awareness of its employees' strengths and weaknesses, attitudes and behaviors, values and ambitions, and skills and competencies, it can effectively use those abilities and place them in appropriate roles (Shashi et al., 2023).

Successful communication at all levels may be done by ensuring that all critical information and facts are shared across the organization. This promotes an organizational culture that prioritizes and improves communication. Furthermore, it allows businesses to retain key personnel by addressing their wants and preferences. The capacity of the organization to obtain all required information from a single individual enables it to make efforts to retain such employees (Bowling & Thieman, 2020).

Career development is a people-oriented process of improving one's talents to achieve one's chosen career. When employees feel disheartened about the job demands in the organization, job positives bring the light of belongingness to the working environment creating a positive attitude in employees. Motivation is one of the tactics used by superiors to encourage subordinates to perform properly and wisely in accordance with what is expected. Motivation may assist managers in understanding each employee's work attitude. Bosses may encourage their staff in many ways depending on the attitude of each individual. As a result, subordinates must be inspired since many subordinates wish to work after being encouraged by their superiors (Setyawati et al., 2022). Employees who do well are frequently recognized and rewarded for their efforts. Recognizing their efforts and accomplishments boosts their confidence, drive, and happiness, pushing them to keep improving their abilities and pursue professional progress.

Opportunities for progression are essential for bolstering professional development. Jobs that provide clear avenues for advancement, promotions, and increasing responsibilities help to foster a sense of accomplishment and purpose. Employees are motivated to improve their abilities and knowledge in order to advance in their professions (Samroodh et al., 2022). Skills, whether technical or soft, are key assets and traits for job success. It is critical to build transferable abilities that will allow you to compete in the job. Understanding key talents, their significance, and how to acquire them will assist you in continuously improving them for personal and professional progress.

Skills development is critical since it allows you to improve talents and attributes that are required for effective work performance. You may also start your path to personal development by mastering these skills, which will allow you to maximize your potential and achieve your career objectives in record time. Jobs sometimes need specialized technical skills relating to the industry or profession in which you work. You may improve your understanding and competence in these technical skills by gaining hands-on experience. If you are a software developer, for example, you may improve your programming languages, database administration, or web development abilities via hands-on experience and exposure to real-world projects (Wang et al., 2020).

Achieving a work-life balance is critical for those who want to live a happy, productive, and profitable life. It has evolved as a popular ambition for individuals wanting a higher quality of life. Employers can benefit greatly from encouraging a good work-life balance. Implementing an efficient work-life balance approach inside a company allows employees to participate in their community while also managing expenses and minimizing attrition, resulting in increased efficiency. Increased devotion and drive to work lead to increased productivity, lower absenteeism, and enhanced physical and emotional well-being. Striving for this balance improves interpersonal connections.

Changes to work hours, altering task duties, and establishing flexible deadlines are all options. This is a major component of job positives in career development (Adnan Bataineh, 2019). Meaningful work is an essential component of a pleasant work environment that has a substantial influence on career development. When employees do work that coincides with their beliefs, interests, and passions, it gives them a feeling of purpose and fulfillment in their careers. Increased job satisfaction is one of the most important advantages of meaningful employment. Employees are more likely to be satisfied and pleased with their jobs when they consider their work has a point and helps to accomplish something meaningful (Dik et al., 2022). This strong emotional attachment to their work can result in greater motivation, engagement, and satisfaction with their work.

Work that is meaningful increases psychological well-being. Individuals who work on projects that connect with their beliefs and passions report higher levels of psychological well-being, notably higher degrees of happiness, self-esteem, and general life satisfaction. This sense of well-being pervades all aspects of their lives, improving their overall quality of life and adding to their total work success and fulfillment. Positive work settings frequently feature supportive leaders and coworkers who offer direction, feedback, and opportunities for cooperation. This promotes collaboration, trust, and camaraderie, which improves job satisfaction and career advancement. Continuous education and development opportunities in positive work settings provide employees with the skills and knowledge required for career advancement.

Implementing a career development program might assist in motivating people to perform at their best. They will work hard to achieve their objectives since they are aware that they have them (Rodríguez-Sánchez et al., 2020). It has been argued that the emotionally draining nature of mental health work raises the risk of burnout, which has also been linked to decreased employee satisfaction and higher rates of desire to quit. Job-life integration is the harmonious merger of job responsibilities with personal commitments and interests, resulting in a balanced and rewarding living. Positive work environments value work-life integration and proactively encourage policies and practices that help employees attain that equilibrium (Lamovšek et al., 2023).

8.4 PERSONAL RESOURCES AND WORK OUTCOMES IN THE CAREER DEVELOPMENT PROCESS

8.4.1 Personal Resources

Personal resources are characteristics of the self that are frequently related to resiliency and hint to people's knowledge of their ability to successfully manage and modify their environment. Personal resources are important in shaping professional growth because they influence an individual's ideas, attitudes, and behaviors. These tools, such as self-efficacy, resilience, and learning orientation, help with professional decisions, adaptation, and ultimately, accomplishment. It has been discovered that self-efficacy is a strong predictor of job interests, goals, exploration, decision-making, performance, and adjustment. Individuals with greater degrees of self-efficacy are more proactive, tenacious, and resilient in their professional endeavors. They are more inclined to establish lofty objectives, take chances, get feedback, and learn from their errors. They are also more likely to handle professional changes, disappointments, and anxieties (Pallavi et al., 2023).

Employees with strong self-efficacy are more likely to set ambitious goals and persevere in the face of hardship, and make aggressive efforts to achieve their professional goals. They have faith in their ability to learn the required skills, overcome obstacles, and achieve in their chosen careers. As a consequence, they are more self-assured and driven to make educated professional decisions and take activities that are in line with their objectives (Burnette et al., 2020). Self-efficacy is a dynamic and situational belief that may be modified by a variety of events. Gender, culture, personality characteristics, environment, social norms, and reinforcement can all have an influence on self-efficacy beliefs in professional advancement. Gender stereotypes and socialization can have an impact on self-efficacy views in specific job fields, while culture might place an emphasis on social support and compliance.

Extraversion, openness, and neuroticism are all personality qualities that might influence self-efficacy (Lee & Brown, 2022). Another key human resource in professional advancement is resilience. It refers to a person's capacity to recover from failures, adapt to changes, and sustain happiness in the face of hardship. Individuals who are resilient cope better with the challenges and uncertainties that arise in their jobs. They exhibit resilience, problem-solving abilities, and the capacity to learn from setbacks, all of which improve their job flexibility. Resilient individuals are more likely to handle job changes, explore new possibilities, and recover from failures by efficiently managing stress and adjusting to changing circumstances, thereby adding to their long-term career accomplishments. Self-efficacy, independence, and risk-taking are all components of professional resilience. Each of the three components adds to one's overall career resilience, both individually and together. For example, self-efficacy, as a conviction in one's own talents in handling career affairs, allows a person to have trust in overcoming career difficulties, and so makes him or her psychologically strong (Jiang et al., 2021; Kodama, 2021).

For modern organizations that are successful, innovation is a favored objective. Because innovation is difficult to attain yet highly prized, it might be thought of as a modern analog of the Greek word 'Eureka' (which means 'I have discovered it'). A growth mindset or learning orientation is a personal resource that promotes professional success by cultivating an attitude of continual learning and development. Individuals that have a high learning orientation welcome challenges, seek criticism, and regard failure as an opportunity for improvement. They are curious and eager to learn new things, which allows them to adapt to changing work needs and industry trends. A learning orientation encourages professional exploration, skill development, and the capacity to grab new possibilities, which improves career flexibility and long-term success (Slåtten et al., 2020).

8.4.2 Work Outcomes in Career Development

Work implications in career development are the outcomes and repercussions of people's work experiences and efforts. Job performance, engagement, and well-being are all important outcomes for professional success and happiness. Understanding the connection between job outcomes and career development is critical for individuals and organizations interested in promoting meaningful and happy careers. Because of the significant shift in focus in psychology and human resources literature, the JD-R model has a significant benefit.

Traditional psychology was essentially a healing and restorative science. At the turn of the century, a group of psychologists advocated for a change in emphasis toward positive aspects of human existence, such as happy sentiments, positive personalities, and positive organizations. Positive psychology viewpoints and concepts are currently being applied to a broad variety of professions and industries, with an emphasis on both the dark and bright sides of employees' functioning being emphasized inside organizations (Luo & Lei, 2021). Job performance is an important job result that has a direct influence on career development. It relates to how well employees accomplish job-related objectives and expectations, display expertise poses, and contribute to organizational success. High job performance is frequently related to professional development, recognition, and growth prospects. According to the JD-R Model, persons with enough job resources and fewer job demands are more likely to attain greater levels of job performance, resulting in better career growth opportunities (Ashwini & Khang, 2023).

In accordance with a Gallup survey, organizations with strong employee engagement enhance output by 17%. High-performing employees are 400% more profitable than ordinary performers, according to research performed by the Society for Human Resource Management (SHRM) (Meng, 2022). The study of occupational happiness stems from the field of positive psychology, where the alleged and measurable characteristics of occupational well-being differ across work situations. Possibilities for career growth are divided into three categories: fulfilling one's career objectives, obtaining professional expertise and skills, and receiving rewards corresponding to those talents and knowledge. Career development programs, in particular, allow employees to advance toward their career goals while also developing professional abilities. Employees who believe they have prospects for advancement within the organization are less inclined to seek opportunities outside.

Employees who are satisfied with their career advancement possibilities are more inclined to stay with the organization, cutting turnover rates and associated costs (Higuchi et al., 2022). Another significant job outcome that promotes career growth is engagement. It refers to an individual's level of passion, devotion, and interest in their profession. Employees who are engaged have a feeling of purpose, a connection to their work, and are driven to give their all. Individuals who are engaged are more inclined to seek out challenges, take initiative, and pursue continuous learning, which can help them advance in their careers. According to the study, engaged employees are more likely to achieve professional success and happiness (Koroglu & Ozmen, 2022).

The current wave of job resignations, dubbed as the Great Resignation, is the consequence of people who do not want to be locked in dead-end employment at firms that show little regard for their employees. This enormous departure is motivated by a desire for a better work-life balance, a higher salary, and better benefits. Furthermore, a lack of professional development opportunities adds to high staff turnover rates. According to a Monster poll performed in the fall of 2021, 45% of questioned employees stated that if they were provided extra training, they would be more inclined to stay in their existing employment. Organizations may improve employee retention and drive staff growth by committing to continual talent development. Employees are more likely to feel loyal and committed to a firm when organizations prioritize career development. They understand the investment made in their progress and are less likely to look for chances elsewhere. This results in higher staff retention rates and lower recruiting and training expenditures (Muhammad et al., 2023).

A technology company may implement a rotation program that exposes staff to diverse departments and projects, building loyalty and lowering the risk of talent turnover (Singh, 2019). Work results and career growth are inextricably linked. Positive work outcomes, such as strong job performance, engagement, and mental wellness, help to improve one's career. Individuals who entertain well, feel engaged, and maintain their well-being are more likely to obtain professional prospects, earn promotions, and experience a fulfilling career path. Career development activities, on the other hand, such as requesting feedback, perpetual learning, and skill development can improve work outcomes by boosting job performance, boosting engagement, and fostering well-being.

8.5   INCORPORATION OF ARTIFICIAL INTELLIGENCE IN THE CAREER DEVELOPMENT PROCESS

AI is the replication of human intelligence in computers that are programed to do activities that would normally need human intellect. It entails creating computer systems capable of learning, reasoning, problem-solving, and making data-driven judgments. The AI wave has seen spectacular peaks and somber valleys throughout the last century. The hype prompted organizations and governments to invest in AI development; however, inadequate results forced many programs to be shelved. The maturity of the methods required to wait for data, as well as the data required to wait for prevalent computational power. AI was the science fiction of the 1920s. By the 1950s, a tiny step had been taken to move AI from hype to reality (Pooja & Khang, 2023).

Alan Turing published a paper titled "Computing Machinery and Intelligence" in which he introduced the Imitation Game (commonly known as the Turing Test) for assessing whether or not a machine is intelligent. The test is simple: if a human judge cannot dependably discern between a machine and a person during contact with a computer, the machine passes the test. The term "Artificial Intelligence" initially emerged as a serious scientific notion in 1956 at the "Dartmouth Conference" organized by Marvin Minsky, John McCarthy, Shannon, and Nathan Rochester. A year after the perceptron, or neural network, was invented, Arthur Samuel, an American computer scientist, coined the word "machine learning." (Surden, 2019).

In the framework of career development, the incorporation of AI technology gained traction in the early 21st century. As breakthroughs in machine learning, NLP, and data analytics advanced, the application of AI in career-related industries grew. AI has been used by organizations and people to improve their career planning, developing skills, job matching, and performance assessment (Murire et al., 2023).

AI-powered career platforms and tools have evolved, with features such as personalized career evaluations, skill training, and job suggestions. These systems analyze massive volumes of data, such as job market trends, skill demand, and individual profiles, to deliver significant insights and career assistance. In accordance with career and fit theories, the more effortlessly a person fits into an organization, the happier he or she is with his or her job. Employee burnout has occurred as a result of a poor person–organizational fit. The incorporation of AI changes the structure and allocation of resources inside organizations, and it has the possibility to disrupt the relationship between people and organizations, hence changing the person–organization fit relationship. As an effect of AI awareness, employees suffer worry and insecurity, which affects their job happiness. Employment and self-efficacy are connected to related self-management (Kong et al., 2021).

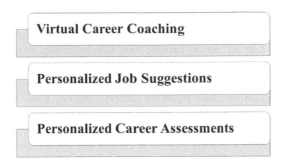

FIGURE 8.3   Significant effect of incorporation of AI in career development.

Source: Author's own framework.

AI in the work environment is critical for increasing efficiency in operations, promoting more rapid decision-making, and driving product and solution development. However, this partnership between humans and machines is proposed to assist workers in enhancing their working talents. It might also progressively establish the assignment (or delegation) of jobs between a worker and their AI helper, in which the AI assistant is assigned organized, repetitive, rules-based activities (Zirar et al., 2023). Some significant effects of AI on career development are described in the Figure 8.3 below.

- **Virtual Career Coaching**: AI chatbots and digital assistants are capable of providing virtual career counseling sessions, advising on career planning, preparation for interviews, and job search techniques. These virtual trainers are available 24 hours a day, 7 days a week to offer assistance and direction. AI chatbots are used by platforms such as "Mya" and "Magnet.me" to give job advice to individuals. These chatbots converse, ask pertinent questions about talents, hobbies, and career objectives and make personalized suggestions for appropriate career pathways and job prospects. Virtual career advisors can help you acquire important abilities for your future job. They can propose suitable online courses, self-learning tools, and insights on in-demand talents for certain sectors or career opportunities (Muthmainnah et al., 2023).
- **Personalized Job Suggestions**: AI algorithms can analyze a person's talents, experience, and career choices to suggest work options that match their profile. This saves time and effort while looking for suitable job vacancies and enhances the likelihood of finding the appropriate fit. LinkedIn's "Jobs You May Be Interested In" function analyzes a user's profile, contacts, and interests to deliver personalized job suggestions based on AI algorithms. To recommend appropriate work prospects, it considers characteristics such as skills, industry, geography, and career record. AI-powered recommendation engines are used in several talent acquisition systems, including Greenhouse and Workday. These platforms examine candidate profiles, talents, and interests in order to match them with appropriate job positions

inside an organization. This improves the recruiting process' efficiency and raises the chance of finding the ideal candidate for a certain post.

- **Personalized Career Assessments**: Individual talents, interests, and preferences may be analyzed by AI-powered technologies to generate personalized job evaluations. This allows people to get insight into their own talents, shortcomings, and potential career options. (2021) is an online portal that provides personalized career evaluations using AI algorithms. Users can take a quiz that evaluates their interests, talents, values, and personality qualities. AIoCF (2021) recommends prospective career pathways that match the individual's profile based on the evaluation findings. CareerExplorer is an AI-powered career evaluation tool that assesses an individual's personality, interests, values, and abilities. The platform analyzes assessment data and generates personalized career suggestions using machine learning techniques.

CIAG services can be developed in a variety of ways, including career counseling, career information, and career learning and matching to relevant occupations. In recent years, AI has been used in industries with comparable societal issues. Given the current absence of CIAG-specific research, experience from similar situations can throw some insight into career-related applications. Therapeutic counseling, library and information services, education, and personnel selection are some of these disciplines.

AI in career development provides individuals with personalized insights, assistance, and tools to help them make educated decisions, improve their abilities, and effectively navigate their professional trajectories. It simplifies several elements of professional growth, allowing individuals to remain competitive and fulfill their professional objectives (Wilson et al., 2022). Career education is another area where AI is anticipated to boost career assistance and advising. AI is already being used to personalize learning paths in educational programs (for example, language learning software) and digital games. It may be used to lead students through educational programs in such a manner that they focus on relevant activities rather than learning about topics they already know (Arpita et al., 2023).

Finally, in the near future, AI applications may be used to boost the provision of career guidance and counseling, for example, by recognizing people with career development needs based on their internet-search behavior and presenting them with specific advertisements for career services that are likely to be of use to them (rather than primarily to those providing the services) (Köchling et al., 2021).

8.6   CONCLUSION

The importance of AI in job development is changing how people manage their professional lives. AI-powered tools and technology provide unparalleled personalized career advising, job matching, and skill development. Individuals may benefit from bespoke career evaluations, personalized job suggestions, and virtual career counseling by employing AI. AI integration in career development enables individuals to make better-educated decisions, uncover appropriate career options, and advance their professional development (Khang & Shah et al., 2023).

To deliver accurate and personalized suggestions, AI systems can analyze massive volumes of data, including talents, experiences, and job inclinations. This saves time and effort during the job search, enhances the likelihood of finding the ideal match, and promotes professional happiness and success. As AI advances, it is critical for career development experts, educators, and regulators to keep current on the newest trends and best practices. Individuals can maneuver the ever-changing job market, make educated decisions, and establish happy and successful careers by integrating AI into career development (Khang & Muthmainnah et al., 2023).

To summarize, the role of AI in career development is an exciting opportunity for individuals to realize their full potential and construct a meaningful and productive professional life. Individuals may use AI to establish a perfect match between their abilities, hobbies, and employment prospects, thereby influencing their future success (Khang & Kali et al., 2023).

REFERENCES

Adnan Bataineh, K. (2019). Impact of work-life balance, happiness at work, on employee performance. *International Business Research*, 12(2), 99–112. https://pdfs.semanticscholar.org/255e/18221ad3601c8d3ac91a74d6c613c58cc6e9.pdf

AIoCF. (2021). AI Skills-based competency ecosystem in digital economy, The AI-oriented Competency Model for Digital Economy 5.0. Retrieved from https://scedex.com/quickstart.htm

Anand, P. (2023). AI tools that can help with career growth. DATAQUEST. Retrieved from https://www.dqindia.com/ai-tools-that-can-help-with-career-growth/

Arpita, N., Ipseeta, S., Patnaik, B. C. M., Baral, S. K., & Khang, A. (2023). Impact of artificial intelligence (AI) on talent management (TM): a futuristic overview. *Designing Workforce Management Systems for Industry 4.0: Data-Centric and AI-Enabled Approaches* (1st ed., pp. 32–50). CRC Press, Boca Raton, FL. https://doi.org/10.1201/9781003357070-9

Ashwini, Y. S. & Khang, A. (2023). Challenges faced by marketers in developing and managing contents in workforce development system. *Designing Workforce Management Systems for Industry 4.0: Data-Centric and AI-Enabled Approaches* (1st ed., pp. 332–359). CRC Press, Boca Raton, FL. https://doi.org/10.1201/9781003357070-18

Babasaheb, J., Sphurti, B., & Khang, A. (2023a). Industry revolution 4.0: workforce competency models and designs. *Designing Workforce Management Systems for Industry 4.0: Data-Centric and AI-Enabled Approaches* (1st ed., pp. 14–31). CRC Press, Boca Raton, FL. https://doi.org/10.1201/9781003357070-2

Babasaheb, J., Sphurti, B., & Khang, A. (2023b). Design of competency models in the human capital management system. *Designing Workforce Management Systems for Industry 4.0: Data-Centric and AI-Enabled Approaches* (1st ed., pp. 32–50). CRC Press, Boca Raton, FL. https://doi.org/10.1201/9781003357070-3

Baker, M. (2022). Gartner HR research finds just 25% of employees are confident about their career at their current organization. Gartner. Retrieved from https://www.gartner.com/en/newsroom/press-releases/2022-09-15-gartner-hr-research-finds-just-25-percent-of-employees-are-confident-about-their-career-at-their-current-organization

Bayireddi, M. (2020). Council post: How AI-powered career pathing refreshes the employee experience. Forbes. Retrieved from https://www.forbes.com/sites/forbestechcouncil/2020/01/23/how-ai-powered-career-pathing-refreshes-the-employee-experience/?sh=17af20df3d2c

Bowling, A. M. & Thieman, E. B. (2020). Exploring stress and recovery among high-achieving career development event teams: a mixed methods study. *Journal of Agricultural Education*, 61(2), 142–161. https://eric.ed.gov/?id=EJ1263511

Briggs, E., Jaramillo, F., & Weeks, W. A. (2012). Perceived barriers to career advancement and organizational commitment in sales. *Journal of Business Research*, 65(7), 937–943. https://www.sciencedirect.com/science/article/pii/S014829631100138X

Brooks, L., Cornelius, A., Greenfield, E., & Joseph, R. (1995). The relation of career-related work or internship experiences to the career development of college seniors. *Journal of Vocational Behavior*, 46(3), 332–349. https://www.sciencedirect.com/science/article/pii/S000187918571024X

Burnette, J. L., Pollack, J. M., Forsyth, R. B., Hoyt, C. L., Babij, A. D., Thomas, F. N., & Coy, A. E. (2020). A growth mindset intervention: enhancing students' entrepreneurial self-efficacy and career development. *Entrepreneurship Theory and Practice*, 44(5), 878–908. https://journals.sagepub.com/doi/abs/10.1177/1042258719864293

Chan, C. D., Hammer, T. R., Richardson, L., & Hughes, C. L. (2022). Through the relational looking glass: applications of relational-cultural theory to career development and mental health. *Journal of Employment Counseling*, 59(4), 168–178. https://onlinelibrary.wiley.com/doi/abs/10.1002/joec.12185

Columbus, L. (2018). 10 charts that will change your perspective on Artificial Intelligence's growth. Forbes. Retrieved from https://www.forbes.com/sites/louiscolumbus/2018/01/12/10-charts-that-will-change-your-perspective-on-artificial-intelligences-growth/?sh=7a19defa4758

Demerouti, E. & Bakker, A. B. (2011). The job demands-resources model: challenges for future research. *SA Journal of Industrial Psychology*, 37(2), 01–09. https://www.scielo.org.za/scielo.php?pid=S2071-07632011000200001&script=sci_arttext

Dik, B. J., Marsh, D. R., & Alayan, A. J. (2022). Career development and mental health assessment, prevention, and intervention for children and adolescents. *Mental Health Assessment, Prevention, and Intervention: Promoting Child and Youth Well-Being*, pp. 391–410. Springer International Publishing, Cham. https://link.springer.com/chapter/10.1007/978-3-030-97208-0_18

Dispenza, F. (2021). Empowering the career development of persons with disabilities (PWD). *Journal of Career Development*, 48(5), 670–685. https://journals.sagepub.com/doi/abs/10.1177/0894845319884636

Gati, I. & Kulcsar, V. (2021). Making better career decisions: from challenges to opportunities. *Journal of Vocational Behavior*, 126, 103545. https://www.sciencedirect.com/science/article/pii/S0001879121000178

Hayati, N., Yusuf, A. M., & Miharja, R. (2022). The relationship between workload and career development on job satisfaction; case study PT XYZ. *HOLISTICA-Journal of Business and Public Administration*, 13(1), 125–132. https://holisticajournal.ro/docs/8b5b58a6423953bf8fdf35d766fdec15.pdf

Herr, E. L. (2001). Career development and its practice: a historical perspective. *The Career Development Quarterly*, 49(3), 196–211. https://onlinelibrary.wiley.com/doi/abs/10.1002/j.2161-0045.2001.tb00562.x

Higuchi, T., Takahashi, K., & Yan, L. (2022). An analysis of work engagement based on the job demands-resources (Jd-R) model: empirical findings under remote work environments in Japan. *Global Business Journal*, 8(1), 24–33. https://www.jstage.jst.go.jp/article/gbj/8/1/8_24/_article/-char/ja/

Jiang, Z., Jiang, Y., & Nielsen, I. (2021). Thriving and career outcomes: the roles of achievement orientation and resilience. *Human Resource Management Journal*, 31(1), 143–164. https://onlinelibrary.wiley.com/doi/abs/10.1111/1748-8583.12287

Khang, A., Kali, C. R., Suresh Kumar, S., Amaresh, K., Sudhansu Ranjan, D., & Manas Ranjan, P. (2023). Enabling the future of manufacturing: integration of robotics and IoT to smart factory infrastructure in industry 4.0. *AI-Based Technologies and Applications in the Era of the Metaverse* (1st ed., pp. 25–50). IGI Global Press, Hershey, PA. https://doi.org/10.4018/978-1-6684-8851-5.ch002

Khang, A., Muthmainnah, M., Seraj, P. M. I., Yakin, A. A., Obaid, A. J., & Panda, M. R. (2023). AI-aided teaching model for the education 5.0 ecosystem. *AI-Based Technologies and Applications in the Era of the Metaverse* (1st ed., pp. 83–104). IGI Global Press, Hershey, PA. https://doi.org/10.4018/978-1-6684-8851-5.ch004

Khang, A., Rani, S., Gujrati, R., Uygun, H., & Gupta, S. K. (2023). *Designing Workforce Management Systems for Industry 4.0: Data-Centric and AI-Enabled Approaches*. CRC Press, Boca Raton, FL. https://doi.org/10.1201/9781003357070

Khang, A., Shah, V., & Rani, S. (2023). *AI-Based Technologies and Applications in the Era of the Metaverse* (1 ed.). IGI Global Press, Hershey, PA. https://doi.org/10.4018/978-1-6684-8851-5

Köchling, A., Wehner, M., & Ruhle, S. A. (2021). Feeling treated fairly? Employee reactions toward AI in career development systems. *Academy of Management Proceedings*, 2021(1), 13802. https://journals.aom.org/doi/abs/10.5465/AMBPP.2021.13802abstract

Kodama, M. (2021). Functions of career resilience against changes during working life in Japan: focus on health condition changes and task or job changes. *Sage Open*, 11(1), 21582440211002182. https://journals.sagepub.com/doi/abs/10.1177/21582440211002182

Kong, H., Yuan, Y., Baruch, Y., Bu, N., Jiang, X., & Wang, K. (2021). Influences of artificial intelligence (AI) awareness on career competency and job burnout. *International Journal of Contemporary Hospitality Management*, 33(2), 717–734. https://www.emerald.com/insight/content/doi/10.1108/IJCHM-07-2020-0789/full/html

Koroglu, Ş. & Ozmen, O. (2022). The mediating effect of work engagement on innovative work behavior and the role of psychological well-being in the job demands-resources (JD-R) model. *Asia-Pacific Journal of Business Administration*, 14(1), 124–144. https://www.emerald.com/insight/content/doi/10.1108/APJBA-09-2020-0326/full/html

Lamovšek, A., Černe, M., Radević, I., & Božič, K. (2023). The key to work-life balance is (enriched) job design? Three-way interaction effects with formalization and adaptive personality characteristics. *Applied Research in Quality of Life*, 18(2), 647–676. https://link.springer.com/article/10.1007/s11482-022-10100-9

Lannin, A. & Townsend, M. A. (2020). Graduate student perspectives: career development through serving as writing-intensive GTAs. Graduate writing across the disciplines: Identifying, teaching, and supporting. University Press of Colorado. The WAC Clearinghouse. https://www.fuctcompany.com/docs/books/graduate/lannin.pdf

Lee, R. & Brown, C. (2022). The relations among career-related self-efficacy, perceived career barriers, and stigma consciousness in men with felony convictions. *Psychological Services*, 29(2), 143–161. https://psycnet.apa.org/record/2022-48318-001

Lee, Y. & Eissenstat, S. J. (2018). An application of work engagement in the job demands-resources model to career development: assessing gender differences. *Human Resource Development Quarterly*, 29(2), 143–161. https://onlinelibrary.wiley.com/doi/abs/10.1002/hrdq.21310

Lent, R. W. & Brown, S. D. (2013). Understanding and facilitating career development in the 21st century. *Career Development and Counseling: Putting Theory and Research to Work*, 2, 1–26. https://www.google.com/books?hl=en&lr=&id=6mOX6XBxrv0C&oi=fnd&pg=PA1&dq=Understanding+and+facilitating+career+development+in+the+21st+century.+Career+development+and+counseling:+Putting+theory+and+research+to+work&ots=oHInijAU8J&sig=sDB70DZVYbNDrcumSnXCnDh8UI0

Luo, M. & Lei, J. (2021). Using the JD-R model to predict the organizational outcomes of social workers in Guangzhou, China. *Journal of the Society for Social Work and Research*, 12(2), 349–369. https://www.journals.uchicago.edu/doi/abs/10.1086/714311

Meng, L., Du, J., & Lin, X. (2022). Surviving bench stress: meaningful work as a personal resource in the expanded job demands-resources model. *Current Psychology*, 42, 1–12. https://link.springer.com/article/10.1007/s12144-022-02956-9

Muhammad, M. W., Sikandar, A., Muthmainnah, M. A. R., & Khang, A. (2023). Unmanned aerial vehicles (UAVS) in modern agriculture: advancements and benefits. *Advanced Technologies and AI-Equipped IoT Applications in High-Tech Agriculture* (1st ed., pp. 105–126). https://doi.org/10.4018/978-1-6684-9231-4.ch006

Murire, O., Cilliers, L., & Chinyamurindi, W. T. (2023). Social media role in addressing employability challenges of graduates: an expert opinion analysis. *SA Journal of Human Resource Management*, 21, 9. https://sajhrm.co.za/index.php/sajhrm/article/view/2047/3251

Muthmainnah, M., Khang, A., Seraj, P. M. I., Yakin, A. A., Oteir, I., & Alotaibi, A. N. (2023). An innovative teaching model - the potential of metaverse for English learning. *AI-Based Technologies and Applications in the Era of the Metaverse* (1st ed., pp. 105–126). IGI Global Press, Hershey, PA. https://doi.org/10.4018/978-1-6684-8851-5.ch005

Omae, Y., Furuya, T., Matsushita, M., Mizukoshi, K., Yatsushiro, K., & Takahashi, H. (2023). Artificial intelligence education in an elementary school and its evaluation by career development, motivation and rubrics. *Information and Technology in Education and Learning*, 3(1), 2. https://www.jstage.jst.go.jp/article/itel/3/1/3_3.1.Trans.p002/_article/-char/ja/

Pallavi, J., Vandana, T., Ravisankar, M., & Khang, A. (2023). Data-driven AI models in the workforce development planning. *Designing Workforce Management Systems for Industry 4.0: Data-Centric and AI-Enabled Approaches* (1st ed., pp. 179–198). CRC Press, Boca Raton, FL. https://doi.org/10.1201/9781003357070-10

Parker, A., Waldstrøm, C., & Shah, N. P. (2023). The coevolution of emotional job demands and work-based social ties and their effect on performance. *Journal of Management*, 49(5), 1601–1632. https://journals.sagepub.com/doi/abs/10.1177/01492063221087636

Pooja, A. & Khang, A. A study on the impact of the industry 4.0 on the employees performance in banking sector. *Designing Workforce Management Systems for Industry 4.0: Data-Centric and AI-Enabled Approaches* (1st ed., pp. 384–400). CRC Press, Boca Raton, FL. https://doi.org/10.1201/9781003357070-20

Rodríguez-Sánchez, J. L., González-Torres, T., Montero-Navarro, A., & Gallego-Losada, R. (2020). Investing time and resources for work-life balance: the effect on talent retention. *International Journal of Environmental Research and Public Health*, 17(6), 1920. https://www.mdpi.com/1660-4601/17/6/1920

Samroodh, M., Anwar, I., Ahmad, A., Akhtar, S., Bino, E., & Ali, M. A. (2022). The indirect effect of job resources on employees' intention to stay: a serial mediation model with psychological capital and work-life balance as the mediators. *Sustainability*, 15(1), 551. https://www.mdpi.com/2071-1050/15/1/551

Santos, G. G. (2016). Career barriers influencing career success: a focus on academics' perceptions and experiences. *Career Development International*, 21(1), 60–84. https://www.emerald.com/insight/content/doi/10.1108/CDI-03-2015-0035/full/html

Saxena, A. C., Ojha, A., Sobti, D., & Khang, A. (2023). Artificial intelligence (AI) centric model in metaverse ecosystem. *AI-Based Technologies and Applications in the Era of the Metaverse* (1 ed., pp. 1–24). IGI Global Press, Hershey, PA. https://doi.org/10.4018/978-1-6684-8851-5.ch001

Saxena, P., Saxena, V., Pandey, A., Flato, U., & Shukla, K. (2023). *Multiple Aspects of Artificial Intelligence.* Book Saga Publications, Thousand Oaks, CA. https://www.google.com/books?hl=en&lr=&id=HBTJEAAAQBAJ&oi=fnd&pg=PP3&dq=Multiple+Aspects +of+Artificial+Intelligence.+Book+Saga+Publications&ots=E-c8mugR_K&sig=95 YtcANTejbOGmUwPoJHNYDH8H8

Setyawati, N. W., PG, D. S. W., & Rianto, M. R. (2022). Career development, motivation and promotion on employee performance. *East Asian Journal of Multidisciplinary Research,* 1(9), 1957–1970. https://journal.formosapublisher.org/index.php/eajmr/article/view/1453

Shashi, K. G., Khang, A., Parin, S., Chandra Kumar, D., Anchal, P. (2023). Data mining processes and decision-making models in personnel management system. *Designing Workforce Management Systems for Industry 4.0: Data-Centric and AI-Enabled Approaches* (1st ed., pp. 89–112). CRC Press, Boca Raton, FL. https://doi.org/10.1201/9781003357070-6

Singh, D. (2019). A literature review on employee retention with focus on recent trends. *International Journal of Scientific Research in Science and Technology,* 6(1), 425–431. https://www.academia.edu/download/60641051/5095_lit_review_paper20190919- 118800-dzni7v.pdf

Slåtten, T., Mutonyi, B. R., & Lien, G. (2020). The impact of individual creativity, psycho-logical capital, and leadership autonomy support on hospital employees' innovative behaviour. *BMC Health Services Research,* 20(1), 1–17. https://bmchealthservres. biomedcentral.com/articles/10.1186/s12913-020-05954-4

Snehal, M., Babasaheb, J., & Khang, A. (2023). Workforce management system: concepts, definitions, principles, and implementation. *Designing Workforce Management Systems for Industry 4.0: Data-Centric and AI-Enabled Approaches* (1st ed., pp. 384–400). CRC Press, Boca Raton, FL. https://doi.org/10.1201/9781003357070-1

Surden, H. (2019). Artificial intelligence and law: an overview. *Georgia State University Law Review,* 35, 19–22. https://papers.ssrn.com/sol3/papers.cfm?abstract_id=3411869

Vondracek, F. W., Lerner, R. M., & Schulenberg, J. E. (2019). *Career Development: A Life-Span Developmental Approach.* Routledge, London. https://www.google.com/books?hl=en&lr=&id=kTiDDwAAQBAJ&oi=fnd&pg=PT7&dq=Career+develo pment:+A+life-span+developmental+approach.+Routledge.&ots=QeWE0HyOzU &sig=Dx5dOLGyNKqRjueWWd2CYT_t9T4

Wang, K. L., Johnson, A., Nguyen, H., Goodwin, R. E., & Groth, M. (2020). The changing value of skill utilisation: interactions with job demands on job satisfaction and absen-teeism. *Applied Psychology,* 69(1), 30–58. https://iaap-journals.onlinelibrary.wiley.com/doi/abs/10.1111/apps.12200

Westman, S., Kauttonen, J., Klemetti, A., Korhonen, N., Manninen, M., Mononen, A., & Paananen, H. (2021). Artificial intelligence for career guidance--current requirements and prospects for the future. *IAFOR Journal of Education,* 9(4), 43–62. https://eric. ed.gov/?id=EJ1318705

Wilson, M., Robertson, P., Cruickshank, P., & Gkatzia, D. (2022). Opportunities and risks in the use of AI in career development practice. *Journal of the National Institute for Career Education and Counselling,* 48(1), 48–57. https://napier-repository. worktribe.com/output/2848811/opportunities-and-risks-in-the-use-of-ai-in-career-development-practice

Wood, S. M. & Lane, E. M. (2023). Supporting career development in advanced programs. *Content-Based Curriculum for Advanced Learners,* pp. 505–522. Routledge, London. https://www.taylorfrancis.com/chapters/edit/10.4324/9781003310426-33/supporting-career-development-advanced-programs-susannah-wood-erin-lane

Yang, F. H. & Shiu, F. J. (2023). Evaluating the impact of workplace friendship on social loafing in long-term care institutions: an empirical study. *Sustainability,* 15(10), 7828. https://www.mdpi.com/2071-1050/15/10/7828

Zekavat, M. (2022). The social, political, and psychological affordances of pandemic humor and satire in the United States of America and Iran. *Amerikastudien-American Studies*, 67(4), 521–540. https://research.rug.nl/files/348755161/102204011.pdf

Zirar, A., Ali, S. I., & Islam, N. (2023). Worker and workplace artificial intelligence (AI) coexistence: emerging themes and research agenda. *Technovation*, 124, 102747. https://www.sciencedirect.com/science/article/pii/S0166497223000585

9 Artificial Intelligence-Powered Competency Framework for Robo-Advisory Services in the Era of Digital Economy

Debanjalee Bose and Sakthi Srinivasan K.

9.1 INTRODUCTION

The digital economy era has a transformative wave of technological innovations, reshaping industries and revolutionizing conventional business models. In the financial sector, the convergence of artificial intelligence (AI) and data analytics has given rise to a paradigm shift in investment advisory services, epitomized by the emergence of AI-powered robo-advisory platforms. These platforms leverage advanced algorithms and data-driven insights to provide automated and personalized investment advice, catering to a diverse spectrum of investors. As the financial landscape evolves under the influence of technology, the competency framework required by finance professionals to excel in AI-powered robo-advisory services becomes a pivotal area of exploration (Khang & Kali et al., 2023).

The convergence of AI and the digital economy has sparked a profound transformation in the way investment advisory services are conceived, delivered, and received. At the heart of this evolution lies the remarkable emergence of AI-powered robo-advisory services – a fusion of cutting-edge technology and financial acumen that has redefined how individuals and institutions manage their financial portfolios. These robo-advisors, driven by advanced algorithms and data-driven insights, hold the promise of democratizing access to professional investment guidance while streamlining investment processes. However, as the influence of robo-advisory services continues to expand, it becomes essential to evaluate the competency framework underpinning their operations, ensuring that the fusion of technology and financial expertise aligns seamlessly in the dynamic digital economy era (Rana et al., 2021).

This chapter embarks on an exploration of the multifaceted dimensions of this competency framework, delving into the technical prowess, regulatory acumen, behavioral understanding, and ethical considerations that collectively shape the future of AI-powered robo-advisory services. As the financial ecosystem evolves,

DOI: 10.1201/9781003440901-9

this evaluation stands as a cornerstone in understanding how these competencies intertwine to unlock the full potential of AI-driven advisory in an era defined by technological innovation and digital interconnectedness (Khanh & Khang, 2021).

The integration of AI into investment advisory has spurred a dynamic interplay between technological prowess and human expertise. In this context, a robust competency framework serves as the cornerstone for successfully navigating the evolving financial ecosystem. This chapter embarks on a comprehensive evaluation of the essential competencies that underpin the proficient functioning of AI-powered robo-advisory services within the contours of the digital economy era. The exploration extends beyond technical skills and delves into regulatory acumen, ethical considerations, and behavioral attributes that collectively shape the role of finance professionals in this technology-driven landscape. The advent of AI-powered robo-advisory services has not only transformed investment strategies but has also catalyzed a significant evolution in the roles and responsibilities of financial professionals. As the digital economy continues to reshape the financial landscape, the competencies required for finance professionals engaging with robo-advisory services have undergone a notable transition (Rani et al., 2021).

9.2   WHY IS AI IMPORTANT?

AI is a skill that can be learned. The need for people with AI abilities has increased enormously as AI technology continues to improve and change several sectors, including banking, healthcare, and technology. Learning AI involves understanding the principles, algorithms, and tools that enable machines to simulate human-like intelligence, such as machine learning, neural networks, natural language processing, and computer vision (Khang & Muthmainnah et al., 2023).

AI holds immense significance as it revolutionizes industries, enhances efficiency, and unlocks unprecedented possibilities. By mimicking human intelligence, AI enables machines to learn, reason, and make decisions, thus enabling the automation of complex tasks and data analysis on a scale beyond human capacity. AI-powered systems are transforming healthcare by aiding diagnostics and personalized treatments, optimizing supply chains, enhancing customer experiences through chatbots, and contributing to scientific discoveries. Its potential to predict trends, optimize operations, and adapt to changing environments has become a strategic advantage in business. As AI continues to advance, its role in reshaping economies, improving lives, and pushing the boundaries of innovation remains pivotal (Muthmainnah et al., 2023).

9.3   ROBO-ADVISORY SERVICES IN AI

Robo-advisory services represent a significant application of AI in the financial domain. These services leverage AI technologies to offer automated and algorithm-driven investment advice and portfolio management. By harnessing sophisticated algorithms, machine learning, and data analytics, robo-advisors analyze investors' financial goals, risk tolerances, and market conditions to provide tailored investment recommendations. This integration of AI enables efficient portfolio diversification, risk assessment, and rebalancing, often at a fraction of the cost of traditional human advisors (Khang & Misra et al., 2023).

As a result, robo-advisory services have democratized access to financial planning, making it accessible to a wider range of individuals and investors. The fusion of robo-advisory services and AI illustrates the transformational potential of technology in reshaping traditional financial practices, optimizing investment strategies, and fostering a more inclusive approach to wealth management.

9.4 FUNCTIONS OF ROBO-ADVISORY SERVICES

- **Risk Assessment and Profiling**: Robo-advisors use questionnaires and algorithms to assess an investor's risk tolerance, financial goals, and investment preferences. This profiling helps determine the most suitable investment strategy.
- **Portfolio Creation and Customization**: Based on the investor's risk profile and financial objectives, robo-advisors create personalized investment portfolios. These portfolios are typically diversified across various asset classes to manage risk and optimize returns.
- **Automated Asset Allocation**: Robo-advisors use sophisticated algorithms to allocate funds across different asset classes such as stocks, bonds, and alternative investments. This allocation is based on the investor's risk tolerance and market conditions.
- **Continuous Monitoring and Rebalancing**: Robo-advisors continuously monitor the performance of the investment portfolio. When the actual allocation deviates from the target allocation due to market fluctuations, the robo-advisor automatically rebalances the portfolio to bring it back in line with the intended distribution.
- **Tax Optimization and Harvesting**: Some robo-advisors offer tax optimization strategies, including tax-loss harvesting. This involves strategically selling losing investments to offset capital gains and reduce tax liabilities.
- **Goal-Based Investing**: Investors can set specific financial goals, such as retirement savings or buying a house. Robo-advisors help create investment plans aligned with these goals and provide projections on whether the goals are achievable based on the current strategy.
- **Dynamic Adjustments**: AI-powered robo-advisors can adapt investment strategies based on changes in the investor's financial situation, goals, and market conditions. This ensures that the investment approach remains relevant over time.
- **Behavioral Finance Insights**: Some robo-advisors incorporate behavioral finance principles to mitigate emotional biases that can impact investment decisions. They may provide guidance to help investors stay disciplined during market volatility.
- **Access to Financial Education**: Many robo-advisory platforms offer educational resources and insights to help investors better understand financial concepts and make informed decisions.

9.5   RISE OF ROBO-ADVISORS

The robo-advisors in financial decision-making have inaugurated a fresh era in investment management. These AI-driven platforms have risen to prominence as revolutionary instruments that democratize and streamline the investment process. Capitalizing on technological strides and data analysis, robo-advisors extend accessible and automated solutions to individuals pursuing their financial aspirations. These platforms offer an array of advantages, including tailored portfolio creation based on individual risk profiles, effective mechanisms for maintaining desired asset allocations through rebalancing, and ongoing vigilance of market trends (Khang & Hajimahmud et al., 2023).

Furthermore, robo-advisors frequently present lower fees in contrast to traditional human advisors, rendering proficient investment management more inclusive for a wider spectrum of investors. By assimilating data insights and principles of behavioral finance, robo-advisors mitigate biases and optimize investment choices. While they excel in convenience, transparency, and cost-effectiveness, investors should still weigh their personal inclinations and the extent of human engagement they seek as shown in Figure 9.1.

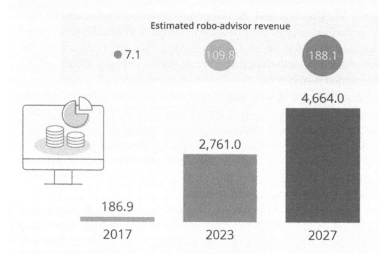

The Rise of the Robo-Advisors

Estimated worldwide assets under management of robo-advisors (in billion U.S. dollars)

Estimated robo-advisor revenue

● 7.1   109.8   188.1

4,664.0

2,761.0

186.9

2017   2023   2027

* Automated online portfolio management of private assets
Source: Statista Market Insights

statista 🗲

FIGURE 9.1   Rise of robo-advisors.

9.6   EFFECT OF AI IN ROBO-ADVISORY SERVICES

The effect of AI-powered robo-advisory services in the digital economy era is profound and far-reaching. These services have revolutionized the landscape of investment advisory by leveraging AI and data analytics to offer automated, personalized, and data-driven investment recommendations. They have democratized access to financial advice, catering to a broader spectrum of investors with lower costs and reduced entry barriers. The integration of AI-driven algorithms has enhanced portfolio optimization, risk assessment, and decision-making processes, leading to more informed and efficient investment strategies. However, this technological advancement also poses challenges such as regulatory compliance, ethical considerations, and the need to balance human expertise with algorithmic insights. Overall, AI-powered robo-advisory services have redefined the way investment advice is delivered, reshaping financial roles and prompting a paradigm shift in how individuals and institutions manage their finances in the digital economy era (Khang & Gupta et al., 2023).

9.7   EFFECT OF AI-POWERED ROBO-ADVISORY SERVICES FOR A BETTER FUTURE

AI-powered robo-advisory services hold the potential to shape a better future by democratizing access to financial advice, enhancing investment strategies, and promoting financial literacy. As these services continue to evolve, they contribute to a more inclusive, efficient, and technologically advanced financial ecosystem that benefits both individual investors and the broader economy.

- **Accessibility and Inclusion**: AI-powered robo-advisory services democratize access to financial advice, making it accessible to a wider demographic, including those with limited financial resources. This inclusivity fosters greater financial literacy and empowerment, ultimately leading to better decision-making and improved financial well-being.
- **Innovation Catalyst**: The growth of AI-powered robo-advisory services stimulates innovation in the financial industry, encouraging traditional financial institutions to adopt technology-driven solutions. This innovation spurs healthy competition, driving the industry toward more efficient and customer-centric approaches.
- **Personalization and Tailored Advice**: AI-powered algorithms analyze vast datasets to provide personalized investment recommendations aligned with individual risk profiles and financial goals. This tailored advice enhances investor engagement and satisfaction, promoting long-term financial planning.
- **Data-Driven Insights**: Robo-advisory services generate valuable insights from data patterns that can inform economic trends and market behaviors. These insights contribute to a better understanding of financial markets and the broader economy, aiding policymakers and researchers.

9.8   IMPACT OF AI IN THE DIGITAL ECONOMY

The impact of AI is profound and far-reaching in the place of the digital economy, touching virtually every facet of our lives and industries. AI's ability to process and analyze massive amounts of data at lightning speed has revolutionized decision-making processes, enhancing efficiency and accuracy across sectors. In customer-centric industries, AI-powered chatbots and personalized recommendations have transformed the way businesses interact with consumers, leading to more tailored and responsive services.

In finance, AI-driven algorithms drive robo-advisory services, automating investment strategies and democratizing financial advice. Moreover, AI's predictive capabilities enable proactive maintenance in manufacturing and healthcare, optimizing operations and reducing downtime. However, this transformative impact also raises ethical considerations related to privacy, bias, and job displacement. Ultimately, AI's integration into the digital economy has redefined how we work, communicate, and innovate, promising both unprecedented opportunities and the need for thoughtful regulation and adaptation to ensure a harmonious and prosperous future (Khang & Rani et al., 2022).

9.9   IMPACT OF ROBO-ADVISORY SERVICES IN THE DIGITAL ECONOMY

The impact of robo-advisory services is significant and far-reaching in the era of the digital economy, fundamentally altering the landscape of investment advisory and financial services. These AI-powered platforms have democratized access to professional investment guidance, catering to a broader range of investors with cost-effective solutions. By leveraging data analytics and algorithms, robo-advisors offer personalized investment strategies aligned with individual risk profiles and financial goals, thereby enhancing investment outcomes. This disruption has prompted traditional financial institutions to adapt their services and adopt technology-driven approaches, fostering healthy competition and innovation.

AI robo-advisory services bring innovation to the forefront by harnessing advanced algorithms and machine-learning techniques. These technologies allow for the seamless analysis of extensive datasets and market trends, enabling the creation of sophisticated investment portfolios tailored to individual investors' risk profiles and financial goals. The integration of AI in robo-advisory services enhances the accessibility of professional financial guidance (Beck, 2021). However, the rise of robo-advisory services also poses challenges, including the need for robust regulatory frameworks, addressing ethical considerations, and ensuring transparent communication of algorithmic decision-making. Despite these challenges, robo-advisory services hold the potential to foster financial literacy, optimize investment strategies, and shape a more accessible and efficient financial ecosystem within the digital economy era.

9.10 THE MULTIFACETED COMPETENCY FRAMEWORK FOR ROBO-ADVISORY SERVICES

Robo-advisory services, powered by AI and data analytics, have revolutionized the landscape of investment advisory. In this era of digital finance, the competency framework required for professionals engaging with robo-advisory services has become increasingly intricate and multifaceted. This chapter delves into the diverse and essential competencies that underpin the effective design, development, implementation, and oversight of robo-advisory services. The competency framework discussed here encompasses technical expertise, regulatory acumen, ethical considerations, and the ability to navigate behavioral intricacies, all of which are crucial for successful robo-advisory operations in the digital economy era (Rani et al., 2022).

- **Technical Prowess in Algorithm Development and Data Analytics**: The foundation of robo-advisory services lies in advanced algorithms that drive portfolio optimization, risk assessment, and investment decision-making. Professionals operating in this domain must possess a comprehensive understanding of machine learning techniques, data analysis, and quantitative modeling. The ability to develop and refine algorithms that align with investors' risk profiles and financial goals is paramount. This involves an exploration of algorithm-driven investment decisions, portfolio optimization techniques, and the integration of data analytics for informed investment strategies (Scholz, 2020). Moreover, proficiency in monitoring algorithmic performance and adapting strategies to evolving market conditions ensures the robustness and effectiveness of robo-advisory platforms.
- **Navigating the Regulatory Landscape**: As financial services increasingly embrace AI-driven solutions, navigating the regulatory environment becomes a critical competency. Professionals must possess a deep comprehension of financial regulations, data privacy laws, and ethical guidelines that govern algorithmic trading and client data management. The competency framework mandates the ability to ensure compliance, safeguard client interests, and uphold transparency in algorithmic recommendations.
- **Behavioral Attributes for Effective Client Interaction**: While robo-advisory services leverage technology, the human element remains pivotal in client interactions. Professionals engaging with robo-advisory clients must exhibit a profound understanding of investor behavior, cognitive biases, and emotional responses. The concept of investment is embedded in the context of consumers' adoption of AI services for personal financial investing; it could analyze whether individuals with a higher level of investment knowledge or a propensity for active investment engagement are more inclined to adopt AI-powered tools (Atwal & Bryson, 2021). The competency framework includes the capacity to build rapport, communicate effectively, and

guide clients through market fluctuations. Cultivating emotional intelligence and empathy is integral to fostering trust and enhancing the overall client experience.

- **Ethics and Transparency in AI-Powered Advisory**: The ethical dimension of robo-advisory services is a cornerstone of the competency framework. Professionals must ensure that algorithmic decisions are devoid of biases and adhere to ethical standards. Transparent communication of the underlying methodologies, potential risks, and expected outcomes is crucial for establishing trust with clients. The ability to navigate ethical dilemmas arising from algorithmic trading and decision-making is a key facet of this competency.

- **Competency Framework for Evolving Roles**: The competency framework for financial professionals engaging with robo-advisory services must adapt to the evolving roles in the digital economy era. Technical skills, such as proficiency in algorithm development and data analytics, are essential for techno-finance experts who drive AI-driven insights. These professionals must possess the ability to design, test, and refine algorithms that align with clients' risk preferences and financial goals. However, technical prowess alone is insufficient. Finance professionals must also excel in client interactions, leveraging behavioral attributes such as emotional intelligence, empathy, and effective communication. This human touch complements AI-driven insights, enhancing client relationships and trust. Moreover, as highlighted by Lim & Yoon (2020), the competency framework must encompass an understanding of personal values and ethical considerations to navigate the ethical implications of algorithmic decision-making.

9.11 LITERATURE REVIEW

The concept of AI-powered robo-advisory services has garnered significant attention in recent literature, reflecting its transformative impact on investment practices and client–advisor interactions. Researchers have acknowledged the potential of robo-advisors to democratize investment advisory by providing cost-effective and accessible solutions for a broader spectrum of investors. As highlighted by Statista (2022), the global robo-advisory market has witnessed exponential growth in assets under management, underscoring the escalating acceptance of these platforms by investors. This growth trajectory underscores the need for a comprehensive evaluation of the competencies that finance professionals require to harness the power of AI in a rapidly evolving financial landscape.

AI robo-advisory introduces a new level of accessibility to investment advice (Bhatia et al., 2021). The automation of investment processes through robo-advisors democratizes access to expert financial advice, making it available to a wider range of clients. This not only transforms the way investment advisory is delivered but also empowers clients with insights and strategies that can contribute to more informed financial decision-making (Dietzmann et al., 2023). AI-driven investment platforms offer investors streamlined access to personalized investment advice and strategies, altering the way investment services are accessed and consumed.

Empirical studies have delved into robo-advisory service adoption and acceptance dynamics. Research by Kim and Park (2020) employs the technology acceptance model (TAM) to dissect investor behavior and acceptance patterns concerning robo-advisory platforms. Their findings emphasize the importance of perceived usefulness and ease of use in influencing investor decisions, reflecting the necessity for finance professionals to possess competencies that align with these user-centric dimensions. In terms of technical competencies, the works of Bouri et al. (2022) emphasize the significance of financial experience in shaping robo-advisory service acceptance.

As algorithmic intricacies drive the functioning of robo-advisory platforms, finance professionals must possess expertise in algorithm development, machine learning, and data analytics. Ethical considerations also feature prominently in the discourse surrounding robo-advisory services. The research by Gupta and Misra (2021) underscores the pivotal role of trust and privacy concerns in shaping investor perceptions of robo-advisors. This highlights the ethical dimensions that finance professionals must navigate in ensuring transparent and ethical algorithmic recommendations.

As the financial landscape evolves, so too do the roles and skillsets require of finance professionals. The shift from traditional to hybrid advisory models necessitates skill transformation, as highlighted by Häcker and Homan (2019), wherein financial advisors transition from transactional roles to advisory roles complemented by AI-driven insights.

This chapter contributes to the existing body of knowledge by undertaking a comprehensive evaluation of the competency framework necessary for finance professionals to excel in the domain of AI-powered robo-advisory services. By synthesizing insights from a wide array of disciplines, including finance, technology, ethics, and behavioral sciences, the chapter seeks to provide a holistic understanding of the multifaceted competencies required to navigate the intricacies of the digital economy era.

9.12 CONCLUSION

The evaluation of the competency framework for AI-powered robo-advisory services in the digital economy era underscores the transformative potential of technology within the financial domain. The seamless integration of advanced algorithms, data analytics, and financial expertise has yielded a novel approach to investment advisory, democratizing access, and streamlining decision-making. The multifaceted competencies encompassing technical proficiency, regulatory adeptness, behavioral understanding, and ethical considerations serve as the pillars supporting the successful implementation of robo-advisory services.

As the digital economy era unfolds, the harmonious fusion of these competencies holds the key to unlocking the full potential of AI-powered advisory services. While challenges persist, including regulatory complexities and ethical dilemmas, the collective pursuit of refining and aligning these competencies promises to elevate the role of robo-advisory services as invaluable tools for investors navigating the complexities of today's financial landscape. As we stand at the nexus of technological advancement and financial innovation, this evaluation paves the way for a future where AI and human expertise synergize to create a more accessible, efficient, and ethically responsible approach to investment advisory (Khang & Rashmi et al., 2024).

9.13   FUTURE SCOPE OF WORK

Further investigations could delve into the ethical dimensions of algorithmic decision-making in robo-advisory services, establishing frameworks to ensure transparency and fairness. Additionally, studying the integration of behavioral finance principles could enhance the effectiveness of advisory experiences. Exploring the role of human interaction and user experience design in AI-driven advisory services offers room for enhancing trust and engagement. Investigating robo-advisors' potential in facilitating sustainable investing aligns with growing interest in socially responsible finance (Khang, 2024).

The evolution of regulatory frameworks to accommodate AI-driven advisory and the adaptation of competency models to different cultural contexts present intriguing research opportunities. Moreover, hybrid advisory models, the long-term performance analysis of AI recommendations, and the educational requirements for professionals in this field warrant in-depth study. As AI continues to evolve, understanding how to enhance algorithm transparency, integrate natural language processing, and navigate the psychological aspects of human–AI interaction remains pivotal. These unexplored areas collectively contribute to the expanding landscape of AI-powered robo-advisory services, shaping its trajectory in the dynamic digital economy era (Khang & Shah et al., 2023).

REFERENCES

Atwal, G. & Bryson, D. (2021). Antecedents of intention to adopt artificial intelligence services by consumers in personal financial investing. *Strategic Change*, 30(3), 293–298. https://onlinelibrary.wiley.com/doi/abs/10.1002/jsc.2412

Beck, A. D. (2021). The role of artificial intelligence in robo-advisory. *Robo-Advisory: Investing in the Digital Age* (pp. 227–243). Palgrave Macmillan, Cham. https://link.springer.com/chapter/10.1007/978-3-030-40818-3_11

Bhatia, A., Chandani, A., Atiq, R., Mehta, M., & Divekar, R. (2021). Artificial intelligence in financial services: qualitative research to discover robo-advisory services. *Qualitative Research in Financial Markets*, 13(5), 632–654. https://www.emerald.com/insight/content/doi/10.1108/QRFM-10-2020-0199/full/html

Bouri, E., Molnár, P., Azzi, G., Roubaud, D., & Hagfors, L. I. (2022). The influence of financial experience on robo-advisory acceptance: an empirical study. *International Journal of Information Management*, 63, 102315. https://www.it-in-industry.org/index.php/itii/article/view/593

Dietzmann, C., Jaeggi, T., & Alt, R. (2023). Implications of AI-based robo-advisory for private banking investment advisory. *Journal of Electronic Business & Digital Economics*, 2(1), 3–23. https://www.emerald.com/insight/content/doi/10.1108/jebde-09-2022-0037

Gupta, A., & Misra, S. (2021). Robo-advisory services: unpacking the role of trust and privacy concerns. *Journal of Retailing and Consumer Services*, 63, 102656.

Häcker, F. T. & Homan, A. C. (2018). A decade of digitalization in the financial services industry: is robo-advisory the end of the financial advisor? *Business Research*, 12(1), 3–41.

Khang, A. (2024). *AI-Oriented Competency Framework for Talent Management in the Digital Economy: Models, Technologies, Applications, and Implementation.* CRC Press, Boca Raton, FL. https://doi.org/10.1201/9781003440901

Khang, A., Gupta, S. K., Rani, S., & Karras, D. A. (2023). *Smart Cities: IoT Technologies, Big Data Solutions, Cloud Platforms, and Cybersecurity Techniques.* CRC Press, Boca Raton, FL. https://doi.org/10.1201/9781003376064

Khang, A., Hajimahmud, V. A., Gupta, S. K., Babasaheb, J., & Morris, G. (2023). *AI-Centric Modelling and Analytics: Concepts, Designs, Technologies, and Applications* (1 ed.). CRC Press, Boca Raton, FL. https://doi.org/10.1201/9781003400110

Khang, A., Kali Charan, R., Surabhika, P., Pokkuluri Kiran, S., & Santosh Kumar, P. (2023). Revolutionizing agriculture: exploring advanced technologies for plant protection in the agriculture sector. *Handbook of Research on AI-Equipped IoT Applications in High-Tech Agriculture* (pp: 1–22). IGI Global Press, Hershey, PA. https://doi.org/10.4018/978-1-6684-9231-4.ch001

Khang, A., Misra, A., Gupta, S. K., & Shah, V. (2023). *AI-aided IoT Technologies and Applications in the Smart Business and Production.* CRC Press, Boca Raton, FL. https://doi.org/10.1201/9781003392224

Khang, A., Muthmainnah, M., Seraj, P. M. I., Yakin, A. A., Obaid, A. J., & Panda, M. R. (2023). AI-aided teaching model for the education 5.0 ecosystem. *AI-Based Technologies and Applications in the Era of the Metaverse* (1 ed., pp. 83–104). IGI Global Press, Hershey, PA. https://doi.org/10.4018/978-1-6684-8851-5.ch004

Khang, A., Rani, S., & Sivaraman, A. K. (2022). *AI-Centric Smart City Ecosystems: Technologies, Design and Implementation* (1st ed.). CRC Press, Boca Raton, FL. https://doi.org/10.1201/9781003252542

Khang, A., Rashmi, G., Hayri, U., Tailor, R. K., & Sanjaya S. G. (2024). *Data-driven Modelling and Predictive Analytics in Business and Finance.* CRC Press, Boca Raton, FL. https://doi.org/10.1201/9781032600628

Khang, A., Shah, V., & Rani, S. (2023). *AI-Based Technologies and Applications in the Era of the Metaverse* (1 ed.). IGI Global Press, Hershey, PA. https://doi.org/10.4018/978-1-6684-8851-5

Khanh, H. H. & Khang, A. (2021). The role of artificial intelligence in blockchain applications. *Reinventing Manufacturing and Business Processes through Artificial Intelligence* (vol. 2, pp. 20–40). CRC Press, Boca Raton, FL. https://doi.org/10.1201/9781003145011-2

Kim, Y. & Park, H. S. (2020). The influence of robo-advisors on the investment behavior of retail investors: a perspective of the technology acceptance model. *Sustainability*, 12(17), 6826. https://www.emerald.com/insight/content/doi/10.1108/JOSM-10-2020-0378/full/html

Lim, S. S. & Yoon, S. J. (2020). Robo-advisors in personal financial management: exploring the role of personal values. *Information Systems Frontiers*, 2020, 1–17. https://www.mdpi.com/1911-8074/15/4/163

Manrai, R. & Gupta, K. P. (2023). Investor's perceptions on artificial intelligence (AI) technology adoption in investment services in India. *Journal of Financial Services Marketing*, 28(1), 1–14. https://link.springer.com/article/10.1057/s41264-021-00134-9

Muthmainnah, M., Khang, A., Seraj, P. M. I., Yakin, A. A., Oteir, I., & Alotaibi, A. N. (2023). An innovative teaching model - the potential of metaverse for English learning. *AI-Based Technologies and Applications in the Era of the Metaverse* (1st ed., pp. 105–126). IGI Global Press, Hershey, PA. https://doi.org/10.4018/978-1-6684-8851-5.ch005

Rana, G., Khang, A., Sharma, R., Goel, A. K., & Dubey, A. K. (2021). *Reinventing Manufacturing and Business Processes through Artificial Intelligence.* CRC Press, Boca Raton, FL. https://doi.org/10.1201/9781003145011

Rani, S., Bhambri, P., Kataria, A, & Khang, A. (2022). Smart city ecosystem: concept, sustainability, design principles and technologies. *AI-Centric Smart City Ecosystems: Technologies, Design and Implementation* (1st ed., pp. 1–20). CRC Press, Boca Raton, FL. https://doi.org/10.1201/9781003252542-1

Rani, S., Chauhan, M., Kataria, A., & Khang, A. (2021). IoT equipped intelligent distributed framework for smart healthcare systems. *Networking and Internet Architecture* (vol. 2, pp. 30). Springer Nature Singapore, Singapore. https://doi.org/10.48550/arXiv.2110.04997

Scholz, P. (Ed.). (2020). *Robo-Advisory: Investing in the Digital Age.* Springer Nature, Singapore. https://link.springer.com/chapter/10.1007/978-3-030-40818-3_1

Statista. (2022). Total assets under management (AUM) in the robo-advisory market worldwide from 2016 to 2022 (in billion U.S. dollars). https://doi.org/10.1201/9781032600628

10 Future-Proofing Talent Management

Anticipating the Evolution of AIoCF Model in the Digital Economy

*Tarun Kumar Vashishth, Vikas Sharma,
Kewal Krishan Sharma, and Sachin Chaudhary*

10.1 INTRODUCTION

The digital economy has witnessed unprecedented growth and disruption, driven by the rapid advancements in artificial intelligence (AI) and its applications. As organizations strive to stay competitive and relevant in this evolving landscape, the role of talent management becomes increasingly crucial. The ability to anticipate and adapt to the evolving AI-oriented competencies is essential for organizations to future-proof their workforce (Khang & Rani et al., 2023).

This book chapter, titled "Future-proofing Talent Management: Anticipating the Evolution of AI-Oriented Competencies in the Digital Economy," aims to explore the challenges and opportunities organizations face in managing talent in the era of AI-driven technologies. By understanding the current state of AI-oriented competencies and anticipating future trends, organizations can proactively develop strategies to attract, develop, and retain talent with the necessary skills to thrive in the digital age (Khang & Shah et al., 2023).

10.1.1 OBJECTIVES

The Impact of AI on the Digital Economy: The digital economy has witnessed a seismic shift with the emergence of AI technologies. AI has become an integral part of various industries, transforming business processes, enhancing efficiency, and enabling new forms of innovation. From predictive analytics to natural language processing and robotic process automation, AI has proven its potential to revolutionize traditional business models. Industries such as finance, healthcare, manufacturing, and customer service have experienced significant AI-driven transformations. Automated financial analysis, AI-assisted diagnosis in healthcare, smart factories,

DOI: 10.1201/9781003440901-10

and personalized customer experiences are just a few examples of AI's impact. As organizations embrace AI, the demand for talent with AI-oriented competencies grows exponentially.

The Current State of AI-Oriented Competencies: To future-proof talent management, organizations must first understand the existing AI-oriented competencies and skill sets required in the digital economy. AI-oriented competencies encompass both technical and non-technical skills. Technical skills include proficiency in programming languages, data analytics, machine learning, and AI algorithms. Non-technical skills encompass critical thinking, creativity, problem-solving, and the ability to work collaboratively with AI systems. However, the current workforce often faces a significant skills gap in AI-oriented competencies. Traditional education systems have struggled to keep pace with the rapid evolution of AI, leading to a scarcity of talent with the necessary skills. Bridging this gap requires a multifaceted approach that includes redefining education and training programs, fostering interdisciplinary collaboration, and promoting lifelong learning (Snehal et al., 2023).

Anticipating the Evolution of AI-Oriented Competencies: As AI technologies continue to advance, organizations need to anticipate the future evolution of AI-oriented competencies. This involves understanding the emerging trends, advancements, and potential disruptions in the AI landscape. For instance, the convergence of AI with other transformative technologies like blockchain (Khanh & Khang, 2021), the Internet of Things (IoT), and augmented reality (AR) presents new opportunities and challenges for talent management. Additionally, the field of AI itself is evolving rapidly. New algorithms, techniques, and frameworks are constantly emerging. Organizations must stay abreast of these developments to ensure their talent management strategies align with the future needs of the digital economy. Continuous monitoring of AI research, industry trends, and collaboration with academia can help organizations anticipate the changing competencies required for AI-driven roles.

Strategies for Future-proofing Talent Management: To effectively future-proof talent management, organizations need to adopt strategies that address the challenges posed by the evolving AI-oriented competencies. These strategies should encompass various stages of talent management, including recruitment and selection, development and training, and retention and motivation. Recruitment and selection practices should be redesigned to identify candidates with the potential to develop AI-oriented competencies. Beyond traditional qualifications, organizations should look for attributes such as adaptability, curiosity, and a passion for lifelong learning. Utilizing AI-driven recruitment tools, like natural language processing-based resume screening or automated candidate assessment, can enhance the efficiency and effectiveness of talent acquisition processes. Developing and training the existing workforce is crucial to bridge the skills gap. Upskilling and reskilling initiatives should be prioritized to equip employees with AI-oriented competencies. Organizations can offer training programs, certifications, and collaborations with educational institutions to facilitate continuous learning and professional development. Creating a culture of innovation, experimentation, and knowledge-sharing encourages employees to embrace AI technologies and continuously enhance their skills (Rana et al., 2021).

10.1.2   OVERVIEW OF THE DIGITAL ECONOMY AND THE ROLE OF AI-ORIENTED COMPETENCIES

10.1.2.1   Overview of the Digital Economy

The digital economy encompasses the economic activities and transactions that are driven by digital technologies and the internet. It has revolutionized industries across the globe, transforming the way businesses operate, interact with customers, and create value. In the digital economy, information flows rapidly, communication is instant, and data is a valuable resource. The proliferation of digital technologies has led to the emergence of new business models, innovative products and services, and unprecedented connectivity. E-commerce platforms, social media networks, cloud computing, and mobile applications have become integral parts of our daily lives, influencing consumer behavior and reshaping industries.

10.1.2.2   The Role of AI-Oriented Competencies in the Digital Economy

Artificial intelligence (AI) is a key driving force behind the digital economy. It refers to the development of computer systems that can perform tasks that typically require human intelligence, such as visual perception, natural language understanding, decision-making, and problem-solving. AI technologies, including machine learning, deep learning, and cognitive computing, have the potential to automate processes, extract insights from vast amounts of data, and enable intelligent decision-making.

In the digital economy, organizations rely on AI-oriented competencies to unlock the full potential of AI technologies. These competencies encompass both technical and non-technical skills that enable individuals to effectively work with AI systems and leverage their capabilities. Technical skills include proficiency in programming languages, data analysis, statistical modeling, and AI algorithms. Non-technical skills encompass critical thinking, creativity, ethical considerations, and the ability to collaborate with AI systems. AI-oriented competencies are vital for organizations to leverage AI technologies in various domains. For instance, in finance, AI-powered algorithms can automate risk assessment and investment strategies, but individuals with AI-oriented competencies are needed to develop and fine-tune these algorithms. In healthcare, AI can assist in diagnosing diseases and recommending treatments, but healthcare professionals with AI-oriented competencies are required to interpret and validate AI-generated insights.

10.1.2.3   Future-proofing Talent Management for the Digital Economy

Future-proofing talent management is imperative for organizations to thrive in the digital economy. It involves anticipating and adapting to the evolving AI-oriented competencies to ensure a skilled workforce that can effectively harness the potential of AI technologies. Organizations need to develop strategies to attract, develop, and retain talent with the necessary AI-oriented competencies. Recruitment and selection practices should be aligned with the requirements of AI-driven roles.

Organizations can employ innovative approaches, such as AI-based resume screening and automated candidate assessment tools, to identify candidates with the potential to develop AI-oriented competencies. Beyond technical qualifications, attributes like adaptability, curiosity, and a passion for lifelong learning should be considered. By future-proofing talent management and nurturing AI-oriented competencies, organizations can position themselves as leaders in the digital economy. They can harness the transformative power of AI technologies, drive innovation, and gain a competitive edge in a rapidly evolving business landscape.

10.1.3 IMPORTANCE OF FUTURE-PROOFING TALENT MANAGEMENT STRATEGIES

The importance of future-proofing talent management strategies in the context of "Future-proofing Talent Management: Anticipating the Evolution of AI-Oriented Competencies in the Digital Economy" is paramount. The rapid advancements in AI and its pervasive influence across industries necessitate a proactive approach to talent management to ensure organizations can thrive in the digital economy. Here are some key reasons why future-proofing talent management strategies are crucial:

- **Meeting Evolving Skill Demands**: The digital economy is characterized by dynamic and evolving skill requirements. AI-oriented competencies, both technical and non-technical, are in high demand. By future-proofing talent management strategies, organizations can ensure they have the right skills within their workforce to leverage AI technologies effectively. It enables them to meet the evolving skill demands of the digital economy and remain competitive.
- **Addressing the Skills Gap**: There is often a significant skills gap between the competencies organizations require and what the current workforce possesses. Future-proofing talent management strategies involve identifying the skills gap and implementing initiatives to bridge it. This may include upskilling and reskilling programs, partnering with educational institutions, and fostering a culture of continuous learning. By addressing the skills gap, organizations can ensure they have a talent pool equipped with the necessary AI-oriented competencies.
- **Attracting and Retaining Top Talent**: In the digital economy, competition for talent with AI-oriented competencies is fierce. Future-proofing talent management strategies involve creating an environment that attracts and retains top talent. This can be achieved through offering opportunities for professional development, providing clear career pathways, and recognizing and rewarding employees for their AI-related skills and contributions. By prioritizing talent management, organizations can build a reputation as an employer of choice and attract high-quality talent.
- **Enhancing Innovation and Agility**: AI technologies are catalysts for innovation and enable organizations to adapt to changing market dynamics with agility. Future-proofing talent management strategies involve fostering a culture of innovation, experimentation, and knowledge-sharing. It encourages employees to explore and apply AI technologies creatively, driving innovation within the organization. By nurturing AI-oriented competencies, organizations can enhance their ability to innovate and adapt to the evolving digital landscape.

- **Mitigating the Risk of Disruption**: Organizations that fail to future-proof their talent management strategies may face the risk of being disrupted by competitors who embrace AI-oriented competencies. Disruption in the digital economy can occur swiftly, and organizations must have the talent in place to leverage AI technologies for competitive advantage. Future-proofing talent management strategies help mitigate the risk of disruption by ensuring the organization has the necessary skills and capabilities to navigate the AI-driven future.

In conclusion, the importance of future-proofing talent management strategies in the context of "Future-proofing Talent Management: Anticipating the Evolution of AI-Oriented Competencies in the Digital Economy" cannot be overstated. It enables organizations to meet evolving skill demands, address the skills gap, attract and retain top talent, enhance innovation and agility, and mitigate the risk of disruption. By proactively preparing their workforce with AI-oriented competencies, organizations can position themselves for success in the digital economy.

The digital economy, driven by AI, presents both challenges and opportunities for talent management. Organizations must adapt their strategies to anticipate and address the evolving AI-oriented competencies. By understanding the impact of AI on the digital economy, assessing the current state of AI-oriented competencies, and anticipating future trends, organizations can future-proof their talent management practices. Embracing strategies that focus on recruitment, development, and retention of talent with AI-oriented competencies enables organizations to thrive in the era of digital transformation. This book chapter will provide insights, best practices, and actionable recommendations to guide organizations on this transformative journey.

10.2 CURRENT STATE OF TALENT MANAGEMENT IN THE DIGITAL ECONOMY

The current state of talent management in the digital economy is undergoing significant transformations due to the pervasive influence of digital technologies and the need for AI-oriented competencies. Here are key aspects characterizing the current state of talent management:

10.2.1 SHIFTING SKILL REQUIREMENTS

The digital economy demands a different set of skills compared to traditional economies. AI-oriented competencies, such as data analysis, machine learning, programming, and digital literacy are in high demand. Organizations are seeking professionals who can leverage digital tools and technologies to drive innovation, make data-driven decisions, and adapt to rapidly changing business environments.

10.2.2   Skills Gap and Talent Scarcity

One of the challenges in talent management is the skills gap between the competencies organizations require and the skills possessed by the existing workforce. The rapid pace of technological advancements, including AI, has outpaced the traditional education system's ability to produce talent with the necessary skills. This has resulted in a scarcity of individuals proficient in AI-oriented competencies, creating fierce competition among organizations to attract and retain top talent.

10.2.3   Rise of Remote and Flexible Work

The digital economy has facilitated the rise of remote and flexible work arrangements. Organizations are increasingly embracing remote work options, allowing them to tap into talent pools globally and access specialized skills. Talent management strategies must adapt to these new work arrangements, focusing on effective virtual collaboration, communication, and performance management to ensure success in a distributed work environment.

10.2.4   Emphasis on Continuous Learning and Upskilling

To keep up with the rapid pace of digital transformation, talent management in the digital economy emphasizes continuous learning and upskilling. Organizations recognize the need to invest in their employees' professional development to foster AI-oriented competencies. This includes providing learning opportunities, offering online courses, partnering with educational institutions, and encouraging employees to engage in self-directed learning.

10.2.5   Data-Driven Talent Analytics

The digital economy generates vast amounts of data, and organizations are increasingly leveraging talent analytics to make data-driven decisions in talent management. By analyzing data on recruitment, performance, engagement, and career progression, organizations can gain insights into their talent pool, identify skills gap, predict future talent needs, and design targeted talent development programs.

10.2.6   Focus on Employer Branding and Employee Experience

In the digital economy, organizations must differentiate themselves as attractive employers to secure top talent. Employer branding efforts highlight the organization's values, culture, commitment to innovation, and opportunities for growth. Creating a positive employee experience, characterized by a supportive work environment, opportunities for collaboration, and recognition for contributions is essential for talent acquisition and retention.

10.2.7 Collaboration between HR and IT Departments

Talent management in the digital economy requires close collaboration between HR and IT departments. HR professionals need a strong understanding of digital technologies and AI-oriented competencies to effectively recruit, develop, and manage talent in the digital era. Collaboration between HR and IT ensures alignment between talent management strategies and technological capabilities, enabling organizations to leverage AI and digital tools effectively.

In summary, talent management in the digital economy is characterized by shifting skill requirements, a skills gap, remote work, continuous learning, data-driven talent analytics, employer branding, and collaboration between HR and IT departments. Organizations must adapt their talent management strategies to attract, develop, and retain talent with AI-oriented competencies to thrive in the digital economy.

10.3 LITERATURE REVIEW

The literature review examines the concept of future-proofing talent management in the context of the evolving digital economy and the increasing influence of AI on organizational processes. The review focuses on exploring the anticipated evolution of AI-oriented competencies and their implications for talent management strategies. By analyzing relevant research articles, reports, and scholarly works, this review provides insights into the challenges and opportunities that organizations may face in preparing their workforce for the AI-driven digital future. Here is a summary of the key findings and contributions of each study:

Döngül and Leonardo Cavaliere (2022) delve into the integration of human–computer interaction (HCI) technology and platform business ecosystems into the AI environment. They recommend the creation of intelligent clothing ecosystems and develop a flexible production planning model using optimization algorithms. The study proposes a human–computer interactive customer trust model and demonstrates its potential for solving strategic management problems within platform business ecosystems (Döngül and Leonardo Cavaliere, 2022).

Ermakova (2021) focuses on the implementation of AI in various areas, including social, economic, educational, and other spheres. The study emphasizes the need for fundamental reconstruction of contemporary economic and political relations to successfully leverage advanced AI technologies. It highlights the potential of AI in areas such as economics, education, agriculture, medicine, and the aerospace industry.

Fernandez et al. (2021) aim to connect NDE (Non-Destructive Evaluation) professionals with supplementary resources that can aid their use of AI technologies for professional and personal development. The study aims to enhance professionals' confidence and understanding of AI's applications in their field (Fernandez et al., 2021)

Gao et al. (2021) propose and categorize the five most concerned topics in AI research: perception intelligence, human mind simulated intelligence, classical model-based machine learning, bio-inspired intelligence, and big-data-based intelligence. The study provides insights into the current trends and areas of focus within the field of AI (Gao et al., 2021).

Huateng et al. (2021) highlight the importance of digital transformation in the manufacturing industry as a means of economic transformation, upgrading, and fostering new drivers for economic growth. The study underscores the potential benefits of embracing digital technologies to enhance the manufacturing sector (Huateng et al., 2021).

Li et al. (2023) investigate the relationship between AI orientation (AIO) and the performance of new ventures. They find that AIO positively affects new venture performance, and this relationship is influenced by factors such as firm growth and regional economic development. The study provides empirical evidence highlighting the significance of AIO for new ventures (Li et al., 2023).

Li et al. (2021) apply upper echelons theory to investigate the influence of Chief Information Officers (CIOs) and boards of directors on the development of AIO within firms. Their findings demonstrate that the presence of a CIO positively affects AIO , and board characteristics such as educational diversity, R&D experience, and AI experience moderate this effect. The study contributes to the understanding of how top management teams and boards can effectively develop AIO (Li et al., 2021).

Nosovo et al. (2021) focus on assisting market economy actors in effectively utilizing AI participation and transitioning from experimental phases to obtaining reliable opportunities. The study emphasizes the importance of AI in providing competitive flexibility to organizations and fostering economic growth at the national level (Nosova et al., 2021).

Thowfeek and Samsudeen (2020) focus on the drivers and barriers to successful AI implementation in the banking sector. Based on interviews with AI experts, the study identifies the significance of AI-oriented role models and process capabilities in achieving AI applications that can operate independently. It emphasizes the need for adequate preparation before AI algorithms can function without human involvement or ethical concerns.

Wamba-Taguimdje et al. (2020) analyze the influence of AI on firm performance, particularly by leveraging the business value of AI-based transformation projects. Through a four-step sequential approach, the study explores AI concepts/technologies, examines case studies across various industrial sectors, collects data from AI-based solution providers, and reviews existing AI literature. The research sheds light on the impact of AI-enabled projects on organizational performance (Wamba-Taguimdje et al., 2020).

Overall, these studies collectively contribute to the understanding of AI's impact on various sectors, the factors influencing AI adoption and implementation, and the potential benefits and challenges associated with leveraging AI technologies. By reviewing the existing literature, this study aims to provide a comprehensive understanding of the challenges and opportunities associated with future-proofing talent management in the context of AI-oriented competencies. It offers valuable insights for HR professionals, organizational leaders, and policymakers to effectively navigate the changing landscape of work in the digital economy.

10.4   EMERGING TRENDS IN AI TECHNOLOGIES

In the context of "Future-proofing Talent Management: Anticipating the Evolution of AI-Oriented Competencies in the Digital Economy," it is crucial to consider the emerging trends in AI technologies that are shaping the talent landscape. These trends directly impact the competencies and skills required in the digital economy. Here are some noteworthy emerging trends in Figure 10.1.

Arrows and lines indicate the flow of information and connections between these concepts, symbolizing the integration and synergy among different AI technologies. The image represents the dynamic nature of AI and its potential to revolutionize industries and transform the future of technology.

- **Augmented Intelligence**: Augmented intelligence, also known as intelligence amplification, focuses on combining human capabilities with AI technologies to enhance decision-making and problem-solving. Rather than replacing humans, augmented intelligence emphasizes collaboration between humans and AI systems. This trend highlights the need for talent with skills in human–AI interaction, ethical considerations, and the ability to leverage AI technologies as tools for amplifying human potential.
- **Natural Language Processing (NLP) and Conversational AI**: NLP and conversational AI are revolutionizing how humans interact with machines. Advanced NLP models and chatbot technologies enable natural and contextual conversations, making user interfaces more intuitive and user-friendly. Talent management in the digital economy should include skills in developing and managing conversational AI systems, including NLP, sentiment analysis, and voice recognition.

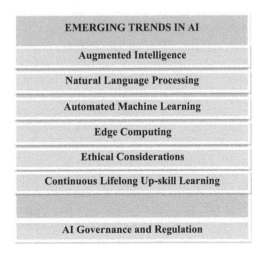

FIGURE 10.1   Component of emerging trends in AI.

- **Automated Machine Learning (AutoML)**: AutoML refers to the automation of the machine learning process, allowing non-experts to build and deploy AI models without extensive programming or data science expertise. This trend streamlines the development and deployment of AI models, making AI more accessible to a broader range of professionals. Talent management strategies should consider providing training and upskilling opportunities for employees to utilize AutoML platforms and tools effectively.

- **Edge Computing and AI at the Edge**: Edge computing involves processing and analyzing data closer to the source, reducing latency and dependency on centralized cloud infrastructures. AI at the edge refers to deploying AI algorithms and models directly on edge devices. This trend is particularly relevant for industries requiring real-time and low-latency AI applications, such as autonomous vehicles, IoT, and smart cities (Khang & Gupta et al., 2023). Talent management strategies should incorporate skills in edge computing, AI deployment on edge devices, and efficient data processing in decentralized environments (Rani et al., 2021).

- **Responsible AI and Ethical Considerations**: As AI becomes more pervasive, ensuring responsible and ethical AI practices is gaining prominence. Talent management strategies must consider the skills needed for ethical AI development, including fairness, accountability, transparency, and bias mitigation. Professionals who can navigate the ethical challenges associated with AI technologies and ensure responsible AI deployment will be in high demand.

- **AI Governance and Regulation**: As AI technologies continue to evolve, there is a growing need for governance and regulation to address potential risks and ensure ethical use. Talent management strategies should account for skills in AI governance, compliance, and regulatory frameworks. Professionals capable of navigating the legal and ethical landscape surrounding AI technologies will play a crucial role in shaping the future of talent management in the digital economy.

- **Continuous Learning and Lifelong Upskilling**: The rapid pace of AI advancements necessitates a culture of continuous learning and upskilling. Talent management strategies should prioritize lifelong learning initiatives, training programs, and partnerships with educational institutions to ensure employees stay abreast of AI-oriented competencies. Encouraging a growth mindset and providing opportunities for professional development will be essential to future-proofing talent management.

In summary, emerging trends in AI technologies have a significant impact on talent management strategies in the digital economy. Organizations must be proactive in identifying and nurturing the AI-oriented competencies required to thrive in this evolving landscape. By understanding and adapting to these trends, organizations can future-proof their talent management practices and ensure they have the skills necessary to harness the full potential of AI technologies.

10.5 ANTICIPATING CHANGES AND ADDITIONS TO AI-ORIENTED COMPETENCY FRAMEWORK

The Chapter emphasizes the need to anticipate changes and additions to the AI-oriented competency framework to stay ahead in talent management. As the field of AI evolves rapidly, new skills and competencies will become essential. Here are some anticipated changes and additions to the AI-oriented competency framework.

10.5.1 INTERDISCIPLINARY SKILLS

The AI-oriented competency framework will increasingly include interdisciplinary skills that combine AI knowledge with expertise from other domains. For example, AI professionals with domain-specific knowledge in healthcare, finance, or marketing will be highly sought after. The ability to apply AI techniques in specific industry contexts and understand the nuances of different domains will become crucial.

10.5.2 EXPLAINABILITY AND ETHICAL COMPETENCIES

As AI technologies become more complex, the competency framework will incorporate skills related to explainability and ethical considerations. Professionals will need to understand how to interpret and explain AI models, algorithms, and decisions to ensure transparency and accountability. Competencies related to bias detection and mitigation, fairness, and privacy will be essential to address ethical concerns associated with AI technologies.

10.5.3 HUMAN-AI INTERACTION

The competency framework will focus on skills related to human–AI interaction and collaboration as in Figure 10.2.

Professionals will need to understand how to work effectively alongside AI systems, leveraging their strengths and compensating for their limitations. Skills such as empathy, emotional intelligence, and effective communication with AI systems will be valuable to optimize human–AI collaboration.

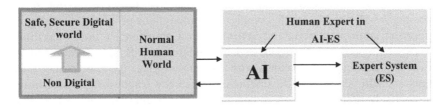

FIGURE 10.2 How AI with the help of an expert system changes common human life.

10.5.4 DATA GOVERNANCE AND MANAGEMENT

With the increasing reliance on data for AI applications, the competency framework will include skills related to data governance and management. Professionals will need to understand data privacy, data quality, data security, and data governance principles to ensure responsible and effective use of data in AI projects. Competencies in data collection, creation, and pre-processing will also be crucial.

10.5.5 LIFELONG LEARNING AND ADAPTABILITY

The competency framework will emphasize the importance of lifelong learning and adaptability in the AI-driven digital economy. Professionals will need to continually update their skills and stay abreast of the latest advancements in AI technologies. Competencies related to self-directed learning, adaptability to new tools and techniques, and the ability to quickly acquire new knowledge will be highly valued.

10.5.6 BUSINESS AND STRATEGY ACUMEN

The competency framework will include skills related to business acumen and strategic thinking in the context of AI. Professionals will need to understand how AI technologies align with organizational goals, identify AI opportunities, and develop AI strategies that drive business value. Competencies in project management, risk assessment, and ROI evaluation in AI initiatives will be necessary.

10.5.7 CULTURAL AND ETHICAL SENSITIVITY

As AI technologies are deployed globally, the competency framework will encompass cultural and ethical sensitivity. Professionals will need to understand the cultural nuances and ethical considerations of different regions and communities. Competencies related to diversity, inclusivity, and global awareness will enable professionals to develop AI solutions that are respectful and relevant across diverse contexts.

10.5.8 ADAPTIVE PROBLEM-SOLVING

The competency framework will focus on adaptive problem-solving skills. Professionals will need to tackle complex and ambiguous problems related to AI development, implementation, and optimization. Competencies in critical thinking, creativity, and the ability to experiment and iterate will be crucial to address the evolving challenges associated with AI technologies. By anticipating these changes and additions to the AI-oriented competency framework, organizations can proactively prepare their talent management strategies to ensure they have the right skills and competencies to navigate the evolving landscape of AI in the digital economy.

10.6   RETAINING AND ENGAGING AI TALENT

Retaining and engaging AI talent is a crucial aspect of future-proofing talent management in the digital economy. As the demand for AI-oriented competencies continues to grow, organizations must implement strategies to attract and retain skilled professionals in this field. Retaining and engaging AI talent requires addressing their unique needs and providing an environment that fosters their professional growth and satisfaction. Here are some key considerations and strategies for retaining and engaging AI talent:

- **Competitive Compensation and Benefits**: Offering competitive salaries and benefits is essential to attract and retain AI professionals. As the demand for these skills increases, organizations need to ensure that their compensation packages are commensurate with the market rates. Additionally, providing benefits such as flexible work arrangements, healthcare coverage, and professional development opportunities can enhance employee satisfaction and loyalty.
- **Career Development and Learning Opportunities**: AI professionals thrive in organizations that offer continuous learning and growth opportunities. Providing access to training programs, workshops, conferences, and certifications related to AI technologies allows employees to enhance their skills and stay updated with the latest advancements. Implementing a clear career development path with opportunities for advancement and promotion also motivates AI talent to stay within the organization.
- **Challenging and Meaningful Projects**: AI professionals are often motivated by intellectually stimulating and impactful work. Assigning them to challenging projects that leverage their skills and expertise keeps them engaged and motivated. Providing opportunities to work on cutting-edge technologies, research initiatives, or innovative AI applications allows AI talent to contribute to the organization's success and make a significant impact.
- **Collaborative and Supportive Culture**: Fostering a collaborative and supportive work culture is crucial for retaining AI talent. Encouraging teamwork, knowledge sharing, and cross-functional collaboration helps AI professionals learn from each other, solve complex problems collectively, and feel valued within the organization. Additionally, creating a supportive environment where employees have access to mentors, coaching, and peer feedback can contribute to their professional growth and job satisfaction.
- **Recognition and Rewards**: Recognizing and rewarding the achievements of AI talent is essential for their engagement and retention. Acknowledging their contributions through public recognition, rewards, bonuses, or promotions reinforces their value to the organization and motivates them to continue delivering high-quality work. Creating a culture of appreciation and celebrating successes can go a long way in retaining AI professionals.

- **Work-Life Balance and Well-being**: AI professionals often work on complex projects that require intense focus and long hours. Ensuring a healthy work-life balance and promoting employee well-being is vital for their long-term retention. Offering flexible work arrangements, promoting work-life balance policies, and providing resources for stress management and mental health support demonstrates an organization's commitment to the well-being of its AI talent.
- **Opportunities for Innovation and Research**: AI professionals are often driven by curiosity and a desire to push the boundaries of technology. Providing opportunities for innovation, research, and experimentation can be a significant draw for these individuals. Allowing them to explore new ideas, collaborate on research projects, and contribute to open-source initiatives can enhance their job satisfaction and commitment to the organization.

In summary, retaining and engaging AI talent requires organizations to prioritize competitive compensation, career development opportunities, challenging projects, a collaborative culture, recognition and rewards, work-life balance, and opportunities for innovation. By implementing these strategies, organizations can create an environment that attracts and retains AI professionals, ensuring their continued contribution to the organization's success in the evolving digital economy.

10.7   THE FUTURE OF TALENT MANAGEMENT IN THE DIGITAL ECONOMY

The future of talent management in the digital economy is characterized by the increasing importance of AI-oriented competencies and the need for organizations to adapt their talent management strategies to thrive in this evolving landscape. As technology continues to advance and reshape industries, talent management practices must also evolve to attract, develop, and retain the right skills for success in the digital era. Here are some key aspects that define the future of talent management in the digital economy:

- **Emphasis on Digital Skills and AI Competencies**: In the digital economy, organizations will place a significant emphasis on recruiting and developing talent with digital skills and AI competencies. These skills include data analytics, machine learning, programming, cybersecurity, and other AI-related proficiencies. Talent management strategies need to align with the demand for these skills, ensuring that organizations have the necessary talent to leverage emerging technologies and stay competitive as in Figure 10.3.
- **Agile Talent Acquisition and Workforce Planning**: The digital economy is characterized by rapid technological advancements and changing business needs. To effectively manage talent, organizations must adopt agile talent acquisition and workforce planning practices. This involves identifying future skills requirements, proactively sourcing talent, and adapting workforce plans to accommodate the evolving needs of the digital landscape.

FIGURE 10.3   Future-proofing talent management: ai-oriented competent digital economy.

Agile talent management allows organizations to stay ahead of the curve and quickly respond to emerging opportunities and challenges.

- **Remote and Flexible Work Arrangements**: The digital economy has accelerated the adoption of remote work and flexible work arrangements. Talent management in the future will need to embrace this shift and provide the necessary infrastructure and policies to support remote collaboration and flexible work hours. This includes leveraging digital communication tools, establishing performance metrics for remote work, and ensuring a work environment that fosters productivity and employee well-being.

- **Continuous Learning and Skill Development**: In the fast-paced digital economy, skills become outdated quickly. To remain competitive, organizations must prioritize continuous learning and skill development for their workforce. Talent management strategies should include ongoing training programs, upskilling and reskilling initiatives, and access to relevant learning resources. This allows employees to stay abreast of technological advancements, acquire new skills, and adapt to changing job requirements.

- **Data-Driven Talent Decisions**: In the digital economy, data plays a crucial role in making informed talent management decisions. Organizations will increasingly rely on data analytics and talent intelligence to drive recruitment, performance management, and career development processes. By leveraging data, organizations can identify talent gaps, optimize talent allocation, personalize employee experiences, and make evidence-based decisions to enhance overall talent management effectiveness.

- **Focus on Diversity and Inclusion**: Diversity and inclusion are key drivers of innovation and organizational success in the digital economy. Talent management strategies of the future will prioritize building diverse teams and fostering inclusive work environments. This includes implementing unbiased recruitment practices, creating inclusive policies, providing equal opportunities for career advancement, and fostering a culture of respect and belonging. Embracing diversity and inclusion enhances creativity, boosts collaboration, and helps organizations tap into a broader talent pool.

- **Talent Analytics and Predictive Insights**: As AI and data analytics capabilities continue to advance, talent management will increasingly leverage predictive analytics to anticipate future talent needs and optimize workforce performance. By analyzing employee data, organizations can identify patterns, predict attrition, assess skills gap, and make informed decisions about talent acquisition, development, and retention strategies. Talent analytics enables organizations to align their talent management practices with business goals and stay proactive in managing their workforce.

In summary, the future of talent management in the digital economy requires organizations to prioritize digital skills, embrace agility, support remote work, foster continuous learning, leverage data-driven insights, promote diversity and inclusion, and harness talent analytics. By adopting talent management practices to the changing demands of the digital era, organizations can build a competitive advantage and attract and retain the talent necessary for success in the digital economy (Khang & Muthmainnah et al., 2023).

10.8 CONCLUSION

In conclusion, future-proofing talent management in the digital economy, particularly in the context of AI-oriented competencies, is crucial for organizations to thrive in the evolving landscape. As AI continues to revolutionize industries and reshape the nature of work, talent management strategies must adapt to attract, develop, and retain the right skills and competencies. The conclusion of this topic emphasizes the following key points:

- **Importance of AI-Oriented Competencies**: AI is becoming increasingly essential in the digital economy, and organizations need to prioritize talent acquisition and development in this area. Skills such as data analytics, machine learning, and programming are in high demand, and talent management must align with these requirements.
- **Agile Talent Acquisition and Workforce Planning**: With technology and business needs rapidly evolving, organizations must adopt agile talent acquisition and workforce planning practices. This ensures that the organization can quickly respond to changing demands and acquire the right talent at the right time.
- **Embracing Remote Work and Flexibility**: The digital economy has accelerated the adoption of remote work and flexible arrangements. Talent management strategies should support these new work models, providing the necessary infrastructure and policies to enable remote collaboration and ensure work-life balance.
- **Continuous Learning and Skill Development**: In a fast-paced digital landscape, continuous learning and skill development are crucial. Organizations must invest in ongoing training programs, upskilling, and reskilling initiatives to keep their workforce updated with the latest technologies and trends.
- **Data-Driven Decision-Making**: Data and analytics play a vital role in talent management. Organizations should leverage data to make informed decisions regarding recruitment, performance management, and career development. By analyzing talent data, organizations can identify trends, make predictions, and optimize talent strategies.

- **Focus on Diversity and Inclusion**: Embracing diversity and fostering an inclusive work environment is essential for innovation and success in the digital economy. Talent management strategies should prioritize diversity in hiring, provide equal opportunities for growth, and create an inclusive culture that values all employees.
- **Continuous Adaptation and Future Readiness**: Future-proofing talent management requires an ongoing commitment to adaptation and readiness. Organizations must stay updated with emerging technologies, industry trends, and evolving talent needs to ensure their talent strategies remain relevant and effective.

By anticipating the evolution of AI-oriented competencies and implementing strategies that address the unique challenges and opportunities of the digital economy, organizations can future-proof their talent management practices and build a workforce that is equipped to thrive in the rapidly changing technological landscape (Khang & Kali et al., 2023).

10.9   KEY TERMS

- **Artificial Intelligence**: Artificial Intelligence (AI) refers to the development of computer systems and algorithms that possess the ability to perform tasks that typically require human intelligence. AI enables machines to perceive, reason, learn, and make decisions, mimicking human cognitive processes. It involves various techniques, such as machine learning, natural language processing, computer vision, and robotics, to process and analyze large amounts of data and derive meaningful insights. The ultimate goal of AI is to create intelligent systems that can solve complex problems, adapt to new situations, and enhance human experiences across various domains.
- **Augmented Intelligence**: refers to the integration of AI technologies with human intelligence to enhance and amplify human capabilities rather than replace them. It focuses on utilizing AI as a tool to assist and collaborate with humans in decision-making, problem-solving, and cognitive tasks. Augmented Intelligence systems leverage AI algorithms, data analysis, and machine learning to process vast amounts of information, extract insights, and present relevant information to humans in a meaningful way. The goal of augmented intelligence is to combine the strengths of both humans and machines to achieve more effective and efficient outcomes in various fields, including healthcare, finance, education, and business.
- **Natural Language Processing (NLP)**: refers to the field of AI that focuses on enabling computers to understand, interpret, and process human language in a way that is similar to how humans do. NLP involves the development of algorithms and techniques to analyze and extract meaning from text or speech data. It encompasses tasks such as language understanding, sentiment analysis, text generation, machine translation, and speech recognition. NLP systems use machine learning and linguistic rules to comprehend

and generate human language, enabling applications such as chatbots, voice assistants, and language translation tools.

- **Automated Machine Learning (AutoML)**: refers to the use of AI techniques and algorithms to automate and streamline the process of building and deploying machine learning models. AutoML systems aim to reduce the manual effort and expertise required in traditional machine learning workflows. They automate tasks such as data preprocessing, feature engineering, model selection, hyperparameter tuning, and model evaluation. AutoML platforms leverage algorithms to automatically search, select, and optimize the best models for a given dataset and problem domain.

REFERENCES

Döngül, E. S. & Leonardo Cavaliere, L. P. (2022). Strategic management of platform business ecosystem using artificial intelligence supported human-computer interaction technology. *Management and Information Technology in the Digital Era: Challenges and Perspectives*, 29, 47–61. https://doi.org/10.1108/S1877-636120220000029004

Ermakova, J. D. (2021). Artificial intelligence in the contemporary digital environment. *Current Achievements, Challenges and Digital Chances of Knowledge Based Economy*, 2021, 355–362. https://doi.org/10.1007/978-3-030-47458-4_42

Fernandez, R. S., Hayes, K., & Gayosso, F. (2021). Artificial intelligence and NDE competencies. *Handbook of Nondestructive Evaluation*, 4.0, 1–53. https://doi.org/10.1007/978-3-030-73206-6_24

Gao, F., Jia, X., Zhao, Z., Chen, C. C., Xu, F., Geng, Z., & Song, X. (2021). Bibliometric analysis on tendency and topics of artificial intelligence over last decade. *Microsystem Technologies*, 27, 1545–1557. https://doi.org/10.1007/s00542-019-04426-y

Huateng, M., Zhaoli, M., Deli, Y., Hualei, W., Huateng, M., Zhaoli, M., & Hualei, W. (2021). Vigorously promoting digital transformation in manufacturing. *The Chinese Digital Economy*, 2021, 73–116. https://doi.org/10.1007/978-981-33-6005-1_6

Khang, A. (2023). *Advanced Technologies and AI-Equipped IoT Applications in High-Tech Agriculture* (1st ed.). IGI Global Press, Hershey, PA. https://doi.org/10.4018/978-1-6684-9231-4

Khang, A., Gupta, S. K., Rani, S., & Karras, D. A. (2023). *Smart Cities: IoT Technologies, Big Data Solutions, Cloud Platforms, and Cybersecurity Techniques*. CRC Press, Boca Raton, FL. https://doi.org/10.1201/9781003376064

Khang, A., Kali, C. R., Suresh Kumar, S., Amaresh, K., Sudhansu Ranjan, D., & Manas Ranjan, P. (2023). Enabling the future of manufacturing: integration of robotics and IoT to smart factory infrastructure in industry 4.0. *AI-Based Technologies and Applications in the Era of the Metaverse* (1st ed., pp. 25–50). IGI Global Press, Hershey, PA. https://doi.org/10.4018/978-1-6684-8851-5.ch002

Khang, A., Muthmainnah, M., Seraj, P. M. I., Yakin, A. A., Obaid, A. J., & Panda, M. R. (2023). AI-aided teaching model for the education 5.0 ecosystem. *AI-Based Technologies and Applications in the Era of the Metaverse* (1 ed., pp. 83–104). IGI Global Press, Hershey, PA. https://doi.org/10.4018/978-1-6684-8851-5.ch004

Khang, A., Rani, S., Gujrati, R., Uygun, H., & Gupta, S. K. (2023). *Designing Workforce Management Systems for Industry 4.0: Data-Centric and AI-Enabled Approaches*. CRC Press, Boca Raton, FL. https://doi.org/10.1201/9781003357070

Khang, A., Shah, V., & Rani, S. (2023). *AI-Based Technologies and Applications in the Era of the Metaverse* (1 ed.). IGI Global Press, Hershey, PA. https://doi.org/10.4018/978-1-6684-8851-5

Khanh, H. H. & Khang, A. (2021). The role of artificial intelligence in blockchain applications. *Reinventing Manufacturing and Business Processes through Artificial Intelligence* (vol. 2, pp. 20–40). CRC Press, Boca Raton, FL. https://doi.org/10.1201/9781003145011-2

Li, D., Pan, Z., Wang, D., & Zhang, L. (2023). A study on artificial intelligence orientation and new venture performance. *Asia Pacific Business Review*, 29, 1–23. https://doi.org/10.10 80/13602381.2023.2188764

Li, J., Li, M., Wang, X., & Thatcher, J. B. (2021). Strategic directions for AI: the role of CIOs and boards of directors. *MIS quarterly*, 45(3), 1603–1644. https://search.ebscohost.com/ login.aspx?direct=true&profile=ehost&scope=site&authtype=crawler&jrnl=0276778 3&AN=152360589&h=dneM1SiRC60V2XsaVmmeF5h1s9OV6t%2F7ARMsVO3V 8idoqDX8pODJni7qzP%2FLd2h6BL2rM%2FpfOeHQ2aLT7Cah4w%3D%3D&crl=c

Nosova, S., Norkina, A., Medvedeva, O., Makar, S., Bondarev, S., Fadeicheva, G., & Khrebtov, A. (2021). The collaborative nature of artificial intelligence as a new trend in economic development. Biologically Inspired Cognitive Architectures Meeting, pp. 367–379. Springer International Publishing, Cham. https://doi.org/10.1007/978-3-030-96993-6_40

Rana, G., Khang, A., Sharma, R., Goel, A. K., & Dubey A. K. (2021). *Reinventing Manufacturing and Business Processes through Artificial Intelligence*. CRC Press, Boca Raton, FL. https://doi.org/10.1201/9781003145011

Rani, S., Chauhan, M., Kataria, A., & Khang, A. (2021). IoT equipped intelligent distributed framework for smart healthcare systems. *Networking and Internet Architecture*, 2, 30. https://doi.org/10.48550/arXiv.2110.04997

Snehal, M., Babasaheb, J., & Khang A. (2023). Workforce management system: concepts, definitions, principles, and implementation. *Designing Workforce Management Systems for Industry 4.0: Data-Centric and AI-Enabled Approaches* (1st ed., pp. 1–13). CRC Press, Boca Raton, FL. https://doi.org/10.1201/9781003357070-1

Thowfeek, M. H., Samsudeen, S. N., & Sanjeetha, M. B. F. (2020). Drivers of artificial intelligence in banking service sectors. *Solid State Technology*, 63(5), 6400–6411. https:// www.researchgate.net/profile/Mb-Fathima-Sanjeetha-2/post/Dissertation_topic_ related_to_Cloud_computing_and_AI/attachment/60a88f126b953100014dc146/AS%3 A1026128093384710%401621659410533/download/288.pdf

Wamba-Taguimdje, S. L., Fosso Wamba, S., Kala Kamdjoug, J. R., & Tchatchouang Wanko, C. E. (2020). Influence of artificial intelligence (AI) on firm performance: the business value of AI-based transformation projects. *Business Process Management Journal*, 26(7), 1893–1924. https://doi.org/10.1108/BPMJ-10-2019-0411

11 Enrich Skills Recommendation Based on Sentiment Analysis Using Ensemble Learning

Nidhya M. S., Vijaya Kumar Guivada, Maniraj S. P., and Neetu Pillai

11.1 INTRODUCTION

When selling goods online, the majority of enterprises have their own recommendation system. Most websites, on the other hand, are created with the company's sales department in mind, which uses irrelevant and meaningless advice to persuade customers to buy. An individual user can utilize a personalized recommendation system (PRS) to choose interesting and useful items from a wide range of products. Customers now have a variety of options for products from e-commerce websites as a result of the expansion of the internet. Customers may struggle to find the ideal items at the ideal time (Khang & Rani et al., 2023).

11.1.1 RECOMMENDATION SYSTEM

A recommendation system, usually referred to as a recommender system, is a specific technique for filtering information that suggests subjects most relevant to a specific user, occasionally replacing "system" with words like "platform" or "engine." Making recommendations that are user-friendly is the aim of a recommender system. The two primary categories of methods available to achieve this goal are collaborative filtering strategies and content-based methods (Khang & Muthmainnah et al., 2023).

11.1.2 SENTIMENT ANALYSIS

The emotional undertone of a document can be identified using the natural language processing (NLP) approach known as sentiment analysis. It is frequently called "opinion mining." Organizations frequently employ this technique to learn more about and obtain feedback on a certain product, service, or idea. Determine whether the message's emotional tone is favorable, negative, or neutral while analyzing digital text for sentiment. Feelings, trends, and value are the three key elements of sentiment analysis (Khang & Rashmi et al., 2024).

DOI: 10.1201/9781003440901-11

11.1.3   ENSEMBLE LEARNING

A broad meta approach to machine learning is ensemble learning that aims to enhance predicted performance by combining the results from many models. The underlying idea of ensemble learning is to combine weak learners to create a strong learner. Using a portion of the training data, train a collection of decision trees. Predict the class that receives the maximum using predictions from individual trees. Ensemble learning method is clearly shown in Figure 11.1.

Sarma et al. (2019) said that a customized recommendation system for online courses, books, news, films, music, and research articles are all available to people seeking assistance. Thanks to online training and learning platforms, users were able to choose their preferred courses from a large range. Consumers may make informed decisions fast and from a vast array of possibilities thanks to expert systems. Recommendation systems emerged as a result, allowing users to customize their search results and choose skills from a wide range of possibilities.

The algorithms for the recommendation system were typically created using collaborative filtering, multi-model ensemble, associative rules, and content-based filtering (Lee et al. 2020). Personalized recommendation systems can use multi-model ensemble techniques; however, for content-based filtering, a significant amount of real-world data is required to create the prediction model. Searching for spatial patterns also uses the multi-model ensemble method. By identifying the spatial anomaly correlation between them, it may put the anomaly correlations in groups. In order to extract patterns from spatial noise, the clustering algorithm serves as a filter (Zhao et al., 2020) (Figure 11.2).

Collaborative filtering allows objects to be filtered based on comparable thoughts. By scanning a huge population, it can identify a smaller group of people who have a passion for collecting items. Collaborative filtering heavily relies on the similarity metric. It can identify user demographics that make specific purchases (Gazdar & Hidri, 2020).

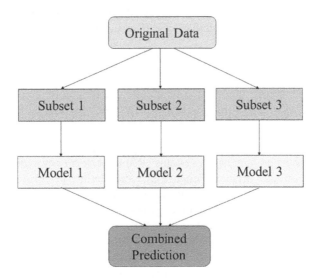

FIGURE 11.1   Ensemble learning.

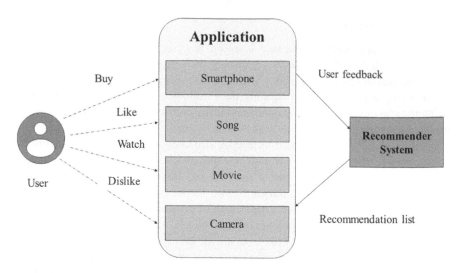

FIGURE 11.2 General recommendation system.

The development of recommendation systems typically employs four basic methods. In this chapter, an ensemble learning approach-based sentiment analysis-based enrich skills recommendation system was proposed. Considering the information for user ratings and preferences, clustering enables grouping of IT skills. A personalized skills recommendation system would benefit greatly from this clustering's impressive prediction abilities. This study's primary objective is to develop a better approach for adjusting the recommendation system. This chapter consists of five parts. Section 11.1 discussed about the introduction of the work. The drawbacks and limitations of present systems are discussed in Section 11.2. Section 11.3 explained about our new suggested system design. Section 11.4 contains our newly constructed system's overall result and discussion. Section 11.5 covers this paper's conclusion.

11.2 LITERATURE REVIEW

In this literature, survey section is discussing about the limitations and drawbacks of the current systems used in enrich skills recommendation process on sentiment analysis. Both private individuals and corporations frequently utilize recommendation algorithms or recommendation systems (RS) to perform news and information searches, pursue online purchases, engage in social dating, perform search engine optimization, etc. (Zou et al., 2017). Skills recommendation systems are increasingly important as the trend of learning courses grows and learner's requests for locating desired skills rise (Lu 2019) while choosing skills.

The majority of systems use artificial intelligence (AI) to search objects based on popularity, correlation, and skill attributes (Cho & Han, 2019). Online search abnormally affects the recommendation algorithm. For instance, clicking on skills with low rankings has a beneficial effect, but clicking on skills with high rankings has no effect (Liu et al., 2019). A personal rank method utilizing a neural network can resolve the classic skill recommendation system's second significant issue with data sparsity (Wen & Li, 2019).

The authors, Hariadi and Nurjanah (2017), discussed about the hybrid approach using an attribute-based system enriched by the personality features. By developing a preference matrix, a user profile is initially created. The title, author, price, genre, and other attributes are among those included in the matrix and change depending on the requirements and goals of the user. In the second phase, building a neighborhood entails figuring out how the active user and the other users are similar to one another. Since the model takes into account how people's interests and rating behavior are related, it is utilized in addition to collaborative filtering. The MSV-MSL (Most Similar Visited Material to the Most Similar Learner) approach is used to determine the skill's score in the third step. The 34 recommendations are generated based on the expected ratings since all the aforementioned processes have been finished.

Mathew et al. (2016) present a hybrid approach that combines filtering based on content, collaboration, keywords, and data mining. The user is prompted to enter passwords that fit his preferences in a database that will be used to provide recommendations in the future for the keyword-based filtering. To find the repeating items in data, this method is called ECLAT (Equivalence class Clustering and bottom-up Lattice Traversal). This technique can parse the items once, from the beginning, to find the most significant and common ones, making it extremely speedy and efficient.

Beleveslis and Tjortjis (2020, June) investigates the research's focus on encouraging the variety of recommendation systems in e-commerce by incorporating LSH (Locality-Sensitive-Hashing), feature weighting, and content-based filtering. The method's architecture is built in three stages. To represent the collection of products, the first one computes a weighted matrix. Then, using the Jaccard 40 formula, the Minhash technique is used to condense representations of the product set and determine which sets share the most characteristics. The algorithm also offers recommendations based on LSH Forest. A pre-possessed version of the already existing Retail Rocket dataset and the best pricing dataset, which was developed from actual e-commerce data, are used to assess the proposed method. It is rapid and effective because there are many frequent ones.

Schoinas and Tjortjis (2019, May), I put forth MuSIF, a new recommender system that incorporates implicit feedback from multiple sources, for product recommendations. The steps are taken to implement the system's design, and the main algorithm uses an IF CF variant with alternative least squares (ALS). After that, the system's performance was assessed when techniques to improve accuracy were used. In the end, four comparisons were built using various techniques. The authors draw the conclusion that once all the approaches were put into practice, the single-source matrix performed better than the multi-source one in every manner examined. Even the multi-source system's highest performance fell short of the single-source system's higher score.

Users may efficiently experiment with ensembles of SVMs thanks to ensemble SVM. The findings of the experiments demonstrate that, while maintaining competitive predictive accuracy, it takes a lot less time to train ensemble models than it does to train ordinary LIBSVM models. The performance of an ensemble may be enhanced by utilizing more complicated aggregation algorithms. Additionally, it enables quick prototyping of new techniques (Claesen et al., 2014).

Due to the present massive increase of comments on computer networks, it is important to analyze any comments that may have embedded attitudes, whether they are good or negative. This process is known as sentiment analysis (Tang et al., 2016). Based on data mining, this classification was created. The two types of sentiment analysis are based on dictionaries and machine learning, respectively. The dictionary base technique is utilized in the preparation process to collect data from each attribute, and then, the machine learning technique is used for sentiment analysis.

The dependence between the base models that ensemble methods have can also be used to distinguish them. Aggregating a group's output from various base learners can be done in a variety of ways. These are most frequently categorized as weighing approaches or meta-learning approaches. "Learning from learners" is referred to as meta-learning. When creating an ensemble, it is also feasible to combine a number of the aforementioned techniques. Sagi and Rokach (2018) use the phrase "ensemble hybridization approach" to describe such a mixture.

11.3 SYSTEM DESIGN

Our novel method is built utilizing a mix of the butterfly optimization technique and the Naive Bayes machine learning algorithm for enriching skills recommendation based on sentiment analysis using ensemble learning (Khang & Shah et al., 2023).

11.3.1 BUTTERFLY OPTIMIZATION ALGORITHM

A recently developed meta-heuristic called the butterfly optimization algorithm (BOA) imitates the butterfly's regular foraging and mating behavior. The primary component of BOA is the scent that butterflies generate, which aids in other butterflies to find food and prospective partners as shown in Figure 11.3.

The butterfly optimization should be initialized before beginning the aforementioned flowchart. First initialize the probability (p), sensory modality (c), and power exponent (alpha). The next step is evaluating the fitness value and also finding the best solution. After that, it checks the relationship between r and p. And also it performs and checks the global search and local search. Once getting the fitness evaluation, update the butterfly values in population. If it is the correct criterion, it delivers the best output solution.

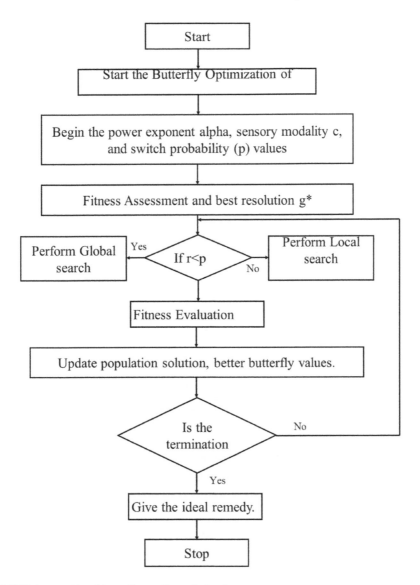

FIGURE 11.3 Algorithm of butterfly optimization.

1. Objective purpose $f(x)$, $x = (x_1, x_2,... x_{dim})$, dim= Dimensions, number
2. The beginning of n butterflies in a population $x_i = (i = 1, 2, 3,...n)$,
3. Stimulus intensity I_i at x_i is determined by $f(x_i)$
4. Describe the sensor modality c, power exponent a, and switch probability p.
5. **while** criterion for stopping not met**do**
6. **for** each butterfly bf in population **do**

7. Determine the scent for *bf* using Equation (11.1)

8. **end for**

9. Choose the best *bf*

10. **for** each butterfly *bf* in population do

11. Create the random number *r* between [0, 1]

12. **If** $r < p$ **then**

13. Utilizing Equation (11.2), proceed to the ideal butterfly or resolution.

14. **else**

15. Using Equation (11.3), move arbitrarily

16. **end if**

17. **end for**

18. Add an update to *a*

19. **end while**

20. Give the best solution you've found.

Butterfly optimization algorithm pseudocode is shown in Figure 11.4. We can clearly understand the working steps of butterfly optimization from the above-mentioned figure. To evaluate the butterfly population value, first find out the

1: objective purpose $f(x), x = (x_1, x_2, \ldots \ldots x_{dim})$, *dim*= Dimensions, number

2: the beginning of *n* butterflies in a population $x_i = (i = 1,2,3, \ldots . n)$

3: Stimulus intensity I_t at x_i is determined by $f(x_i)$

4: Describe the sensor modality *c*, power exponent *a* and switch probability *p*.

5: **while** criterion for stopping not met**do**

6:      **for** each butterfly *bf* in population **do**

7:          Determine the scent for *bf* using Eq. (1)

8:      **end for**

9:      Choose the best *bf*

10:      **for** each butterfly *bf* in population do

11:          Create the random number r between [0, 1]

12:          **If** $r < p$ **then**

13:              Utilising Eq. (2), proceed to the ideal butterfly or resolution.

14:          **else**

15:              Using Eq. (3), move arbitrarily

16:          **end if**

17:      **end for**

18:      Add an update to *a*

19: **end while**

20: Give the best solution you've found.

FIGURE 11.4   Pseudocode for butterfly optimization.

relationship between "*r*" and "*p*." From this relationship, if $r < p$ means that it move toward the best butterfly. Suppose if $r > p$ means that it move toward randomly to the next. And update the value of "*a*" and get the best output solution (Khang, 2023).

11.3.2 NAIVE BAYES

A supervised learning method for classification issues, the Naive Bayes algorithm, is based on the Bayes theorem. The key tool we use for text categorization is a sizable training set (Figures 11.5 and 11.6).

Input:
>   *T*, the practice set
> $F = (f1, f2, f3,...fn)$

Output:
>   a portion of the test data

Step:
1. Study the practice dataset *T*
2. Identify the predictor variables' mean and standard deviation for each class
3. Repeat
    Find the probability of using the Gauss density equation to *fi* for each class
    Until the probability of each predicting factor $(f1, f2, f3,...fn)$ has been calculated
4. Determine each class's probability
5. Obtain the greatest likelihood

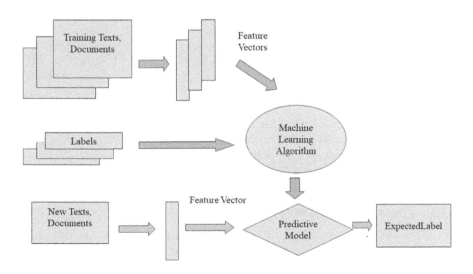

FIGURE 11.5   Text classifications using Naive Bayes.

Input:
T, the practice set
F= $(f1, f2, f3, \ldots, fn)$
Output:
a portion of the test data.
Step:
1. Study the practise dataset T;
2. Identify the predictor variables' mean and standard deviation for each class;
3. Repeat
Find the probability of using the gauss density equation to fi for each class;
Until the probability of each predicting factor $(f1, f2, f3, \ldots, fn$ has been calculated
4. Determine each class's probability;
5. Obtain the greatest likelihood;

FIGURE 11.6　Pseudocode of Naive Bayes algorithm.

11.3.3　BUTTERFLY OPTIMIZATION ALGORITHM WITH NAÏVE BAYES

The optimization algorithm is combined with machine learning technique which is explained below and having the following benefits. Increasing the model's accuracy while lowering the likelihood of mistakes or losses brought on by these predictions is the aim of optimization. Machine learning algorithms are commonly trained on offline or local datasets that are static. Through optimization, predictions and classifications become more accurate, and error is decreased (Khang & Muthmainnah et al., 2023).

The method of optimization involves identifying a set of inputs to an objective function that results in an evaluation of the function that is maximal or minimal. It is the difficult problem that many machine learning methods are based on. By adjusting model hyperparameters, algorithm optimization is the process of enhancing a machine learning model's efficacy and accuracy. A loss function is used in machine learning optimization to calculate the discrepancy between output data's actual and expected values (Khang & Kali et al., 2023).

11.4　RESULTS AND DISCUSSIONS

Enrich skills recommendation using sentiment analysis and UCI, Kaggle, the newly created categorization algorithm is validated using a variety of datasets. Training and testing phases of sentiment analysis are carried out at the second level using ensemble learning. The results of this level are transferred for clustering process. The proposed study employs MATLAB 2013a to assess a classification's efficacy. The newly developed butterfly optimization algorithm with Naïve Bayes technique is contrasted with the current systems of SVM, CNN, ANN, LSTM, and BI-LSTM in terms of accuracy, precision, recall, and F1-measure performance metrics.

11.4.1　PERFORMANCE EVALUATION MATRIX

You can evaluate the performance of your trained machine learning models using performance evaluation metrics. Then, by giving your machine learning model a dataset that it has never seen before, you can assess how effectively it will function as shown in Figure 11.7.

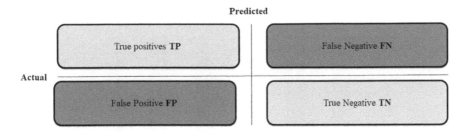

FIGURE 11.7   Performance evaluation matrix.

TABLE 11.1
Accuracy Table

Algorithm	Accuracy
SVM	83.02
CNN	84.12
ANN	91.16
LSTM	93.01
BI-LSTM	94.20
BOA+NB	95.77

11.4.2   ACCURACY

According to the accuracy formula, accuracy is defined as the error rate's departure from 100%. To ascertain accuracy, we must first ascertain the error rate. Additionally, by dividing the observed value by the actual value, the error rate is determined as a percentage as shown in Equation (11.1).

$$\text{Accuracy} = \frac{TP + TN}{TP + TN + FN + FP} \tag{11.1}$$

Table 11.1 and Figure 11.8 show accuracy of the proposed approach comparison with existing approach, which provides a better maximum outcome than the present system, is explained in detail in Table 11.2. It states that the accuracy results of the SVM, CNN, ANN, LSTM, and BI-LSTM are 83.02%, 84.12%, 91.16%, 93.01%, and 94.20%, respectively.

11.4.3   PRECISION

Precision is a number that indicates the number's value and the number's quantity of information digits.

Table 11.2 and Figure 11.9 show detailed explanation of the precision of the proposed approach comparison with existing approaches. The stated precision outcomes for SVM, CNN, ANN, LATM, and BI-LSTM are 83.95%, 85.37%, 92.78%, 92.31%, and 93.45%, respectively.

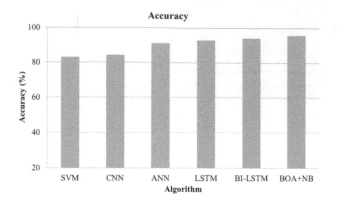

FIGURE 11.8   Accuracy graph

TABLE 11.2
Precision Table

Algorithm	Precision
SVM	83.95
CNN	85.37
ANN	92.78
LSTM	92.31
BI-LSTM	93.45
BOA+NB	96.26

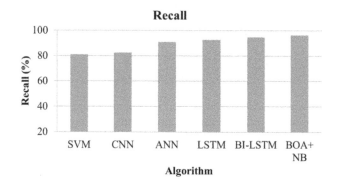

FIGURE 11.9   Precision graph.

11.4.4 RECALL

Recall reveals details on the model's capacity to recognize real positives. There are no true positives in this case to indicate how many people with the illness were actually detected. Two false negatives exist as shown in Equation (11.2).

$$Re\,call = (TP)/(TP + FN) \qquad (11.2)$$

TABLE 11.3

Recall Table

Algorithm	Recall
SVM	80.95
CNN	82.35
ANN	90.91
LSTM	92.71
BI-LSTM	94.62
BOA+NB	96.26

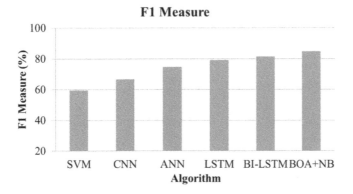

FIGURE 11.10   Recall graph.

Table 11.3 and Figure 11.10 show detailed explanation of the recall of the proposed approach comparison with existing approaches, which offers a superior maximum result than the existing system. The SVM, CNN, ANN, LSTM, and BI-LSTM recall outcomes are given as 80.95%, 82.35%, 90.91%, 92.72%, and 94.62%, respectively.

11.4.5 F1 MEASURE

It is difficult to compare two models that have a high recall but a low precision. Therefore, we want F1-Measure in order to compare them. The F1-Measure is a tool for testing recall and precision at the same time. It replaces the arithmetic mean with the harmonic mean by heavily penalizing the extreme values as shown in Equation (11.3).

$$F1 - Measure = \frac{2*recall*precision}{recall + precision} \tag{11.3}$$

Table 11.4 and Figure 11.11 show detailed explanation of the recall of the proposed approach comparison with existing approaches, which offers a superior maximum

TABLE 11.4

F1 Measure Table

Algorithm	F1 Measure
SVM	59.44
CNN	66.72
ANN	74.69
LSTM	79.22
BI-LSTM	81.36
BOA+NB	84.67

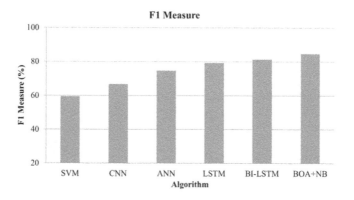

FIGURE 11.11   F1-measure graph.

result than the existing system. The SVM, CNN, ANN, LSTM, and BI-LSTM from F1 measure outcomes are given as 59.44%, 66.72%, 74.69%, 79.22%, and 81.36%, respectively.

11.5   CONCLUSION

In this study, clustering techniques were utilized to improve the ensemble learning-based recommendation system's ability for prediction. We use machine learning (ML) technique proposed based on IT skills for an employee needs to be learned based on competency or information technology. Datasets were gathered from the Kaggle good read skills repository or Skills Taxonomy (SCEDEX, 2021). Using machine learning methods, the total skill ratings of some specific enrich IT skills were processed. For the suggested model's optimization procedures, accuracy, precision, recall, and F1 were evaluated. 95.77%, 96.26%, 96.26%, and 84.67%, respectively, are the overall accuracy, precision, recall, and F1 measure. These findings demonstrate how much more effectively we can eliminate uninteresting skills from the suggested list using our hybrid BOA with NB method (Snehal et al., 2023).

The accuracy, precision, recall, and F1 measure performance graphs were then displayed, and they demonstrate that the majority of datasets maintain a tight relationship with the butterfly optimization classifier. In our upcoming study, we will offer an ensemble learning technique that combines machine learning with optimization to create a suggestion system for enrich skills based on sentiment analysis (Babasaheb et al., 2023).

REFERENCES

Babasaheb, J., Sphurti, B., & Khang, A. (2023). Industry revolution 4.0: workforce competency models and designs. *Designing Workforce Management Systems for Industry 4.0: Data-Centric and AI-Enabled Approaches* (1st ed., pp. 14–31). CRC Press, Boca Raton, FL. https://doi.org/10.1201/9781003357070-2

Beleveslis, D. & Tjortjis, C. (2020). Promoting diversity in content based recommendation using feature weighting and LSH. *IFIP International Conference on Artificial Intelligence Applications and Innovations*, pp. 452–461. Springer, Cham. https://link.springer.com/chapter/10.1007/978-3-030-49161-1_38

Cho, E. & Han, M. (2019). AI powered book recommendation system. *Proceedings of the 2019 ACM Southeast Conference*, pp. 230–232. Kennesaw, GA, USA. https://dl.acm.org/doi/abs/10.1145/3299815.3314465

Claesen, M., De Smet, F., Suykens, J., & De Moor, B. (2014). EnsembleSVM: A Library for Ensemble Learning using Support Vector Machines. arXiv preprint arXiv:1403.0745. https://www.jmlr.org/papers/volume15/claesen14a/claesen14a.pdf

Gazdar, A. & Hidri, L. (2020). A new similarity measure for collaborative filtering based recommender systems. *Knowledge-Based Systems*, 188, 105058. https://www.sciencedirect.com/science/article/pii/S0950705119304484

Hariadi, A. I. & Nurjanah, D. (2017). Hybrid attribute and personality based recommender system for book recommendation. *2017 International Conference on Data and Software Engineering (ICoDSE)*, pp. 1–5. IEEE, Piscataway, NJ. https://ieeexplore.ieee.org/abstract/document/8285874/

Khang, A. (2023). *Advanced Technologies and AI-Equipped IoT Applications in High-Tech Agriculture* (1st ed.). IGI Global Press, Hershey, PA. https://doi.org/10.4018/978-1-6684-9231-4

Khang, A., Kali, C. R., Suresh Kumar, S., Amaresh, K., Sudhansu Ranjan, D., & Manas Ranjan, P. (2023). Enabling the future of manufacturing: integration of robotics and IoT to smart factory infrastructure in industry 4.0. *AI-Based Technologies and Applications in the Era of the Metaverse* (1st ed., pp. 25–50). IGI Global Press, Hershey, PA. https://doi.org/10.4018/978-1-6684-8851-5.ch002

Khang, A., Muthmainnah, M., Seraj, P. M. I., Yakin, A. A., Obaid, A. J., & Panda, M. R. (2023). AI-aided teaching model for the education 5.0 ecosystem. *AI-Based Technologies and Applications in the Era of the Metaverse* (1st ed., pp. 83–104). IGI Global Press, Hershey, PA. https://doi.org/10.4018/978-1-6684-8851-5.ch004

Khang, A., Rani, S., Gujrati, R., Uygun, H., & Gupta, S. K. (2023). *Designing Workforce Management Systems for Industry 4.0: Data-Centric and AI-Enabled Approaches*. CRC Press, Boca Raton, FL. https://doi.org/10.1201/9781003357070

Khang, A., Rashmi, G., Hayri, U., Tailor, R. K., & Sanjaya, S. G. (2024). *Data-driven Modelling and Predictive Analytics in Business and Finance*. CRC Press, Boca Raton, FL. https://doi.org/10.1201/9781032600628

Khang, A., Shah, V., & Rani, S. (2023). *AI-Based Technologies and Applications in the Era of the Metaverse* (1 ed.). IGI Global Press, Hershey, PA. https://doi.org/10.4018/978-1-6684-8851-5

Lee, Y., Wei, C., Hu, P., Cheng, T., & Lan, C. (2020). Small clues tell: a collaborative expansion approach for effective content-based recommendations. *Journal of Organizational Computing and Electronic Commerce*, 30, 1–18. https://www.tandfonline.com/doi/abs/1 0.1080/10919392.2020.1718056

Liu, Q., Zhang, L., & Zhao, Y. (2019). The interaction effects of information cascades, word of mouth and recommendation systems on online reading behavior: an empirical investigation. *Electronic Commerce Research*, 19(3), 521–547. https://link.springer.com/article/10.1007/s10660-018-9312-0

Lu, K. (2019). Research on intelligent detection system for information abnormal defect based on personalized recommendation of E-book. *The International Conference on Cyber Security Intelligence and Analytics,* pp. 1110–1117. Springer, Cham. https://link.springer.com/chapter/10.1007/978-3-030-15235-2_147

Mathew, P., Kuriakose, B., & Hegde, V. (2016). Book recommendation system through content based and collaborative filtering method. *2016 International conference on data mining and advanced computing (SAPIENCE)*, pp. 47–52. IEEE, Piscataway, NJ. https://ieeexplore.ieee.org/abstract/document/7684166/

Sagi, O. & Rokach, L. (2018). Ensemble learning: a survey. *Wiley Interdisciplinary Reviews: Data Mining and Knowledge Discovery*, 8(4), 1–18. https://inria.hal.science/hal-00758208/preview/dmkd_fl_2011_main.pdf

Sarma, D., Mittra, T., & Hossain, M. S. (2021). Personalized book recommendation system using machine learning algorithm. *International Journal of Advanced Computer Science and Applications*, 12(1). https://pdfs.semanticscholar.org/d53f/32726dd28cd19a7b5837 12d33198561b7e09.pdf

SCEDEX. (2021). Skills-based competency ecosystem in digital economy. *The AI-oriented Competency Model for Digital Economy 5.0*. Retrieved from https://scedex.com/skills-framework.htm

Schoinas, I. & Tjortjis, C. (2019). MuSIF: a product recommendation system based on multi-source implicit feedback. *IFIP International Conference on Artificial Intelligence Applications and Innovations*, pp. 660–672. Springer, Cham. https://link.springer.com/chapter/10.1007/978-3-030-19823-7_55

Snehal, M., Babasaheb, J., & Khang, A. (2023). Workforce management system: concepts, definitions, principles, and implementation. *Designing Workforce Management Systems for Industry 4.0: Data-Centric and AI-Enabled Approaches* (1st ed., pp. 1–13). CRC Press, Boca Raton, FL. https://doi.org/10.1201/9781003357070-1

Tang, D., et al. (2016). Sentiment embeddings with applications to sentiment analysis. *IEEE Transactions on Knowledge and Data Engineering*, 28(2), 496–509. https://ieeexplore.ieee.org/abstract/document/7296633/

Wen, G. & Li, C. (2019). Research on hybrid recommendation model based on personrank algorithm and tensorflow platform. *Journal of Physics: Conference Series*, 1187(4), 042086. https://iopscience.iop.org/article/10.1088/1742-6596/1187/4/042086/meta

Zhao, T., Zhang, W., Zhang, Y., & Liu, Z. (2020). Significant spatial patterns from the GCM seasonal forecasts of global precipitation. *Hydrology and Earth System Sciences*, 24(1), 1–16. https://hess.copernicus.org/articles/24/1/2020/

Zou, C., Zhang, D., Wan, J., Hassan, M. M., & Lloret, J. (2017). Using concept lattice for personalized recommendation system design. *IEEE Systems Journal*, 11(1), 305–314. https://doi.org/10.1109/JSYST.2015.2457244

12 Revamping the Hiring Process using WebRTC, AWS Cloud, and Gaze-Tracking Application

*Megharani Patil, Rohan Rajendra Dalvi,
Faraz Hussain, and Suyog Gupta*

12.1 INTRODUCTION

The market is filled with various WebRTC applications (Salvatore & Romano, 2012). Most of them are not suited for taking online interviews. Online interviews often fall prey to malpractices which can mislead recruiters in their recruitment process. The paper provides an implementation path for building a product that aims at filling this gap in the industry and stands out as the only viable option for firms in their online interview process. The paper discusses the development of a browser-based WebRTC platform that helps in detecting these malpractices and gives a versatile environment for taking interviews. The product is built using MVC (Dragos-Paul & Adam, 2014) architecture.

In this chapter, we discuss the implementation and working of WebRTC, the storage of data in a non-relational format, and AI-based verification systems. The platform comes filled with pre-designed question banks based on testing the candidates in a whole spectrum of technical and non-technical topics. We have directed our research in creating a safe environment using several anti-cheat techniques such as encryption (Ondrej, 2015), browser monitoring (Ondrej, 2015), gaze-tracking (Nimesha et al., 2021), etc. The product can take several rounds of interviews at once and allows the human interviewer to intervene and assist wherever required during the interview process.

12.2 BACKGROUND

Our use of star topology for our WebRTC-based interview system stems from the need of having one-to-many bidirectional video connections as stated in the paper (Nakal et al., 2017) "WebNSM: A Novel WebRTC Signaling Mechanism for

DOI: 10.1201/9781003440901-12

One-to-Many Bi-directional Video Conferencing" – "star topology provides better performance for different users, while each participant does not need a high capacity of CPU or high bandwidth."

The paper (Nimesha et al., 2021) "Cheating Detection in Browser-based Online Exams through Eye Gaze Tracking" discusses the idea of a browser-based gaze-tracking cheating detection system – "we generate an eye gaze data set while a student faces an online examination. We then process and analyze this data set to detect any misbehavior during an online examination." Our system uses a similar technique which tracks the candidate's gaze from the candidate's browser, thus reducing the load on the server. For conducting multi-party video conferencing on a large scale, the bandwidth requirement increases accordingly for both the server and the clients. This paper presented the impact of bandwidth limitations and video resolution size on QoE for WebRTC-based mobile multi-party video conferencing that discusses how different bandwidth limitations impact QoE (Quality of Experience) for different video resolutions.

For increasing the flexibility of the system, it is important that we operate the system in a cloud-based environment, and hence, we should know the techniques for cost estimation used in such environments as described in the publication-compostable cost estimation and monitoring for computational applications in cloud-computing environments. We have to estimate the operational requirements using the determined workflows and dataflow.

12.3   PROPOSED SYSTEM

12.3.1   Access Control Modules

12.3.1.1   Model of Access Controls

The product is made to follow a B2B model of operations where the firm delivers services of the software to companies who require support in conducting large-scale online interviews. We have used an RBAC (Role-Based Access Control model) (David et al., 1997) for access control inspired by a hierarchical system of control as shown in Figure 12.1.

12.3.1.2   Roles

The owner and the administrator roles have effectively the same permissions over the software. They can examine the dashboard, add job posts, add exams, add candidates, perform screening, schedule interviews, review the evaluations, and add content in the question banks. The owner has precedence over the administrator as the former can remove the latter. The administrator is responsible for assigning people the role of operators and can remove or add operators. The operator has effectively the same control to create job posts, and exams, add candidates, perform screening, view the dashboard, and schedule and evaluate interviews. But the operator cannot remove job posts, exams, questions, and candidates added by the other operators and neither can the operator schedule interviews for job posts created by the other operators.

This hierarchical structure of management provides effective and fast customer support, reduces redundancy, and makes scheduling and creation of exams faster

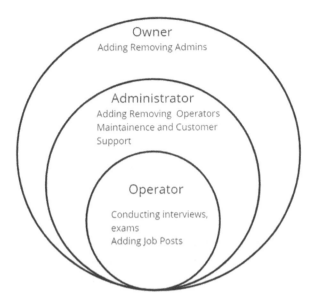

FIGURE 12.1   Role-based hierarchy.

(David et al., 1997), as the question banks are reusable throughout the system and the candidate information is readily available to larger groups of companies. This system of controls provides flexibility and ensures security for the smooth operation of the software (David et al., 1997).

12.3.1.3   Authentication
The process of adding a new operator requires the administrator to enter the email address of the operator into the system. The operator is then sent a mail from the system through which they receive a link to a form from which they can set up their security details. The passwords are then hashed using the SHA-256 algorithm and stored in cloud storage. Thus, each time when the operator wants to access the system, they can do it via the website by just entering the security details. This system ensures that the software is highly accessible, robust, and mobile (Azat, 2014).

12.3.2   System Modules
The modules made for the smooth scheduling of the interview process are the QA bank, exams, job posts, candidates, and screening modules. These modules function as the primary points of interaction for the operator to add questions, candidates, exams, and job posts.

12.3.2.1   QA Bank Module
The QA bank exists as a repository for technical and non-technical questions of several domains which might be required while taking the exams/interview. The QA bank module allows the operator to add questions to a selected domain, such that the

operator can specify its underlined topics, difficulty level, time limit, question type, and scores. Tags can also be added to the questions to make them easily searchable. The system contains customized options for various question types. These are coding questions, SQL coding questions, multiple-choice questions, descriptive questions, and fill-in-the-blanks-type questions. For multiple choice, descriptive, and fill-in-the-blank-type questions, the operator will provide an answer set so that it can be checked against the answers given by the candidate during evaluation. For coding and SQL coding question, the output should be checked against the test cases provided by the operator along with appropriate time-out limitations. A sample code can also be given by the operator such that the candidate has a head start on attempting the question. Our system has used a type of online compiler which can run code in one of Python, Java, C/C++, CPP, JavaScript, Rust, Golang, Ruby, and C# programming language.

12.3.2.2   Exam and Job-Post Modules

The exams and job posts module serves a similar purpose as they represent an opening or a role set up by the client (operator) on the behalf of a company. The job post module allows the operator to create a job post with attributes like job title, company name, company logo, contact details, location, and job details, such as skills, openings, salary range, and job description. The exam module allows the operator to create an exam with attributes like exam title, company name, exam logo, and exam description.

12.3.2.3   Candidate Module

The candidate module contains information of thousands of candidates. This module allows operators to add new candidate information to the software including the candidate's personal details such as gender, date of birth, work experience, full name, email address, salary expectation, and skills. The candidate module also saves documents related to the candidates such as the candidate's profile picture, resume, identification file, voice sample, and other files. The candidate module saves candidates' contact information like their address and phone number along with other details from their social media profile, visa validity, higher education, and other jobs the candidate has applied to.

12.3.2.4   Resume-Screening Module

The resume screening is another crucial part of the software which in turn helps to shortlist candidates from the wider pool into a group of people with the list of skills required in the job. The resume screening module uses data from the candidate module and the exams/job post module to get resumes and required skills, respectively.

12.3.3   Databases

Before continuing with further modules and the functioning of the software, it is crucial for us to understand the structure of the database and its underlined components and to see the dataflow between the modules. We are using a non-relational document database (Nishtha et al., 2012) that provides support for JSON-like storage.

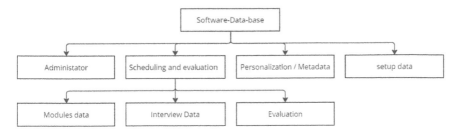

FIGURE 12.2   Database structure.

This storage has a flexible data model that enables users to store unstructured data as it provides full indexing support with replication with rich and intuitive APIs as shown in Figure 12.2.

12.3.3.1   Administrative Database

The first one of these databases is the administration database which contains information about the clients which mainly consist of colleges and companies. This database also contains customer counters, customer manuals, website domain data, and data about the various users (operators, owners, and administrators). The administration database contains four sub-sections. The website sub-section contains information pertaining to the website attributes such as domain data, hosting data front-end and back-end files, and style sheets. The customer sub-section contains data of the customer such as the colleges and companies which have attributes such as ID, name, locations, requirements, openings, etc.

Customers also have their unique IDs and credentials which are stored in the next sub-section which is authorization. The authorization subsection contains IDs, email addresses, and hashed passwords of the various users such as the owners, administrators, and customers. Finally, the last sub-section of the administrator database contains a manual, descriptions, and walkthroughs for the maintenance and administration of the software.

12.3.3.2   Scheduling and Evaluation Database

The second database is for scheduling and evaluation. This is the largest database in the software which contains data collected from different modules, the interview data, and the evaluation data. The data from the modules have information about the candidates, the job posts, the exams, the QA banks, and the resume screening.

The interview data is a collection of files, readings, and recordings taken during the interview/exam process. This includes the interview video media file, the email data, the session data, the candidate user preferences, and the interview answers. The evaluation section of the scheduling and evaluation database contains files for AI face evaluation which contains gaze-tracking and expression evaluation. Along with this, we have a sub-section for AI speech evaluation which contains the evaluated data for the speech rate and speech-to-text function. The evaluation section also stores data for the weighted and grammatical evaluation of the descriptive answers and data collected from the anti-cheat window detection system.

12.3.3.3   Personalization Database

The next database is for the generated data from user interactions during their subsequent sessions. This database contains the personalization and metadata about candidate settings and preferences, user settings and preferences, user-sign-up timings and session information.

12.3.4   SCHEDULING AND INTERVIEW PROCESS

The scheduling part of the scheduling and interview process is handled by the scheduling module. The scheduling module allows the operator to schedule an interview/exam for the selected job post or exam. The operator is required to fill in a due date and time and specify the nature of the interview. The screened and shortlisted candidates can then be notified and added for the scheduled interview. The operator also has to add the required interview/exam questions from the QA bank. The scheduling module then proceeds to email the candidates a notification regarding the interview. By following the link, the candidate can then proceed to join the exam/interview.

The exam/interview takes place on the browser in a room generated using peer-to-peer connections between the hosted server and the candidate's device. The system only allows connection once via the link (Nakal et al., 2017), through a set of secure browsers like Google Chrome which allows for the use of our anti-cheat software (Ondrej, 2015). This method is similar to the rooms created in popular services like Google meet. Once the candidate has joined, they undergo a set of protocols which checks and validates their hardware (Ondrej, 2015), in order to fulfil the requirements posed by the system for the exam/interview. These protocols include the validation of their camera, microphone, screen-sharing, window-lock, facial image capture (Ondrej, 2015), and voice sample confirmation (Widodo et al., 2015).

The candidate can also confirm the language settings for the rest of the interview. They are also requested to open their task manager and end all the other application tasks such that the exam/interview process is more efficient and free of any malware interference. Once the candidate has performed all these validations, they are allowed to start their exam/interview. The timer turns on once the interview/ exam is started. The candidate has to then complete their interviews while being monitored by the gaze-tracking and expression-detection software (Ondrej, 2015). A human proctor can also enter the interview at any time to add new questions or ask other questions while the scheduled interview is taking place. The human proctor is mainly the operator who scheduled the interview, and they can join by clicking the link for the interview/exam in the scheduled section of the website. Once they click, they join the interview via the server and generate another peer-to-peer connection on the server side.

Our interview/examination module is equipped with an online compiler that runs the coding questions with regard to the set timeouts and test cases. These compilers handle several programming languages such as Python, Java, and C++, and can also run SQL queries. The exam/interview is concluded once the time runs out or if the candidate answers all questions and ends the process or if the human proctor closes the interview.

12.3.5 EVALUATION

The evaluation module displays the information collected during the interview about the completion of the exam/interview (Ondrej, 2015). It gives technical feedback on the candidate's performance and evaluates their score for every question. The evaluation module comes with several features containing artificial intelligence-powered evaluation techniques. It matches the candidate's photo ID with their on-camera face. It also compares the candidate's pre-recorded voice with the interview voice (Widodo et al., 2015). The evaluation module displays details about any suspicious software detected while conducting an interview/exam. Suspicious software includes any software that alters the transmission of the video feed, or records the screen or another window/tab opened while taking the interview. The interview module displays the result from gaze-tracking (Ondrej, 2015) done during the interview. It performs post-interview processing of the interview/exam video feed and evaluates the candidate's facial expression to track their confidence and focus (Ondrej, 2015).

12.4 RESULTS AND DISCUSSION

12.4.1 WEBRTC

WebRTC comprises a set of novel technologies which allow the embedding of audio and video communications between applications. It allows browsers to exchange media information in a peer-to-peer fashion (Salvatore & Romano, 2012).

12.4.1.1 WebRTC Architecture

For the implementation of the WebRTC Trapezoid concept, a web application that is downloaded from a separate web server is running in both browsers. Communications are started and stopped using signalling messages. They are sent across web servers using the HTTP or WebSocket protocol, where they can be edited, translated, or managed as necessary. It is important to note that WebRTC does not standardize browser-to-server signalling because it is viewed as an integral component of the application. A peer connection, which affects the data stream, enables media to move between browsers without the need for intermediary servers as shown in Figure 12.3.

12.4.1.2 Implementation of WebRTC for an Interview
Web Application (Topology)

For the implementation of an online interview system, a peer-to-peer connection has to be established between the interviewees and interviewers, and the audio and video streams generated also have to be stored on the server. Using the traditional topology of peer-to-peer connections in a room, we end up with a mesh topology in which the servers as well as the users have to sustain large bandwidth while the interview is taking place. We assume that the devices of the users (candidates and proctors) cannot handle large bandwidth. Thus, we adopt a star topology (Salvatore & Romano, 2012) with the server at its centre and peer-to-peer connections between every user and the server so that the burden of streaming large amounts of data is passed on to the server rather than the user (Figure 12.4).

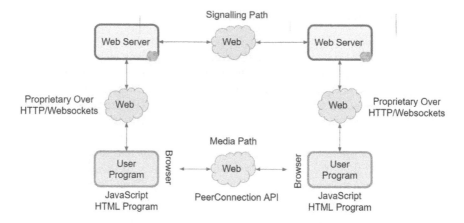

FIGURE 12.3 WebRTC architecture.

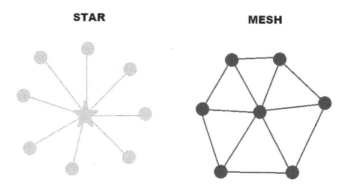

FIGURE 12.4 Star topology and mesh topology.

12.4.1.3 Bandwidth Requirement for the Video Stream

Thus, we get a UDP (Bart et al., 2018) transfer network in which a large amount of data is transferred to and from the server. This data usually consist of two video streams, one from the webcam of the candidate/proctor and one from the shared screen of the user. For our software, we have assumed that these streams have a video quality of 720p and a refresh rate of 30 fps. For two such streams, this results in a bandwidth requirement of 5.36 Mbps and 7.0 Mbps for uploads.

12.4.1.4 Maximum Scenario

Depending on the number of connections between the users and the likelihood of the proctors sharing their screen and video, we obtain 3 possible scenarios. A maximum scenario in which each of the proctors is sharing their screens and webcam feed with each of the other proctors and the candidate as shown in Figure 12.5.

As we can see the below chart, the magenta arrows indicate the UDP streams coming from the candidate to the server, the orange streams coming from the proctors

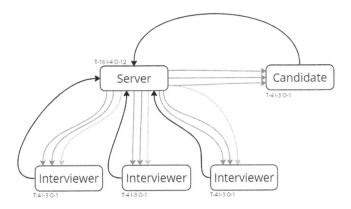

FIGURE 12.5   Maximum scenario with all proctors sharing their screens.

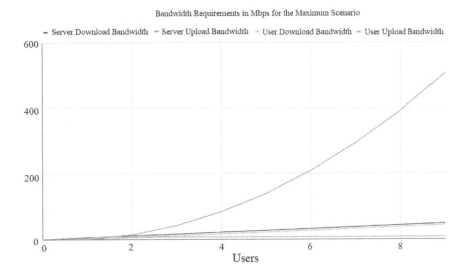

FIGURE 12.6   Graph showing the number of users vs. bandwidth requirement for the maximum scenario.

to the server, and the green and blue arrows indicate the output streams relaying the video streams to the users from the proctors and candidates, respectively, as shown in Figure 12.6.

$$2 \times \text{UPBW} \times \left(N^2 - N \right) \tag{12.1}$$

where

- N, number of users
- UPBW, upload bandwidth (3.5 Mbps for a stream of 720p 30 fps)

Similarly, the server download bandwidth can be shown using Equation (12.2)

$$2 \times \text{DLBW} \times N \tag{12.2}$$

where

- N, number of users;
- DLBW, download bandwidth (2.69 Mbps for a stream of 720p 30 fps);
- the user upload bandwidth remains constant, at the value equivalent to $2 \times$ UPBW, as each user only uploads two video streams each but the user download bandwidth increases with a rate of $\text{DLBW} \times 2 \times (N-1)$ as each user will download the stream of every other user.

12.4.1.5 Minimum Scenario

Likewise maximum scenario, a minimum scenario in which each of the proctors is sharing their screens and webcam feed with each of the other proctors and the candidate as shown in Figure 12.7.

In minimum scenario, the magenta arrows indicate the UDP streams coming from the candidate to the server, the orange streams coming from the proctors to the server, and the green and blue arrows indicate the output streams relaying the video streams to the users from the proctors and candidates, respectively, as shown in Figure 12.8.

$$2 \times \text{UPBW} \times (N-1) \tag{12.3}$$

The total user (proctor and candidate included) upload bandwidth per user decreases as each new user joins as only the candidate is sharing the stream. However, the total download bandwidth per user remains mostly the same.

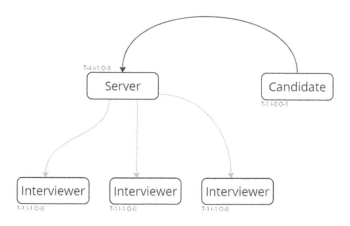

FIGURE 12.7 Minimum scenario with only candidate sharing video stream.

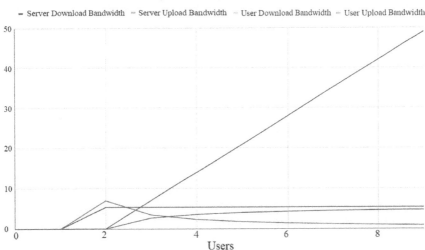

FIGURE 12.8   Graph showing the number of users vs. bandwidth requirement for minimum scenario.

12.4.1.6   Average Scenario

Both of these scenarios are extreme cases and extreme examples of user behaviour. In reality, the proctors would sometimes share their webcam feed and screens. Thus, the mean or average of both of these extreme scenarios is used to stimulate the average behaviour during the interview/exam process as shown in Figure 12.9.

In average scenario, the magenta arrows indicate the UDP streams coming from the candidate to the server, the orange streams coming from the proctors to the server, and the green and blue arrows indicate the output streams relaying the video streams to the users from the proctors and candidates, respectively, as shown in Figure 12.10.

$$(MaxBW + MinBW) / 2 \qquad\qquad (12.4)$$

where

- MaxBW, maximum bandwidth
- MinBW, minimum bandwidth

12.4.2   Rate of Monitoring

In the real world, the system should be capable of conducting multiple interviews and exams simultaneously. Each one of those interviews/exams would have one candidate and their connection to the server. The number of interviewers/proctors joining the interviews may vary according to the organization's requirements. Our system

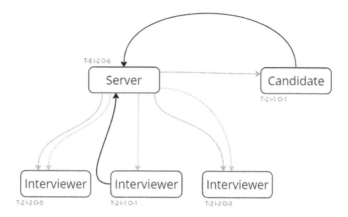

FIGURE 12.9 Average scenario with some interviewers sharing their video streams.

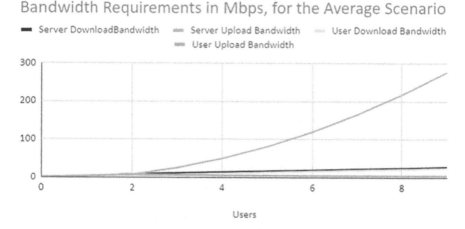

FIGURE 12.10 Average scenario with some interviewers sharing their video streams.

is built as such to allow for a single proctor to join any interview as needed. There can also be multiple proctors joining a single interview to monitor the candidate's conduct. Therefore, to calculate bandwidth requirements for an average interview in the system, we must know the number of proctors joining each interview. This can be done by simply calculating the rate of monitoring. The rate of monitoring (Rm) is the number of proctors over the number of on-going interviews. It is important because it directly affects the server download and upload speeds as shown in Figure 12.11.

Here, the x-axis is the rate of monitoring and the y-axis is the average bandwidth requirement per interview (maximum of 8 proctors) in Mbps. The average bandwidth requirement per interview for a maximum of 8 proctors is calculated using Equation (12.5),

$$\sum_{n=0}^{n}\left(Rm^n \times Bn\right) / \sum_{n=0}^{n}\left(Rm^n\right) \qquad (12.5)$$

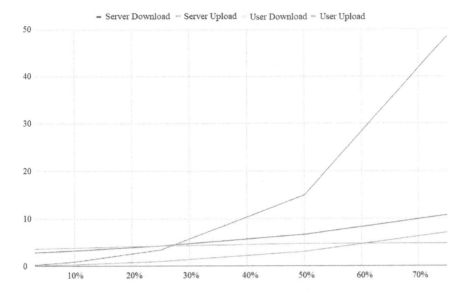

FIGURE 12.11   Graph showing the rate of monitoring vs. average bandwidth requirement.

where

- n, number of proctors,
- Rm, rate of monitoring,
- Bn, bandwidth required for "n" number of proctors.

As the server upload bandwidth increases at the rate of $x \times (x - 1)$ and is not convergent for higher rate of monitoring, one can adjust the values of the rate of monitoring and the number of users to find the optimum mix. For our calculations, we have taken the value of a 5% rate of monitoring and a maximum limit of 8 proctors (Bart et al., 2018). The system can still work fairly well up to a 25% rate of monitoring. Beyond that, the bandwidth requirement rises exponentially.

12.4.3   Memory Requirement and Costing

Our system is designed to store designated data of the interviews/exams in cloud storage. We have used the model of S3 storage given by Amazon AWS to determine the memory costs of running this system at scale. In the previous passage, we used the monitoring rate to calculate the average bandwidth requirement of the interviews. Using this, we can estimate the total memory required to store the video footage for the interviews in our system. Our system uses three types of storage:

- The first is Amazon S3 Standard infrequent access – For long-lived but infrequently accessed data that needs millisecond access which has a per month cost of $0.0125/GB.

- The second one, Amazon's S3 Glacier Flexible Retrieval (Formerly S3 Glacier) – For long-term backups and archives with a retrieval option from 1 minute to 12 hours which has a per month cost of $0.0036/GB.
- Finally, the last type of storage used is Amazon's S3 Glacier Deep Archive – For long-term data archiving that is accessed once or twice in a year and can be restored within 12 hours with a cost of $0.00099/GB.

For the 720p 30 fps footage of webcam and screen recording, for an average of 5% rate of monitoring with a maximum of 8 proctors per interview, we get the per hour memory collection of 10.388 GB using Equation (12.6).

$$\text{AvgIHR} = \text{Average DLBW} \times 60 \times 60 / 1,024 \qquad (12.6)$$

where the average DLBW for the server is equal to 2.95 Mbps and the average UPBW for the server is equal to 0.397 Mbps. We are using the standard infrequent access memory to store the data of the interview for 1 day during which it will be processed to check for expressions tracking, facial recognition, gaze-tracking, and screen monitoring. Then the data would be transferred to the Glacier flexible retrieval storage so that the user can access it within the set period of 1 month. This would result in a cost of $0.04172 per hour of an interview using Equation (12.7),

$$\text{Memory Cost} = \text{AvgIHR} \times ((\text{SIAM} / 30) + \text{GFRM}) \qquad (12.7)$$

Example: $10.388 \times ((0.0125 / 30) + 0.0036)$

where

- AvgIHR, the net data consumption for an average interview for a specific rate of monitoring (here 5%);
- SIAM, Standard Infrequent access per GB per month;
- GFRM, Glacial Flexible Retrieval per GB per month.

We can also store data in the Glacier Deep Archive for each additional month at a cost of $0.010284 per hour of interview.

12.4.4 DATA TRANSFER AND PROCESSING COST

While the data incoming from the internet to EC2 are free, we sustain a $0.05 per GB cost for sending data out from Amazon EC2 to the internet. This totals up to $0.0697 per interview hour for our system. Along with this, we must also account for the cost of running the instances. Each instance of m6in.32xlarge in Amazon EC2 provides the user with 128 core vCPU, 512 GB of RAM, a bandwidth of 200 GBps, and instance storage of elastic capacity with a bandwidth of 80 GBps.

For each interview, we estimate to require up to 0.008 of such instances when we account for the processing requirements of the various AI tools, video streams, and other processing tasks. The cost of one of these instances is equal to $8.91072 per hour. This amount totals $0.00718 per hour of interview. The total cumulative technical running costs of taking these interviews through our system amount to a sum total of $0.11856 per interview hour at a 5% rate of monitoring using the Equation (12.8):

$$\text{Total Cost} = \text{Memory cost} + \text{Data} - \text{transfer cost} + \text{Instance cost} \qquad (12.8)$$

To understand how this might be beneficial to a company transitioning to our AI-based interview system, we need to focus on finding if this switch would result in a more efficient hiring process. To measure success, the company would be using the number of hours saved as the main indicator. Thus, a cost-benefit analysis could be used to prove that our system is better than the traditional method (Khang, 2023).

To do this, we will use the example of company X wishing to interview 5,000 applicants for an opening of 50 posts. Our AI-based resume screening system could first help them shortlist about 4,000 out of those 5,000 candidates to be taken in for the interview. Of these 4,000 candidates, 100 could be shortlisted after the first round of technical (online) interviews to fill in for the final round of in-person interviews lasting about 1 hour. Let us suppose that the company has a team of 20 staff members to conduct interviews. If each primary interview takes 4 hours of time, then based on this calculation, while using our system at a 5% rate of monitoring, it would take about 10 batches to complete the interview process and about 16,000 hours of total interview time. If we multiply it by the system's cumulative operational costs of $ 0.11856, we get a total cost of $1896.973 for the interviews plus $164.55 cost of each additional month of storage. While this might seem expensive, we should also take into consideration the total amount of working hours saved by the recruitment team (Khang & Muthmainnah et al., 2023).

Using our system with a 5% rate of monitoring, it will take the interview team 10 batches (4,000*0.05/20) of 4-hour interviews in the primary round and use face-to-face interviews for the next 5 batches (100/20) of the 1 hour of final round interviews; it would take the interview team of 20 people a total of 45 working hours to complete the interviews, which is approximately 5.625 working days. While using the traditional method of only taking one-on-one interviews would take the same team, 200 batches (4,000/20) of 4 hours to take the primary round and 5 batches (100/20) of 1 hour final round, resulting in a total of 805 working hours or 100.625 working days to complete the task. Thus, in this given example, the total resultant time saved is equal to 15,200 working hours which is equal to 760 hours for each member of the 20-member recruitment team, which is about 95 working days (each of 8 hours) saved per team member.

Thus, even while considering the additional operating costs, support costs, and maintenance costs and time required for scheduling the interviews (maximum 5 hours), we should be able to at least see 93%+ time-saving in the whole interview process for the recruitment teams for a maximum cost of approx. $0.45 per interview (4 hours each) in the given hypothetical example (Khang & Rani et al., 2023).

12.5   CONCLUSION

From this research, we have not only provided the reader with a complete description of our software along with its modules and databases, but we have also given a detailed summary of the mechanism, topologies, and technologies used in building this system. We have done a comprehensive study of the costs that go into operating these mechanisms on a per-interview/exam basis, and we have compared it with the traditional methods to obtain meaningful insights about the effectiveness of using our system (Khang & Shah et al., 2023).

The descriptions included in this chapter cover browser-based gaze-tracking and facial recognition, memory calculations for big data models (Rani et al., 2021), star topology-based study of UDP peer-to-peer connection bandwidth requirements, calculations related to the rate of monitoring, etc. We have also uncovered the mechanisms around one of the most crucial technologies, that is WebRTC, and have provided a summary of a hypothetical example of how such a technology can provide a cost and time-efficient mechanism for conducting interviews and exams in the real world while reducing the risk and cheating discrepancies associated with web browser-based online interviews and exams (Khang & Kali et al., 2023).

ACKNOWLEDGEMENTS

This chapter would not have been possible without the unwavering support from Hiringtek.inc and Thakur College of Engineering and Technology's Computer Engineering Department. We would like to give special thanks to our guide and mentor Dr. Megharani Patil (Associate Professor – Computer Engineering Department and I/C HOD of AI & ML) and Director of Development and Quality Assurance at HiringTek Pvt. Ltd Mr S.P. Laxman. Special thanks to Prof. Dr. Alex Khang[PH], Professor in Information Technology, D.Sc. D.Litt., AI and Data Scientist, Department of AI and Data Science, Global Research Institute of Technology and Engineering, North Carolina, United States.

REFERENCES

Azat, M. (2014). *Practical Node.js: Building Real-World Scalable Web Apps*. APress, Japan. https://doi.org/10.1201/9781003440901

Bart, J., Timothy, G., Varun, G., Fernando, K., & Gill, Z. (2018). Performance evaluation of WebRTC-based video conferencing. *ACM SIGMETRICS Performance Evaluation Review*, 45, 56–68. https://dl.acm.org/doi/abs/10.1145/3199524.3199534

David, F. F., Janet, A. C., & Richard, K. (1997). Role-based access control (RBAC). National Institutes of Standards and Technology U.S Department of Commerce, Gaithersburg, MD. https://dl.acm.org/doi/pdf/10.1145/266741.266759

Dragos-Paul, P. & Adam, A. (2014). Designing an MVC model for rapid application development. *Procedia Engineering*, 69, 1172–1179. https://www.sciencedirect.com/science/article/pii/S187770581400352X

Khang, A. (2023). *Advanced Technologies and AI-Equipped IoT Applications in High-Tech Agriculture* (1 ed.). IGI Global Press, Hershey, PA. https://doi.org/10.4018/978-1-6684-9231-4

Khang, A., Kali, C. R., Suresh Kumar, S., Amaresh, K., Sudhansu Ranjan, D., & Manas Ranjan, P. (2023). Enabling the future of manufacturing: integration of robotics and IoT to smart factory infrastructure in industry 4.0. *AI-Based Technologies and Applications in the Era of the Metaverse* (1st ed., pp. 25–50). IGI Global Press, Hershey, PA. https://doi.org/10.4018/978-1-6684-8851-5.ch002

Khang, A., Muthmainnah, M., Seraj, P. M. I., Yakin, A. A., Obaid, A. J., & Panda, M. R. (2023). AI-aided teaching model for the education 5.0 ecosystem. *AI-Based Technologies and Applications in the Era of the Metaverse* (1st ed., pp. 83–104). IGI Global Press, Hershey, PA. https://doi.org/10.4018/978-1-6684-8851-5.ch004

Khang, A., Rani, S., Gujrati, R., Uygun, H., & Gupta, S. K. (2023). *Designing Workforce Management Systems for Industry 4.0: Data-Centric and AI-Enabled Approaches*. CRC Press, Boca Raton, FL. https://doi.org/10.1201/9781003357070

Khang, A., Shah, V., & Rani, S. (2023). *AI-Based Technologies and Applications in the Era of the Metaverse* (1 ed.). IGI Global Press, Hershey, PA. https://doi.org/10.4018/978-1-6684-8851-5

Nakal, M. E., Al-Sherbaz, A., & Turner, S. J. (2017). WebNSM: a novel WebRTC signalling mechanism for one-to-many bi-directional video conferencing. *IEEE Computing Conference*. IEEE, Piscataway, NJ. https://link.springer.com/chapter/10.1007/978-3-030-01177-2_40

Nimesha, D., Asara, S., Tharindu, Y., & Nalin, W. (2021). Leelanga seneviratne, cheating detection in browser-based online exams through eye gaze tracking. *2021 6th International Conference on Information Technology Research (ICITR)*. IEEE, Moratuwa, Sri Lanka. https://ieeexplore.ieee.org/abstract/document/9657277/

Nishtha, J., Sahil, P., Mehak, A., Ishita, K., & Dishant, G. (2012). A survey and comparison of relational and non-relational database. *International Journal of Engineering Research & Technology (IJERT)*, 1(6), 598–601. https://www.academia.edu/download/76957411/a-survey-and-comparison-of-relational-and-non-relational-database.pdf

Ondrej, K. (2015). Dávid Cymbalák, and František Jakab Adaptive Web-Based System for Examination with Cheating Prevention Mechanism. https://www.researchgate.net/profile/Frantisek-Jakab-2/publication/269516806_Adaptive_Web-Based_System_for_Examination_with_Cheating_Prevention_Mechanism/links/548d66770cf2d1800d80db14/Adaptive-Web-Based-System-for-Examination-with-Cheating-Prevention-Mechanism.pdf

Rani, S., Chauhan, M., Kataria, A., & Khang, A. (2021). IoT equipped intelligent distributed framework for smart healthcare systems. *Networking and Internet Architecture*, 2, 30. https://doi.org/10.48550/arXiv.2110.04997

Salvatore, L. & Romano, S. P. (2012). *Real-Time Communication with WebRTC* (1st ed., pp. 68–73). IEEE, Piscataway, NJ. https://ieeexplore.ieee.org/abstract/document/6319299/

Widodo, B., Alexander Agung, S. G., Sari, A. C., & Ngarianto, H. (2015). Designing of humanoid robot with voice recognition capability. *The3rd IIAE International Conference on Intelligent Systems and Image Processing 2015*. IEEE, Piscataway, NJ. https://pdfs.semanticscholar.org/0d28/f7faf2eef2e9474527551656ab28756bbd45.pdf

13 Conceptual Analysis Study
A Paradigm Shift Transformation in Human Resource Management

Muna Fathmath and Albattat Ahmad

13.1 INTRODUCTION

Human resource management (HRM), which is addressed in both theoretical and practical strategies used to manage the workforce, combines abstract theories with organizational practice. Today's Human Resource Management (HRM) paradigm has evolved into a collection of innovation and transitional strategies for new business models and revenue streams, built on the three main pillars of automation, process improvement in manufacturing, and production optimization. There is a new resolution for the industrial digital revolution (Khang & Shah et al., 2023). At the same time, new trendy human resource tools are emerging. Yet, paradigm transformations in human resource management are as crucial to the industrial revolution (Tyagi et al., 2023). Hybrid models have replaced traditional methods as the norm. Internet services are expanding. Communications and digital connections are developing. Digital work and paperless transactions have evolved into new workplace practices. Despite this, the paradigm shift will have a significant impact on the ability to deliver higher-quality services within no time (Xu, 2023).

Many models and tactics for recruitment and selection have been designed. To submit a form for a job application or any other service, people no longer actually need to go anyplace physically. Internet portals make it easier for applicants to submit their applications and allow them to see their status in real time. Interviews can be conducted online using platforms like Zoom, Google Meet, and other cutting-edge technology. This led to a new level of organizational paradigm shift transformation. Yet, the essential tenet of the market-oriented approach to human resource management modifies in response to external environments, particularly to market demand (Khang & Muthmainnah et al., 2023).

Since the 1980s, employee turnover has dominated discussion about job performance and has emerged as an illusive phenomenon. Staffing affects organizational performance and results. Therefore, it's important to practice in an organization the three phases of HRM acquisition, HRM development, and employee termination

DOI: 10.1201/9781003440901-13

in order to attain human resource management. The pre-hiring phase, the training phase, and the post-hiring phase are the names of these stages.

Yet, it is necessary to use a data-driven system while performing HRM responsibilities, particularly Recruitment and Selections (RS), which are the major factors that might affect any organization's effectiveness. As a result, the initial step is the recruitment and selection process. Positioning potential candidates requires consideration of recruitment and selection theory, model, and practice. Models are based on the practical aspects of a theory. It appears that if an organization uses effective recruitment and selection techniques along with theoretical models, it will have an integrated, long-term people management strategy that will produce long-term organizational performance and employee advantages (Muthmainnah et al., 2023).

To ensure that all candidates were treated equally and fairly, laws, regulations, and a code of conduct were used to guide the recruitment and selection process. Political pressure had to be avoided at all times throughout the process. For people with the necessary skills and knowledge, HRM is a highly specialized talent. According to the 2019 Making People Count study, HR professionals think that public companies have poor analytical skills. Organizations require employees with analytical skills, according to this report. However, in a competitive job market, candidates sometimes struggle to find a position that is ideal for their career (Khang & Kali et al., 2023).

Every firm aspires to have effective recruiting and selection methods to retain well-qualified, experienced, and skilled personnel for certain positions. It is crucial for them to comprehend the knowledge of their sector without failing to understand its new directions if they want to manage their business successfully over the long term in a competitive market. Also, for organizations to function more effectively in the future, competent staff members with advanced knowledge and broad perspectives are required. If it doesn't a paradigm change will affect an organization. As a consequence, this study aims to analyze the real-world effects of a paradigm shift in human resource management. Determining the practical effects of a paradigm change in human resource management is the goal of this study. The objective of this study is to understand how profoundly evolving technology is affecting organizational human resource management (Khang & Rani et al., 2023).

13.2 LITERATURE REVIEW

13.2.1 RECRUITMENT AND SELECTION TECHNIQUES

According to Khang et al. (2023), any recruitment and selection process should be built upon the three core concepts of effectiveness, efficiency, and fairness. Khang and AIoCF (2024) proposed a number of indicators that may be used to assess the efficacy of recruiting and selection, including retention rates, promotion rates, job satisfaction, and organizational success. An organization's overall performance and ongoing development are directly correlated with its human capital. Recruitment and selection procedures are the most crucial component of resource allocation and they are related to workplace productivity (Pooja et al., 2023).

New research has identified the true strategies to improve the number of qualified candidates. Nonetheless, it has been suggested that to write a job advertisement that speaks to applicants, use the necessary key words in the Ad. at the right places. It is essential to organize opportunities in a way that is simple for job seekers to grasp if you want to draw in applicants who are the best fit for the organization. Online job applications, especially those that use a URL, ought to be simpler to submit. In today's world, social media usage is essential, and attractive tools are essential to advertise job positions. Significant changes are being made in the paradigm shift of human resource management (Babasaheb et al., 2023a).

13.2.2 HUMAN CAPITAL TRENDS TRANSFORMATION

The term "HR transformation" refers to how the HR function has developed. It is the process of adapting and seamlessly incorporating personnel, technology, and service delivery into HR strategy to increase business value. The all-digital world is changing how we live and work, and HR can assist business leaders and employees in adopting a digital mind-set, it is a digital way of managing, organizing, and leading change. HR processes, systems, and organizations can also adopt new digital platforms, apps, and ways of providing HR services. This new organizational model—a network of teams has a high degree of empowerment, strong communication, and rapid information flow (Xu, 2023).

A professional HR competency framework is urgently needed in today's society due to the paradigm shift. First and foremost, data-driven systems are crucial since organizations often employ evidence-based HR. So, for the purposes of carrying out everyday tasks, validated assumptions based on data, research, and data analysis are important. Second, understanding the internal structure of the organization, including its policies and principles. Finally, it's critical to establish a culture that optimizes the employee experience and the workforce's digital skills by integrating digital technology and comprehending various technical tools. People's advocacy is the fourth and final key factor, fostering a positive culture through training people to be effective communicators (Lashkari & Cheng, 2023).

13.2.3 HUMAN CAPITAL TRENDS TRANSFORMATION IN NEXT 10 YEARS

Verlinden (2022), a digital HR expert from 2022, stated that every HR leader will need to have HR skills in order to manage the organization and give better higher-quality services. These abilities include the ability to analyze people, organizational design, risk management, ethics and data privacy, change management consulting, resilience, and safe human resources. A competency study conducted by Academic Innovative Human Research (AIHR) found that just 41% of HR professionals are proficient in using technology and data to increase productivity and create corporate value. Additionally, 85% of firms have accelerated their digitization since the epidemic began, and by 2025, roughly 40% of work-related skills will have changed. It is evident that compared to a few years ago, HR is already expected to do a lot (more). Yet, the nature of HR roles and the skills needed to fill them have evolved dramatically and quickly, and this trend is expected to continue (Verlinden, 2022).

According to the most recent study, in 2023 human resource management would aid in increasing employee retention by continuously securing and enhancing organizational performance. As a result, HR must be aware of the following: HR should begin actively and holistically managing its contingent workforce; HR must help to increase retention by ensuring continuity; and HR must improve organizational performance by dealing with HR compliance by dividing compliance responsibilities within the HR department. Being aware of legal requirements, adhering to corporate standards consistently, and periodically auditing the policies and procedures are also important. The four areas with the highest levels of digital transformation within an organization are as follows. They include the following: domain transformation, business model transformation, process transformation, and cultural transformation (Sharma et al., 2023).

Apparently, scenario planning is a method that is frequently used in the process of strategic planning. Consequently, it's important to use the data to define strategic decisions and develop potential future scenarios. Yet, by using data and developing scenarios, one may comprehend the outside forces that might affect an organizational operation. HR also plays a significant role in this process, particularly in managing the dynamics of the scenario team as part of the scenario planning stage. Discover how to make sure HR complains at the organization, stay out of trouble, and develop trustworthy employees. With the uncertain future, scenario planning is becoming more and more crucial. Hence, two ideas that HR practitioners need to learn are the scenario planning process and how it relates to organizations and HR. The strategic HR professional's toolkit should include scenario planning. It offers the chance to adapt the organization and its executives to handle the uncertainty of the new workplace (Veldsman, 2022a).

13.3 RESEARCH METHODOLOGY

The secondary source of data was examined in this qualitative research in order to reach a conclusion. For this basic fundamental study data was gathered from peer-reviewed journal articles, Reports, and other conference proceedings. Data was gathered using exploratory research questions, then it was analyzed, and interpreted using diagrams with relevant information. The practical Consequences and the peer-reviewed articles were examined in this conceptual background study (Babasaheb et al., 2202b).

13.4 RESULTS AND DISCUSSIONS

13.4.1 STRATEGIC WORKFORCE PLANNING & HR ANALYTICS

The strategy for human resource management has transformed, and workforce planning now depends heavily on HR analytical skills. There are four main factors in strategic workforce planning such as; the right size, shape, time/cost, and agility of the workforce as shown in Figure 13.1.

FIGURE 13.1   HR analytics needed to understand strategic workforce planning.

Source: Veldsman (2022a).

13.4.2   HR Analytics Skills HR Leaders and Practitioners Need to Build

The most recent study, however, has revealed that HR Analytics Skills are important for everyone to understand, especially those who work as HR Leaders, HR Practitioners, or in the HR Team. They must provide unique HR analytical skills such as IT, data analysis, marketing, HR, and business skills. In addition to these abilities, they must have a deep grasp of psychological knowledge and be skilled storytellers. These abilities are crucial for a paradigm shift in human resource management in this growing society as shown in Figure 13.2.

> *You can think, dream, create, design, built, and even select the most wonderful workplace, but it requires you to make the dream a reality.*

Olden days strategy has changed to controller to consultant & assistant	*In the era of digital economy, HR manager's strategically guide and identified about execution, monitoring, coaching, and proposal of opportunities to the people and proactive to focus on strategy development and are accountable to offering effective services for both employers and workers and other stakeholders (Prof. Dr. Alex Khang).*

13.4.3   Roles of HRM in Recruitment Selection Process

The recruitment and selection process would not need to be carried out if the position was not required by the organization. Job analysis is therefore the first step in the recruitment and selection process, and it's crucial to undertake a job analysis, including a job description and job specification, taking into account the job's primary competencies, in order to have the right selection criteria as shown in Figure 13.3.

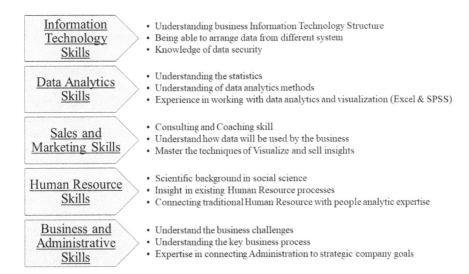

FIGURE 13.2   HR analytical skills for HR leaders and HR practitioners.

Source: Veldsman (2022b).

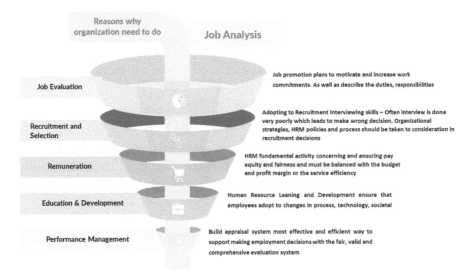

FIGURE 13.3   Roles of HRM in recruitment & selection process.

Source: Jay (2022).

13.4.4   Recruitment and Selection Methods and Their Validity

The recruiting and selection team for the organization must use fair recruitment and selection practices, and adhere to guidelines, procedures, and rules when choosing the best applicant. In recruitment, it needs to attract candidates who are the best matches for the organization and then write a job advertisement to speak to the applicants. In reality, if the proper selection process isn't used, it won't be fair to advance and recruit qualified candidates. First and foremost, the interview panel should be knowledgeable about their obligations and demonstrate utmost integrity when choosing the best candidate (Khang & Rashmi et al., 2024).

Yet, having a better technique is crucial when choosing interview panel members. The panel should consist of a minimum of four members who have been designated as Detailed, Hostile, Loving, and Technical skilled Examiners (Distributed hyper ledger technology and AI for IoT-based FinTech platform (DHLT)). In order to pick the most qualified candidates for the organization, the panelists must then behave with self-integrity by avoiding the Halo/Horns effect, prototype effect, and stereotype selection methods as shown in Figure 13.4.

In addition to the interview, every level should conduct a written exam to choose the best candidates. Instead of having multiple-choice questions, the exam paper should cover scenario-based questions. Only then could the examiner or the panel member determine if the candidate would be a good fit for the organization or not. Indeed, all interview questions should adhere to the following guidelines: they should be open-ended rather than closed-ended, they should focus on organizational objectives, and they should collect performance data and assess the personality of the applicant. The reason for hiring the candidate would be strengthened by reference checking and validating the candidate's information. Eventually, the selection method shall vary in organizational validity as shown in Figure 13.5.

- Have a systematic recruitment and selection practice
- Design the selection process
- Select the candidates based on their job specification within the procedure and strategic guideline
- Make hiring decisions based on factual evidence
- Select the candidates with the writing skills and abilities

There is a post-hiring phase after the hiring procedure. It's a crucial time to officially welcome and onboard newly hired candidates to the organization. The ideal methods for performing an induction program are as;

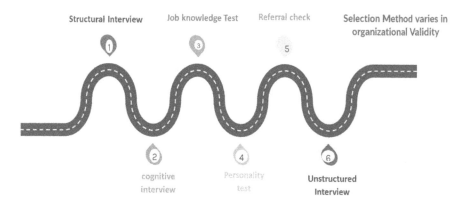

FIGURE 13.4   Recruitment & selection methods and their validity for the selection process.
Source: Jay (2022).

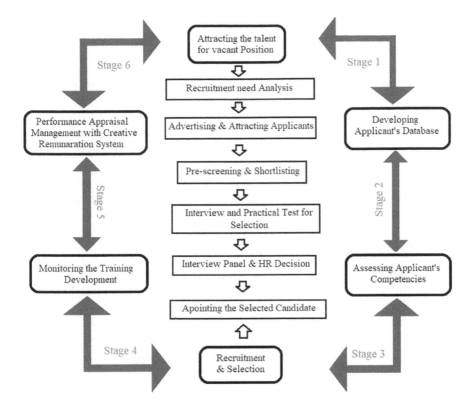

FIGURE 13.5 Six stages of the recruitment and selection process.

Source: Muna (2022).

- Begin communication & start welcoming the new employee.
- Communicate necessary details & documents (policies, procedures, organizational values, etc.)
- Keep their workstation & gadgets fully prepared
- Create a work plan for new employees
- Set up a meeting with their boss
- Introduce new hires & colleagues with each other
- Take them on a full office tour
- Give new employees a guide
- Take feedback from the new employees & have a nice talk

The second element in the post-hiring phase is HR KPIs, which are important for measuring organizational-level dimensions and evaluating the performance of recruiting and selection practices.

- Employees with high performance
- Scope of Leadership Programs
- Scope and use of technology platforms as a source of innovation

- Positive assessment of the work characteristics
- Positive assessment of the organization's selection, procedures, and technologies

But if the organization doesn't track and measure the hiring efforts and consistently spot and fix issues, and recruitment and selection plan becomes ineffective. The correct recruiting KPIs may help continually streamline and enhance the strategy and are a priceless asset both during and after the hiring process. Recruitment KPIs help employees, managers, and business leaders understand where they currently are, what needs to improve, and the exact steps to take to reach those goals. Also, it's important to ensure that the recruiting efforts are continuously assessed and strengthened, it is crucial to select the right KPIs for the organization, monitor them, and make adjustments as needed.

13.4.5 Cost of Poor Recruitment and Selection Decision

Poor recruitment and selection practices can lead to issues including; high turnover, poor performance, weak culture, absenteeism, and are time consuming. Therefore, adequate recruitment and selection procedures must be practiced in an organization in order to reduce the impact of these problems as shown in Figure 13.6.

13.4.6 Benefits of an HRIS (Human Resources Information System)

Veldsman (2022a) a European HR psychologist, stated that technology has a significant impact on HR. Technology developments are to be expected to lead to new and

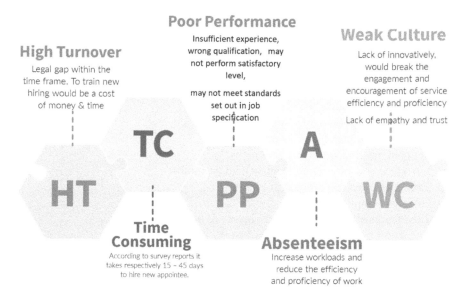

High Turnover
Legal gap within the time frame. To train new hiring would be a cost of money & time

Poor Performance
Insufficient experience, wrong qualification, may not perform satisfactory level,
may not meet standards set out in job specification

Weak Culture
Lack of innovatively, would break the engagement and encouragement of service efficiency and proficiency
Lack of empathy and trust

Time Consuming
According to survey reports it takes respectively 15 – 45 days to hire new appointee.

Absenteeism
Increase workloads and reduce the efficiency and proficiency of work

TC   A
HT   PP   WC

FIGURE 13.6 Cost of poor recruitment and selection decisions.

Source: Papapolychroniadis et al. (2017).

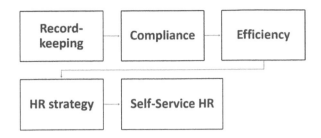

FIGURE 13.7 Human resource information system (HRIS) benefits.

Source: Satispi et al. (2023) and Sharma et al. (2023).

evolving jobs, just like during the Industrial Revolution. With these developments, we have the opportunity to improve previous shortcomings as shown in Figure 13.7. Ultimately, HR must always grow, reskill, upskill, and keep updated on the latest advancements. HR practitioners will need to develop their technological leadership abilities or manage the implementation of automation (Veldsman, 2022b).

13.5 HR LEADERS – PEOPLE MANAGEMENT SKILLS

13.5.1 HR LEADERS – UNDERSTAND 6 DEEP PSYCHOLOGICAL PATTERNS

Every HR professional, HR leader, and HR practitioner should be aware of six psychological patterns. It's important to understand people's communication patterns and behavioral tendencies. The six essential patterns to be aware of are Proactive and Reactive, Generic and Particular, Internal and External, Options and Processes, Toward and Away from, and finally, Things and Person as shown in Figure 13.8.

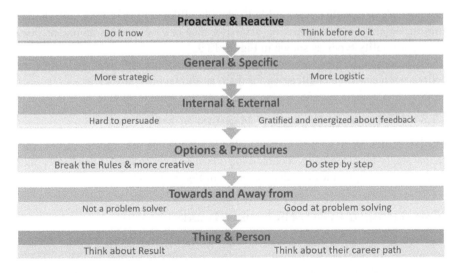

FIGURE 13.8 Six psychological patterns every HR leader should understand.

Source: Charvet (2022) and Verlinden (2022).

- Proactive individuals are first and foremost those who take the initiative to act immediately and insist on taking action. Reactive people would assess the context, analyze it, and weigh what they understood before acting.
- Yet, General people are more strategic and won't provide enough information in a detailed manner. They will have a broad perspective but won't be able to easily provide details. Where else Specific people are the people who need more details and need all the information more logically.
- Others are classified as Internal and External. Internal type people are more difficult to convince and persuade. They do what they do because they love doing it (Charvet, 2022). On the other hand, receiving comments from co-workers and superiors makes external individuals happier.
- **Options**: People are known as those who break the rules yet are more creative. They have a ton of innovative concepts and strategies for breaking the law. Nonetheless, procedural: persons tend to finish the work they started since they follow a step-by-step procedure.
- **Toward**: people need to be motivated. They are better organized to get the job done, but they lack problem-solving skills. Away from: people always believe that something will inevitably go wrong. So, they are adept at solving issues.
- **Things**: People always consider the outcomes and how they affect the business, while **Persons**: people constantly consider their career path and never give consideration to the success of the organization (Charvet, 2022).

13.5.2   Leaders should Understand the Characteristics of People's Personality Type!

Every person's personality type and individual qualities will vary from one another within an organization. Aggressive, assertive, and passive are the names of three ifferent personality types as shown in Figure 13.9.

If HR leaders lack these qualities and skills, managing employees and providing excellent service will be challenging for them as shown in Figure 13.10.

13.5.3   HR Leaders – Knowledge Management Skills

Every HR leader needs to be knowledgeable in knowledge management techniques such as single-loop learning, double-loop learning, procedural knowledge, tacit knowledge, and implicit knowledge as shown in Figure 13.11.

13.5.4   Becoming a Great HR Leader – Leaders Qualities

To become a successful HR leader or HR practitioner they must possess positive traits in terms of offering high-quality services as shown in Figure 13.12.

Thus, every HR practitioner needs to be familiar with fundamental principles and best practices as shown in Figure 13.13.

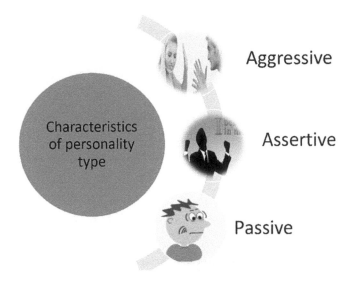

FIGURE 13.9 Characteristics of personality type HR leader must understand.

Source: Verlinden (2022).

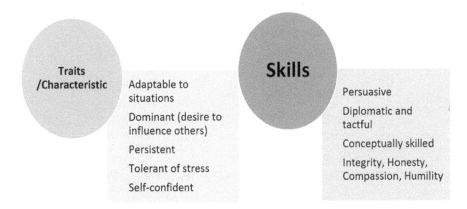

FIGURE 13.10 HR leadership personality traits/characteristics.

Source: Verlinden (2022).

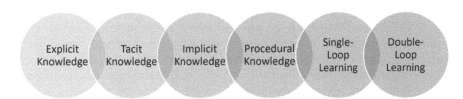

FIGURE 13.11 HR leaders' knowledge management skills.

Source: Charvet (2022).

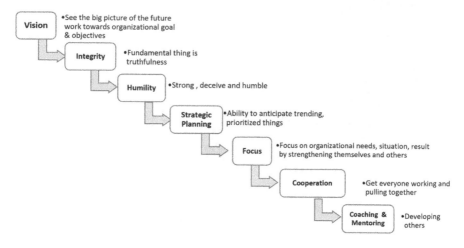

FIGURE 13.12 HR leaders' qualities to become great leaders.

Source: Verlinden (2022).

Best Practices of HR Practitioners

Best Practice of HR Leaders	Balance your work, personal and professional life
	Manage any kind of conflict or grievance
	Build trust among your subordinates and co-workers
	Focus on everyone's wellbeing
	Communication
	Ethical Decision making style
	Lead by example
	Sharpen your knowledge
	Know how to prioritize your work and plan in advance
	Mange data with technology advancement

FIGURE 13.13 HR leaders' qualities to become great leaders.

13.6   CONCLUSION

In the process of human resource management, the effective tool of digital technology and advancement of new technology are significant for an organization's steady stream power of growth in terms of providing quality service and financial efficiency. When the organization has poor recruitment and selection practices the organization is more likely to have a weak organizational culture, most of the employees will be demotivated, the organization will have less skilled employees, where there will be no innovation, and finally, poor relationship coordination and all these factors would lead to high employee turnover rate (Shah & Khang, 2023).

The chapter indicated that to be practically usable, automated recruitment tools should focus on addressing cognitive bias in the recruitment process, supporting team decision-making, facilitating (rather than performing) decision-making, and empowering job seekers while supporting recruiters. Our work provides valuable information for researchers and practitioners who focus on automated recruitment tools and other related decision-support systems. However, for recruitment and selection organizations must employ applicant software and advanced technology which will enable them to select the best suitable candidates. Indeed, the Appraisal system must be implemented in the most effective and efficient way to support making employment decisions with a fair, valid, and comprehensive evaluation system.

REFERENCES

Babasaheb, J., Sphurti, B., & Khang, A. (2023a). Industry revolution 4.0: workforce competency models and designs. *Designing Workforce Management Systems for Industry 4.0: Data-Centric and AI-Enabled Approaches* (1st ed., pp. 14–31). CRC Press, Boca Raton, FL. https://doi.org/10.1201/9781003357070-2

Babasaheb, J., Sphurti, B., & Khang, A. (2023b). Design of competency models in the human capital management system. *Designing Workforce Management Systems for Industry 4.0: Data-Centric and AI-Enabled Approaches* (1st ed., pp. 32–50). CRC Press, Boca Raton, FL. https://doi.org/10.1201/9781003357070-3

Charvet, S. R. (2022). *Motivational Traits for Recruiting Team Member.* Institute for Influence, Canada.

Khang, A., Kali, C. R., Suresh Kumar, S., Amaresh, K., Sudhansu Ranjan, D., & Manas Ranjan, P. (2023). Enabling the future of manufacturing: integration of robotics and IoT to smart factory infrastructure in industry 4.0. *AI-Based Technologies and Applications in the Era of the Metaverse* (1st ed., pp. 25–50). IGI Global Press, Hershey, PA. https://doi.org/10.4018/978-1-6684-8851-5.ch002

Khang, A., Muthmainnah, M., Seraj, P. M. I., Yakin, A. A., Obaid, A. J., & Panda, M. R. (2023). AI-aided teaching model for the education 5.0 ecosystem. *AI-Based Technologies and Applications in the Era of the Metaverse* (1st ed., pp. 83–104). IGI Global Press, Hershey, PA. https://doi.org/10.4018/978-1-6684-8851-5.ch004

Khang, A., Rani, S., Gujrati, R., Uygun, H., & Gupta, S. K. (2023). *Designing Workforce Management Systems for Industry 4.0: Data-Centric and AI-Enabled Approaches.* CRC Press, Boca Raton, FL. https://doi.org/10.1201/9781003357070

Khang, A., Shah, V., & Rani, S. (2023). *AI-Based Technologies and Applications in the Era of the Metaverse* (1 ed.). IGI Global Press, Hershey, PA. https://doi.org/10.4018/978-1-6684-8851-5

Khang, A. & AIoCF. (2024). *AI-Oriented Competency Framework for Talent Management in the Digital Economy: Models, Technologies, Applications, and Implementation.* CRC Press, Boca Raton, FL. https://doi.org/10.1201/9781003440901

Khang, A., Gujrati, R., Uygun, H., Tailor, R. K., & Gaur, S. S. (2024). *Data-driven Modelling and Predictive Analytics in Business and Finance* (1st ed.). CRC Press, Boca Raton, FL. https://doi.org/10.1201/9781032600628

Lashkari, M. & Cheng, J. (2023). "" Finding the Magic Sauce": Exploring Perspectives of Recruiters and Job Seekers on Recruitment Bias and Automated Tools." arXiv preprint arXiv:2301.11958.

Muna, F. (2022). *Analyse the Recruitment and Selection Theories and Practical Implication of Civil Service Sector,* pp. 125–133. International Hellenic University, Thessaloniki, Greece. https://doi.org/10.1201/9781003440901

Muthmainnah, M., Khang, A., Seraj, P. M. I., Yakin, A. A., Oteir, I., & Alotaibi, A. N. (2023). An innovative teaching model - the potential of metaverse for English learning. *AI-Based Technologies and Applications in the Era of the Metaverse* (1st ed., pp. 105–126). IGI Global Press, Hershey, PA. https://doi.org/10.4018/978-1-6684-8851-5.ch005

Papapolychroniadis, I., et al. (2017). Comparative analysis of recruitment systems in the public sector in Greece and Europe: trends and outlook for staff selection systems in the greek public sector. *Academic Journal of Interdisciplinary Studies*, 6(1), 21–30.

Pooja, K., Babasaheb, J., Ashish, K., Khang, A., & Sagar, K. (2023). The role of blockchain technology in metaverse ecosystem. *AI-Based Technologies and Applications in the Era of the Metaverse* (1st ed., pp. 228–236). IGI Global Press, Hershey, PA. https://doi.org/10.4018/978-1-6684-8851-5.ch011

Satispi, E., et al. (2023). Human resources information system (HRIS) to enhance civil servants' innovation outcomes: compulsory or complimentary? *Administrative Sciences*, 13(2), 32.

Shah, V. & Khang, A. (2023). Metaverse-enabling IoT technology for a futuristic healthcare system. *AI-Based Technologies and Applications in the Era of the Metaverse* (1st ed., pp. 165–173). IGI Global Press, Hershey, PA. https://doi.org/10.4018/978-1-6684-8851-5.ch008

Sharma, C., et al. (2023). Role and impact of human resource information system (HRIS) on organizational activities. *AIP Conference Proceedings,* AIP Publishing LLC, New York.

Tyagi, A. K., et al. (2023). Blockchain-internet of things applications: opportunities and challenges for industry 4.0 and society 5.0. *Sensors* 23(2), 947.

Veldsman, D. (2022a). Scenario planning: What HR needs to know?. Accadamy to Innovate HR. https://www.aihr.com/blog/scenario-planning/

Verlinden, N. (2022b). 13 future HR skills you need to start building now. Co Innovate HR. https://www.aihr.com/blog/future-hr-skills/

Xu, X. (2023). Research on the digital development of human resource management in large enterprises in the post-epidemic period. *Proceedings of the 4th International Conference on Economic Management and Model Engineering, ICEMME 2022*, Nanjing, China.

14 Shaping Artificial Intelligence-Perceived Hybrid Learning Environment at University Toward the Global Talent Development Strategy

*Muthmainnah M, Besse Darmawati,
Alex Khang, Ahmad Al Yakin,
Ahmed J. Obaid, and Ahmed A. Elngar*

14.1 INTRODUCTION

The rapidly developing field of artificial intelligence (AI) has been widely used in the classrooms of art. To effectively perform their teaching activities and contribute to the development of society, teachers need to be educated in the practices, strategies, and policies that provide them with the professional knowledge, teaching skills, evaluation techniques, and ethical orientations they need to do so Salas-Pilco et al. (2022). It is commonly agreed that there are three distinct but interconnected stages of teacher education (pre-service, induction, and in-service) Dunkin (1987). Therefore, teacher education includes both initial preparation programs and ongoing professional development programs already in the classroom.

Most educators now understand the value that technology can add to classroom instruction. Therefore, other forms of technology, such as social media, blogs, web conferences, and discussion forums, are incorporated into traditional and online teacher preparation programs. Several issues exist, including the school culture, the availability of resources, teachers' attitudes, expertise, and skills. Farjon et al. (2019) and Menabò et al. (2021) make it challenging to integrate technology into classes. Despite this, governments around the world are passing legislation to incorporate technological tools into the classroom (Butler et al. (2018) and Stokes-Beverley and Simoy (2016)). Therefore, it is crucial that instructors receive training in the proper use of technology in the classroom.

The term AI is used to describe computer programs that can predict outcomes, provide advice, or make decisions that have consequences in the real or virtual worlds based on a set of human-defined objectives. AI systems have direct and indirect effects on the world around us. They frequently act independently and show signs of learning and adaptability in response to their environment.

Although the way we live and conduct our daily business has changed rapidly, traditional methods are still widely used in the education sector. These days, most kids would rather be doing anything but coloring in coloring books and playing with board games. The majority of their time is spent engrossed in digital displays, such as those found in laptops and smartphones. Despite this, modern educational systems have not evolved much and instead remain static and conventional. Therefore, the educational paradigm needs to shift to accommodate students' evolving expectations and ways of life. Education systems in developing nations like India, Swaziland, and South Africa are slow to adopt new forms of technology, but this is good news for students because they will have a head start on realizing their full potential as unique individuals. It will also benefit teachers because they will be freed from some of the more tedious tasks inherent in the current model of instruction.

According to Xin (2021) and Alam (2021), learning how to use a computer is not the same as learning about AI, nor is it digital education. It is possible to instruct students in AI concepts and procedures without resorting to technology. AI can help schools just as much as it can be taught in schools, and this must be recognized. AI is reshaping the underlying mechanisms of learning and teaching. As was previously mentioned, AI has an effect on human activities, including pedagogical practices. It is possible that the use of machine learning in the classroom will allow us to abandon the "one-size-fits-all" mentality and instead create individualized learning plans and pedagogical resources (Pedro et al., 2019). "AI-based learning systems that would be able to give instructors important information on their student's learning styles, talents, and progress and provide advice on how to customize their teaching techniques to individual needs", writes (Ramu et al., 2022; Devi et al., 2020). It would make it possible for children who are having trouble in learning to get supplemental tutoring that is tailored to their individual needs.

A student's level of knowledge can be assessed with the help of AI, and a study plan can be tailored to each individual learner based on the gaps in their knowledge (Hager et al., 2019). In this way, AI can help students maximize their study time by adapting to their unique requirements (Prentzas, 2013). Using the knowledge space principle, many businesses educate their AIs to spot information gaps by weighing the relative complexity of scientific topics (for instance, one can stimulate the training of another or become the basis for remedial training).

Even if the way we live and conduct our daily business has changed at a breakneck pace, old approaches are still widely used in the educational system. Undergraduate students today rarely use printed books, play board games, or engage in other more conventional forms of undergraduate students' entertainment. Their lives are mostly governed by electronic displays, such as those found in laptops, cellphones, and tablets. However, modern educational systems are much more rooted in the past (Hannafin & Land, 1997). Therefore, a shift in the educational paradigm is required to accommodate students' evolving needs and preferences.

Experts believe that by combining AI with traditional teaching methods, students will be better able to acquire and address 21st century abilities such as self-evaluation, confidence, teamwork, and others (Churchill, 2020; Seldon et al., 2020; Popkova & Gulzat, 2019; Wogu et al., 2018; Venugopal et al., 2021). There is always a chasm between courses and learning in traditional pedagogy because lecturers rarely realize the weaknesses of their lectures. When a significant number of students in a course submit different questions and incorrect work, it becomes impractical for a single mentor to respond to each and every one of them. This issue can be resolved using AI. Already, Coursera, one of the largest online course providers, is incorporating this technology into its lessons. AI can also be used to track students' development and alert teachers to any problems it detects.

The use of AI in education has the potential to reshape the profession. As AI becomes more pervasive in the classroom, educators will need to refocus their attention from administrative tasks like grading to more substantive matters of student learning. Using AI, educators may provide a resource for students to ask questions and get answers. The pace of the course is a frequent source of frustration for educators who are responsible for instructing multiple pupils at once. When using AI, this issue will essentially disappear because students will be able to progress through the material at their own pace. Therefore, human teachers will provide support alongside the AI lessons, allowing for more meaningful student-teacher interactions.

Traditional and online news outlets alike are focusing their coverage on the ChatGPT trend in 2022, when the technology first appears. The unpredictable behavior of ChatGPT makes it a potentially "black swan" event. Public interest in the positive and negative impacts of AI on society has resurfaced with the presence of ChatGPT today, although AI has been covered in the past (Sołtysik et al., 2023). Some early adopters and inventors were excited about ChatGPT's accessibility and convenience, while others were concerned about the possible downsides it might cause, as depicted in the Terminator movie. According to Ariyaratne et al. (2023), ChatGPT can also produce written text that closely resembles human writing; its ability raises questions about its implications for education and academia, which is one of the main areas of discussion around implementing ChatGPT in the education arena.

What we see today is technology that has evolved from a harmless novelty to a game changer. Its success will be determined by a number of circumstances, including the motivation for its use, and the effect will be small if it is banned or rejected. Understanding this technology, which is based on a very large language model, is critical to recognizing its possible benefits, drawbacks, and limitations (Javaid et al., 2023). The implications of ChatGPT and related technologies in certain industries, such as education or the digital economy, must also be understood. The potential benefits of these instruments for education and the drawbacks they may present must be carefully considered.

Educators and other stakeholders will be better able to weigh the pros and cons of implementing new technologies into classroom teaching and make informed decisions about how best to implement them (Gill et al., 2024). Once someone has learned what ChatGPT is and what it can do, they can decide whether to use it or not, weighing the potential advantages and disadvantages. To do this, people may need to change habits they have developed over time, which may be difficult because of their natural aversion to change.

AI is an increasingly popular subfield of computer science concerned with building robots with human-like intelligence aimed at interacting with humans. Many fields have found uses for AI, including healthcare, transportation, education, and training. In addition, AI and IoT can be combined to create a new technology called AIoT. One of the most promising AI technologies is ChatGPT (Hariri, 2023), a natural language processing (NLP) system that can replicate human-like conversations and is used for language acquisition classes in English education, which will be studied in this study (Khang & Vugar et al., 2024).

Learning English in Indonesia is now seen as less to an end (liberal arts education) and more to an end (creating a workforce for international trade, human development, and cultural interaction), as has been the case in many other non-English speaking nations in recent years (Dardjowidjojo, 2000; Lowenberg, 1991; Gustine, 2018). Moreover, English is used as a common language between many nations, and recent technological developments have accelerated globalization, increasing the demand for communication in both the corporate and public sectors. Considering the importance of being able to communicate in English in a global context. For these reasons, English language education in Indonesia has changed from years of learning grammar and translation-based curricula for whole skills or a combination of English, with a larger focus on communication today. Despite the necessity of English, most Indonesian colleges do not yet include communication lessons taught by native English teachers as part of their required core English curriculum, and most English lecturers at the undergraduate level have not seen any change in the content of the courses; they teach in English one of them by incorporating media and technology like AI (Rana et al., 2021).

Using relevant theories of curriculum theory, literature analysis, and field investigation, this paper investigates the application of AI in English teaching in middle schools. Several studies have demonstrated the benefits of using AI to learn English, including Bin and Mandal (2019). It is proposed that an AI-powered system aiding college-level English instruction be put into place. Some aspects of the English education system are enhanced and softened when combined with English instruction. To boost the efficiency and effectiveness of English instruction, researchers are looking at using AI technology in the classroom. In this article, we introduce a deep learning-assisted, online intelligent English teaching system that may be used to build a cutting-edge platform for assisting students in becoming more proficient in English in accordance with their own individual levels of knowledge and character development. An English teaching assessment implementation model using decision tree technologies has been developed with the help of the decision tree algorithm and neural networks. It summarizes rules and facts, draws useful material from a wealth of sources, and aids educators in raising student achievement in English.

The AI expert system's line of thinking has been incorporated into this system. The pilot program shows that the method can boost students' productivity in the classroom and make course materials more relevant. Additionally, the system defines a reference model and provides a reference model with comparable procedures. Several factors, including AI-based college English listening instruction, robotic college English oral instruction, AI-based college English writing instruction volume, and cloud service-based AI-college English translation instruction, should inform the mode of college English instruction against the backdrop of AI. Both Y. and Liu were authors on this study (2019).

This study investigates the relationship between student attitudes and learning outcomes toward AI in improving English learning for undergraduate students in Universitas Al Asyariah Mandar. The arguments and solutions given in this study are different from those given in previous studies.

14.2   ARTIFICIAL INTELLIGENCE IN LANGUAGE LEARNING

AI is an emerging trend in language classroom technology, and in the wake of the pandemic, this application is growing more and more as a support for TEFL (Nagro, 2021). One of the most important learning strategies in the modern era is the use of AI. Understanding how AI systems perform tasks that humans are now more proficient at is the study of AI (Marcus & Davis, 2019). It is assumed by Dong et al. (2020) that "AI" means the simulation of human intelligence and logical abilities. This time frame is important because it marks the beginning of the application of AI in EFL and hybrid environments. Having AI assist in language learning will completely change the educational landscape for both students and teachers (Roll & Wylie, 2016).

With the emergence of the new millennium and the advancement of computer technology, AI is starting to show its significance in dealing with huge amounts of data, and in the context of this study, about learners' language acquisition data (Granger, 2014). Advances in NLP methods have also resulted in some exciting new opportunities for language learning. The goal of NLP is to enable computers to analyze, understand, and generate spoken and written language in a human-like manner (Juhn & Liu (2020) and Stone et al. (2022)), with the help of these advances, AI-enabled gadgets can now interact with students in a variety of ways, including by conversing with them, evaluating their written and oral work, and providing feedback on both Robaldo et al. (2019).

Similarly, Kim (2019) looks at researching the effects of AI chatbots on students' understanding of English grammar. The research was conducted in Korea. Seventy students took part in this study. Individuals also have varying levels of linguistic competence. There are three levels: beginner, intermediate, and advanced. The research aims to determine whether and how AI-powered chatbots can help students learn foreign languages. In addition, this study is of various quantitative types. Two groups were created from the participants. Replica, a messaging software, was used to teach grammar to an experimental group. Another group learns grammar through pair discussion. Furthermore, pre-test and post-test data were used to construct this analysis. Both groups improved their grammar skills because of the research, but those who interacted with the chatbots saw the biggest gains. The study concluded that using AI chatbots to teach Korean students a foreign language is effective.

Soliman (2016) conducted research on the topic of using an AI-powered virtual learning environment to teach medical students in English. The city of Bisha, Saudi Arabia, is the research location. The main claim of this study is that the ineffective teaching methods used in medical English materials are largely to blame for students' inability to understand complex concepts.

There were also 29 medical faculty professors from Bisha University who took part in the study. The aim of this research is to create an AI-powered 3D virtual classroom for medical students to practice their English skills. In addition, this research is quantitative in nature. Information was collected through a questionnaire. There are 40 questions covering topics such as technology, education, and AI. Most professors surveyed strongly support student exposure to an AI-based learning environment for English. The study reported here aims to investigate the current state of teaching English to medical students at Bisha University by implementing a proposed virtual learning environment based on AI.

Obari and Lambacher (2019) conducted another investigation on the topic of teaching native Japanese speakers English with AI, Tokyo, Japan, became the research location. During the study, 47 new students took part. Moreover, these students are first-language speakers. Their language proficiency is somewhere between beginner and advanced. The research aims to determine whether AI can help native English speakers become more proficient in the language. English studies should also cover the acquisition of AI and other 21st century skills. The study authors also mention incorporating AI into modern classrooms to help students acquire the skills needed for today and tomorrow. Moreover, it is a mixed-methods study. The research subjects were divided into two groups.

- The AI experiment included grouping participants into eight different categories. The first class spent the semester learning with the help of Google Home Mini. The main aim of the first set of lessons is to cultivate better communication skills through attentive listening and clear expression.
- The second group also took advantage of Alexa throughout the semester to improve their vocabulary and understand complex texts better. While using the AI for training purposes, the researcher performed additional tasks for both teams to complete. Participant responsibilities include working with AI speakers, watching videos, and summarizing them in 300 words, developing presentations, learning about worldview theories and philosophies underlying 21st century skills and, finally, participating in conversations. The research findings indicate the growth of both groups' linguistic competence in English. The study authors found that AI helps students acquire 21st century and English skills.

However, researchers and educators have mixed feelings about the integration of AI technologies into language teaching. Many studies extol the advantages of AI in language learning because it promises personalized learning. AI systems, for example, have been found to process students' language input Sun et al. (2021) and provide more effective grammatical feedback Chen et al. (2022). Murphy (2019) studies of AI in the past focused mainly on grammar, but more recent studies, made possible by advances in computer technology, have revealed that AI has much more to give. In language schools, for example, AI applications have been found to improve students' abilities to communicate and collaborate (Vincent-Lancrin & Van der Vlies, 2020), as well as speaking and listening skills (Sun, 2021; El Shazly, 2021), motivation and reading comprehension (Sharma et al., 2020; Keerthiwansha, 2018), 21st skills, and critical thinking (Muthmainnah et al., 2022).

14.3 RELATED WORK

14.3.1 METHOD

This research is based on quantitative methods. The research design that will be used in this study is an experiment on third-year public health faculty and government science faculty students, and the control group is computer science faculty. In this research, after obtaining the data, it will be collected, processed, and analyzed; after that, it will be submitted objectively, which will be adjusted to reflect actual reality. The objectives of this research are clear about the effectiveness of using the AI application in English. Quantitative data was collected using a test and a closed questionnaire, as shown in Figure 14.1.

- E: experiment class
- K: control class
- O 1: Pre-test in experiment class
- O 2: Post-test in experiment class
- O 3: Pre-test in control class
- O 4: Post-test in control class
- X E: Using AI apps
- X K: Use textbook (Hello America Level III)

The researcher designed a closed-ended questionnaire containing 21 statements using a 5-point Likert scale (strongly agree, agree, neutral, disagree, and strongly disagree) to find out the perceptions of teachers and students to using AI to learn English. Questionnaire adapted from Obari and Lambacher (2019) and student perceptions and real-life use of AI in language learning.

Questionnaires were distributed to students of the English Sunset program course level 3 as a first step to gain a clear understanding of the arrangement of EFL classrooms at the university. It is assumed that not all EFL classrooms share this feature based on a list of common characteristics of EFL classrooms. The three basic elements of education (a) curriculum, (b) teaching and learning, and (c) assessment are then emphasized to talk about how much AI can be used in EFL classrooms. The idea of how AIEd is used in EFL classes at the university level was established because of this study.

14.3.2 PARTICIPANT

A total of 141 EFL student teachers participated in this study. All students took part in this study voluntarily. The majority of EFL students are those who never use smartphones for their EFL purposes. The participants in this study were students at Universitas Al Asyariah Mandar, West Sulawesi, Indonesia. The participants'

E	O 1	X E	O 2
K	O 3	X C	O 4

FIGURE 14.1 Two-group pre-test and post-test design.

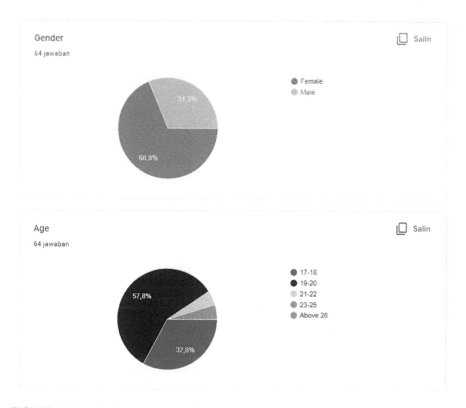

FIGURE 14.2 Students' demographic in the experiment classexperof Students' demographic average age 19–20 years old.

English proficiency level is intermediate for students in the program of English Sunset there are six faculty in this university teacher training and education faculty, government science faculty, public health faculty, agriculture faculty, public health faculty, and Islamic religion faculty (Khang, 2023).

The random sampling technique was used with the reasoning that each group has an equal chance of being used as a research sample. The sample chosen was public health and government science faculty with a total of 64 students who were used as the experimental group, and the second sample was computer science faculty with a total of 77 students who were used as the control group. Then the researcher conducted a pretest so that the students' initial abilities were known. This experimental research aims to compare English skills using AI as the medium in learning English, as shown in Figure 14.2.

14.3.3 Research Procedure

In this design, a researcher gives the text twice, namely before the treatment called the pre-test, which is given to the experimental class. After that, the researcher gave some learning material using an AI application. After that, entering the final stage, the researcher gave assignments to students to write conversation scripts about travel, food, work, and sport with John English boot apps, which they then practiced in the form of role play.

14.3.3.1 Pretest

A pretest is an initial test given to students by researchers prior to the start of the lesson, with the goal of gathering preliminary data and in this first stage, the researcher gave an initial test once in the English sunset program before starting the program.

14.3.3.2 Treatment

At this stage, after carrying out the initial stage (pre-test) on students, the next stage is using treatment. At this stage, the researcher wants to know the students' understanding of the effects of using John English Boot applications in the English sunset program. The experimental class and control class were chosen at the treatment stage. Treatment was given to the experimental class using AI media, and the control class was treated using book media with the same material, as shown in Figure 14.3.

- The lecturer starts the lesson by saying hello, praying, and checking student attendance.
- Lecturers deliver learning objectives and convey inspiration to students by explaining the importance of the material to be studied, namely, travel (simple future will and going to), ordering food (talking about food), work (asking about education and experiences), and sport (telling people to do something) with AI apps. All students are directed to open their gadgets (mobile phones) and look for material about food and news on YouTube that has been prepared by researchers using the following link: https://www.youtube.com/watch?v=lvw_ht29FXE
- The lecturer explains about travel (simple future will and going to), ordering food (talking about food), work (asking about education and

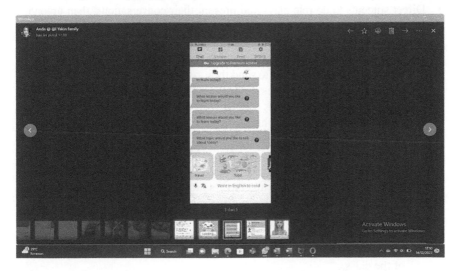

FIGURE 14.3 AI features.

FIGURE 14.4   Students explore AI applications and doing interaction with AI.

experiences), and sport (telling people to do something) on YouTube-based writing script, as shown in Figure 14.4. Then students are asked to install the AI-John English Boot application through the Play Store, and instructors explain how to use the AI-John English Boot (AI) application.

- The task assigned by the teacher was to interact with the John English Boot apps while carefully observing and listening to John English Boot. Then students are asked to write conversation text using John's English. Students are directed to check grammar and pronunciation using the Cambridge online dictionary application.
- Direct students to do role plays and present their conversations, then upload them on their respective YouTube accounts. The lecturer asks students' knowledge about describing food and how to order food. Students are given the opportunity to answer and then respond to the answers and questions asked.
- The teacher instructs students to send their respective YouTube links regarding their English performances to the class WhatsApp group. The learning steps in the control class are as follows:
  - Start the lesson by saying hello, praying, and mark student attendance.
  - delivering learning objectives.
  - convey inspiration to students by explaining the importance of the material to be studied.
  - Students were asked to write conversation scripts.
  - Give examples of conversation texts using textbooks at Hello America level 3.
  - Students are directed to write conversation scripts in books.
  - Several people were appointed to read the conversation script in front of the class.
  - The teacher informs students of the next meeting's lesson plan.

Researchers at the same time made observations during the learning process, and the implementation of this John English Boot application was applied three times. After that, enter the final stage, namely, the post-test.

14.3.3.3 Post-Test

Enter the post-test after completing the pre-tests and treatments. The post-test is the opposite of the pre-test, where the pre-test is given before the process of giving material, and the post-test is given after the process of giving material using John English Boot application.

14.4 DATA ANALYSIS TECHNIQUE

Descriptive statistical analysis and inferential statistical analysis were used in this study. The data analysis technique used by researchers is IBM SPSS Statistics Version 22 for Windows. Inferential statistics are used to analyze data with the aim of determining whether the null hypothesis is accepted or rejected. For the purposes of this inferential statistical analysis, a normality test, a homogeneity test, and a hypothesis test were first carried out.

14.4.1 NORMALITY TEST

Researchers used SPSS to test the normality of research data to find out whether the sample under study was normally distributed or not. With the provisions of the alpha level of 0.05. To carry out the normality test, the statistical hypothesis is as follows:

- H0 = The data is from a typical population.
- H1 = The data comes from a non-normal population.

Therefore, if the data normality test results obtained are significant 0.05, then H0 is accepted, and if the data normality test results obtained are significant 0.05, then H0 is rejected.

14.4.2 HOMOGENEITY TEST

Samples taken from the population that have the same variance and do not show any significant differences from one another can be tested through a homogeneity test. The statistical hypothesis formulation used is:

- H0 = variance of a homogeneous data group
- H1 = variance of group data is not homogeneous

The tests are based on significance, namely: (a) if the significance is 0.05, then H0 is accepted; (b) if the significance is 0.05, then H1 is rejected.

14.4.3 STATISTICAL HYPOTHESIS

The *t*-test (*t*-test) was used in testing the hypothesis of this study using the SPSS Statistics 22 for Windows program. The research to be tested is formulated by:

- If tcount > ttable, then H0 is accepted, and Ha is rejected.
- If tcount > table, then H0 is rejected, and Ha is accepted.
- H0 = There is no difference in writing results after applying AI apps.

There are differences in learning outcomes following the use of AI apps. The *t*-table is determined at a significance level of 0.05; 2 = 0.025 (2-sided test) with df (degrees of freedom) *n*−2 test characteristics, which are as follows: (a) If tcount ttable, then H0 is accepted; and (b) If count ttable, then H0 is rejected. If it is based on significance, namely: (a) if the significance is 0.05, then H0 is accepted; and (b) if the significance is 0.05, then H0 is rejected.

14.5   RESULTS AND DISCUSSION

Participants reported having little or no difficulty learning languages when using AI applications. Some of their thoughts on the value of AI are included here, as shown in Figure 14.5.

To a large extent, students agree with the third-place assertion that AI accelerates language learning and inspires them to learn languages efficiently since it is aimed at language learners at their level as if chatting to their teacher both in and out of the classroom. Most students "agreed" or "strongly agreed" with the fourth statement, which states that the adaptability and user-friendliness of AI make it suitable for students of varying ages and speed up their capacity to learn English (Khang & Muthmainnah et al., 2023).

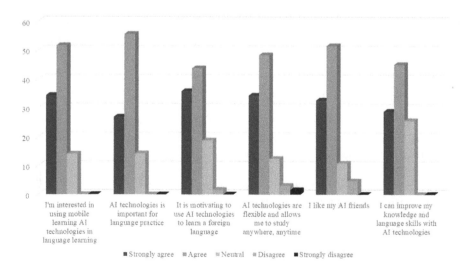

FIGURE 14.5   Students' attitudes toward AI in language learning.

In addition, AI helps teachers identify areas where their students can make the most progress in terms of language competency, allowing them to spend less time on topics in which they are already well-versed. Students agreed or strongly agreed with the fifth statement that AI clarified many points that the explanation teacher could not discuss because the teacher acknowledged that some students in the class were embarrassed to ask about some points that they didn't understand; therefore, students will discuss the system as if they asked a teacher comfortably, and they will also find and search more, and they claim to really like their AI friends (John English Boot).

Students understand that AI can fill in the gaps that their English teacher may have in the classroom, as stated in the sixth statement, with AI fulfilling and complementing all language learning needs of students, not only increasing their knowledge of language and language skills but also improving their skills in using technology, as shown in Figure 14.6.

The strengths of concrete learners are as follows: intuitive thinking and improvisation; collaborative problem solving; various creative pursuits; avoiding monotonous or uninteresting tasks; and the need for a learning environment that incorporates human and social content and the cultural relevance of technology-based learning English. Most respondents strongly agreed 31.3% and agree 43.8 on the statement that AI helps subjects develop 21st century skills. On the statement, AI technologies can help create real-life language learning experiences 37.5% on strongly agree and 50% chose to agree. Respondents felt they could learn more better through AI 34.9% took strongly agree and 42.9 agree, as shown in Figure 14.7.

43.5% of students who responded with "agreed" and 35.5% who took "strongly agreed" to the last statement had their learning facilitated by the statements.

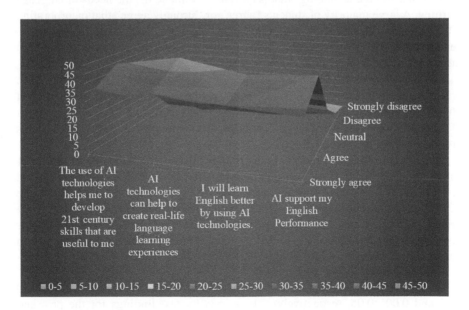

FIGURE 14.6   AI and students 21st century skills.

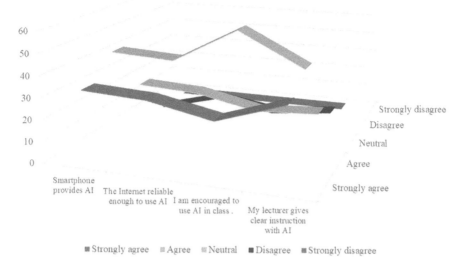

60
50
40
30
20
10
0

Strongly disagree
Disagree
Neutral
Agree
Strongly agree

Smartphone
provides AI   The Internet reliable
enough to use AI   I am encouraged to
use AI in class .   My lecturer gives
clear instruction
with AI

■ Strongly agree   ▨ Agree   ▧ Neutral   ■ Disagree   ■ Strongly disagree

FIGURE 14.7   Students' opinions about AI feature and instructional teaching model.

Lesson explanations are just one area that can benefit from AI simplification and clarification, as students can be simulated and trained with listening and reading and can receive an explanation from AI. All students' language learning demands can be met with the incorporation of AI in this study, as it can be utilized anywhere and at any time, and the technology that is currently available fits the needs of language learning. A large majority of students "agree" or "strongly agree" that incorporating AI into the learning process makes learning a new language easier than the conventional method. This is because it gives students a new way to learn that is both interesting and effective, as shown in Figure 14.8.

The application was updated to version 28 (2021) in order to analyze the data. A statistical significance test was used to analyze the data collected in the pre-test phase. However, previous to employing parametric tests like the t test, examinations of normality were conducted ad hoc. Descriptive data for pre-test and post-test scores are summarized in Table 14.1.

Students learning outcomes for both group is different on the pre-test. Table 14.1 shows the mean score of the control group pre-test score of 55.95 and the post-test score of 64.94. Language learning class using AI shows students a pre-test score of 54.25 and the post-test score of 71.38. With a maximum score of 72, both study groups performed well on the pretest. Considering their starting skill level, their good performance was to be expected. Finally, there was little or no difference in performance between the two groups, indicating that their English skills were comparable, as shown in Table 14.2.

Analyses of both sets of data showed significance levels of less than 0.05 (0.000 0.05 and 0.001 0.05, respectively). The normality test findings for the pre- and post-test data indicated that the data were not normally distributed, hence the

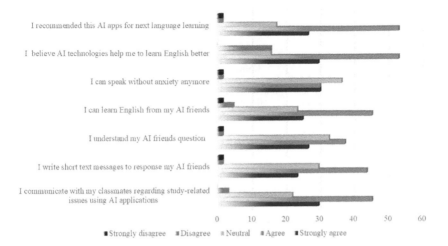

FIGURE 14.8   Teaching model through AI in language learning.

TABLE 14.1
Students Learning Outcomes

Statistics of Control Group		Pre-test of Control	Post-test of Control
1. N	Valid	77	77
	Missing	0	0
Mean		55.95	64.94
Median		56.00	64.00
Mode		52	64
Std. Deviation		7.441	6.118
Variance		55.366	37.430
Range		28	32
Minimum		44	44
Maximum		72	76
Sum		4,308	5,000
	Statistics of Experiment Group		
		Pre-test Experiments	**Post-test Experiments**
N	Valid	64	64
	Missing	0	0
Mean		54.25	71.38
Median		52.00	72.00
Mode		52	72
Std. Deviation		10.616	8.007
Variance		112.698	64.111
Range		32	32
Minimum		40	56
Maximum		72	88
Sum		3,472	4,568

TABLE 14.2
Tests of Normality

Class		Kolmogorov-Smirnova			Shapiro-Wilk		
		Statistic	df	Sig.	Statistic	df	Sig.
Learning	Pre-test experiment	0.225	64	0.000	0.852	64	0.000
outcomes	Post-test experiment	0.187	64	0.000	0.947	64	0.008
	Pre-test control	0.276	77	0.000	0.854	77	0.000
	Post-test control	0.218	77	0.000	0.855	77	0.000

Lilliefors significance correction. The results of the Kolmogorov-Smirnov test at a significance level $= 0.05$ were calculated to yield a significance level of $0.000c$ for pre-test data and $.000c$ for post-test data, as shown in the table above.

TABLE 14.3
Wilcoxon Test

Ranks		N	Mean Rank	Sum of Ranks
Post-test experiment –	Negative Ranks	2[a]	3.50	7.00
Pre-test experiment	Positive Ranks	52[b]	28.42	1,478.00
	Ties	10[c]		
	Total	64		
Postcontrol – precontrol	Negative Ranks	11[d]	15.95	175.50
	Positive Ranks	58[e]	38.61	2,239.50
	Ties	8[f]		
	Total	77		

[a] Post experiment < pre-experiment.
[b] Post experiment > pre-experiment.
[c] Post experiment = pre-experiment.
[d] Post control < pre control.
[e] Post control > pre control.
[f] Post control = pre control.

Wilcoxon test was utilized in this study rather than the paired sample t test, as shown in Table 14.3.

With a mean rank of 3.50 and a sum rank of 7.00, the difference (negative) between the pre- and post-test results for authoring editorial content is 2. If the value drops by 2, it means 2 persons less in the post-test than in the pre-test. There were as many as 52 people with positive data (N), meaning that 52 students saw an improvement in their editorial writing scores between the pre- and post-test, as shown in Table 14.4.

TABLE 14.4
Asymp. Sig. (2-Tailed)

Test Statistics[a]	Post Experiments – Pre Experiment	Post Control – Pre Control
2. Z	−6.358[b]	−6.212[b]
Asymp. Sig. (2-tailed)	0.000	0.000

[a] Wilcoxon signed ranks test.
[b] Based on negative ranks.

Where,

a. Wilcoxon Signed Ranks Test
b. Based on negative ranks.

Asymp. Sig. (2-tailed) has a value of 0.000, as seen in the "test statistics" report. Because 0.000 is smaller than 0.05, the statement "Ha is accepted" can be made. As a result, the pre-test and post-test results of the experimental class were different from the control class in terms of learning English using AI media and technology, as shown in Table 14.5.

According to the data in the table above, there were 64 participants in the post-test experimental group, with a mean rank of 89.53 and a sum of 5,730.00, and 77 participants in the post-test control group, with a mean rank of 55.60 and a sum of 4,281.00, as shown in Table 14.6.

- H0 is rejected and Ha is approved if the Asymp value, significance (2-tailed) is less than 0.05. A two-tailed significance level (Asymp) > 0.05 means that Ho is accepted, while Ha is rejected.
- H0: employing AI medium does not provide noticeably different English outcomes. Really, the media and AI technology make a huge difference in students' ability to learn.

TABLE 14.5
Man-Whitney *U* Test

Ranks	Group	3. N	Mean Rank	Sum of Ranks
Learning Outcomes	Experiment	64	89.53	5,730.00
	Control	77	55.60	4,281.00
	Total	141		

TABLE 14.6

Hypotheses Test Statistics

	Learning Outcomes
Mann-Whitney U	1,278.000
Wilcoxon W	4,281.000
Z	−4.989
Asymp. Sig. (2-tailed)	0.000

Grouping Variable: class

Decision-making basics for the Mann-Whitney test:

- H0 is rejected and Ha is approved if the Asymp value, significance (2-tailed) is less than 0.05. Two-tailed significance level (Asymp) > 0.05 means that Ho is accepted, while Ha is rejected.
- H0: employing AI medium does not provide noticeably different English outcomes. Really, the media and AI technology makes a huge difference in students' ability to learn.

From the results of the Mann-Whitney test, it can be shown that the Asymp. Sig. (2-tailed) of 0.000 is less than the probability value of 0.05. Consequently, "Ha is acceptable" as the basis for a conclusion based on the Mann-Whitney test presented above.

14.6   LIMITATION AND DISCUSSION

The limitation of the AI application used in this study is that there is no feature to correct language errors in the experimental class. Spada and Lightbown (2022) acknowledges, however, that being aware of one's language errors is the most typical form of negative evidence in SLA. There are many approaches and degrees to this phenomenon. For example, some interactionists advocate reshuffling as a kind of remediation, arguing that it is the least disruptive cognitive attack that is likely to have the desired remedial effect Zhao and Ellis (2022). However, Chekili (2022) argues that rearranging does not force students to self-correct while providing important negative evidence. This study also assumes that language errors made using the John Boot English application cannot be corrected by this app when they interact with the AI device. However, the feature shown is that the AI application can communicate like a native, which can help undergraduate students correct their language mistakes independently, as described in Figures 14.7 and 14.8. Apart from the reform, comparisons between research and theory about the types of corrections and their effectiveness look ambiguous. These types of corrections include stating relevant linguistic rules, providing error type indicators without rearranging, using simple underlines, and providing simple error counts per line.

Preferences for the type of correction among students also appear to vary, and whether a given student's preferred method is the most effective or not is uncertain. However, it is very clear that students crave feedback on their mistakes, and this can be realized when interacting with the AI. Although until now researchers have not reached a consensus on the best way to correct language errors, which may be caused by differences between students (Papadakis et al., 2022), several studies lend credence to the idea that correcting mistakes is useful and important and can even result in learning (Patra et al., 2022). These studies find that AI is especially helpful for EFL.

Integrating AI technology in language learning and innovative teaching models in this study, as shown in Figures 14.4–14.6 and 14.8, shows that the instructional teaching model used in this study is focused and systematic, working alone or with a compatible partner, setting goals and directing learning, and pursuing limited examples, trial and error, rules, and definitions. Concrete or "relational" learners use several realistic instructional strategies at once, collaborate with others to achieve common goals, and support structure, modeling, guidance, and explicit feedback figures. The analytical student's self-esteem is less-dependent on peer approval; he/she is task-oriented and inattentive to emotional cues in interpersonal encounters; and he/she is dependent on self-imposed personal identity and social roles when interacting with others. He/she is person-focused and attuned to verbal and nonverbal cues in interpersonal interactions; he/she looks to groups for definitions of identity and roles; learning achievement increases when an authoritative group or figure gives praise and encouragement (Khang & Muthmainnah et al., 2023).

This study investigates a learner trait known as language anxiety, Horwitz et al. (2010) depicted in Figure 14.8, describe how the benefits of AI can reduce anxiety in learning. The impact of anxiety is not constant. Anxiety can sometimes help others but cripple others; consequently, we cannot see a direct or linear relationship between the two. Learning anxiety that is overcome with the help of AI technology, as shown by this study, can overcome self-doubt, the anxiety of speaking in public, and the difficulty of expressing the ideas of undergraduate students in speech (Figure 14.4 and Table 14.6). Our EFL undergraduate students will not be immune to anxiety or its potentially damaging effects. Therefore, it is very important that the learning tools we provide (AI) also help students be less stressed.

In the field of personality studies, we talk about characteristics such as risk taking on the one hand and social preferences such as introversion and extroversion on the other (Zafar and Meenakshi (2012)). Galmiche (2018) identified readiness to guess, seem stupid to communicate, and construct unique utterances with up-to-date knowledge as risk-taking behaviors in language learning. Risk-taking behaviors, such as anxiety, can benefit or harm students, depending on the situation. Kim (2019) found that students' willingness and motivation to take risks was a good indicator of whether they would use grammatically accurate spelling when speaking in class, and this study noted that the questionnaire study showed excellent results in this regard with AI (Khang & Santosh et al., 2023).

14.7 CONCLUSION

Our findings show that AI can help EFL undergraduate students' English performance become better. Even though this AI application does not yet have a feature to correct students' language errors, this result does not reduce the credibility of the findings. This research cannot be considered a full study; it suffers from certain pitfalls, and any result-based judgment on the ultimate efficacy of strategy training for student learning in game environments would be premature. This research sample was also conducted at one university due to logistical and financial limitations. The findings of this study may or may not be supported by future research into the potential contribution of smartphone-based AI applications to the learning of a larger sample of EFL students. Researchers did not examine this in this study, but it will be equally interesting to see what, if any, lasting effects AI has on students' ability to learn (Muthmainnah & Yakin et al., 2023).

In addition, gender demographics were not considered as a confounding factor. There were far more women than men in the study group. Gender can interact with teaching methods in ways that unfairly favor one gender in terms of test scores. Taking this into account will allow future replications of the current study to investigate the main effects and interactions of AI-based gender more thoroughly. An additional avenue of investigation could be to survey how useful AI is seen by EFL students, where this interaction is their first interaction and getting to know AI in language learning makes them more interested in learning languages than before. Researchers seeking verifiability (when the two datasets agree) or initiation may find that quantitative method studies produce quantitative data. Alternatively, triangulation investigations in future tests will allow for more definitive conclusions about the efficacy of AI (Khang & Shah et al., 2023).

Researchers with various levels of experience in data analysis can triangulate investigators; this research could be replicated in different locations with different groups of language learners, and data triangulation can be used to gather information from multiple sources (e.g., test scores, questionnaire responses, interview responses, etc.). Finally, a significant limitation of this study is that the conclusions can not only be applied to EFL students at the undergraduate level but also to EFL teachers at all levels or in informal language learning situations with the teaching model applied in this study. To investigate how language proficiency as a potential covariate can reduce the putative influence of teaching mode, future replications of this study could recruit cohorts of mixed-proficiency or mixed-ability learners on different AI platforms (Khang & Kali et al., 2023).

REFERENCES

Alam, A. (2021). Should robots replace teachers? Mobilisation of AI and learning analytics in education. *2021 International Conference on Advances in Computing, Communication, and Control (ICAC3)*, pp. 1–12. IEEE, Piscataway, NJ. https://ieeexplore.ieee.org/abstract/document/9697300/

Ariyaratne, S., Iyengar, K. P., Nischal, N., Chitti Babu, N., & Botchu, R. (2023). A comparison of ChatGPT-generated articles with human-written articles. *Skeletal Radiology*, 37(3), 1–4. https://link.springer.com/article/10.1007/s00256-023-04340-5

Bin, Y. & Mandal, D. (2019). English teaching practice based on artificial intelligence technology. *Journal of Intelligent & Fuzzy Systems*, 37(3), 3381–3391. https://content.iospress.com/articles/journal-of-intelligent-and-fuzzy-systems/ifs179141

Butler, D., Leahy, M., Twining, P., Akoh, B., Chtouki, Y., Farshadnia, S., & Valtonen, T. (2018). Education systems in the digital age: the need for alignment. *Technology, Knowledge and Learning*, 23(3), 473–494. https://link.springer.com/article/10.1007/s10758-018-9388-6

Chekili, F. (2022). Teaching English to Arab learners. *Pedagogical Linguistics*, 4(1), 98–117. https://eric.ed.gov/?id=ED473079

Chen, X., Zou, D., Xie, H., Cheng, G., & Liu, C. (2022). Two decades of artificial intelligence in education. *Educational Technology & Society*, 25(1), 28–47. https://www.jstor.org/stable/48647028

Churchill, N. (2020). Development of students' digital literacy skills through digital story-telling with mobile devices. *Educational Media International*, 57(3), 271–284. https://www.tandfonline.com/doi/abs/10.1080/09523987.2020.1833680

Dardjowidjojo, S. (2000). English teaching in Indonesia. *EA Journal*, 18(1), 22–30. https://search.informit.org/doi/abs/10.3316/AEIPT.113365

Devi, J. S., Sreedhar, M. B., Arulprakash, P., Kazi, K., & Radhakrishnan, R. (2022). A path towards child-centric artificial intelligence based education. *International Journal of Early Childhood*, 14(03), 2022. https://www.researchgate.net/profile/Burada-Sreedhar/publication/361227279_A_Path_Towards_Child-Centric_Artificial_Intelligence_based_Education/links/62a428cfc660ab61f872a70a/A-Path-Towards-Child-Centric-Artificial-Intelligence-based-Education.pdf

Dong, Y., Hou, J., Zhang, N., & Zhang, M. (2020). Research on how human intelligence, consciousness, and cognitive computing affect the development of artificial intelligence. *Complexity*, 2020(1), 1–10. https://www.hindawi.com/journals/complexity/2020/1680845/

Dunkin, M. J. (1987). *The International Encyclopedia of Teaching and Teacher Education*. Pergamon Press, Oxford, UK. https://korthagen.nl/en/wp-content/uploads/2018/07/The-relationship-between-theory-and-practice-in-teacher-education.pdf

El Shazly, R. (2021). Effects of artificial intelligence on English speaking anxiety and speaking performance: a case study. *Expert Systems*, 38(3), e12667. https://onlinelibrary.wiley.com/doi/abs/10.1111/exsy.12667

Farjon, D., Smits, A., & Voogt, J. (2019). Technology integration of pre-service teachers explained by attitudes and beliefs, competency, access, and experience. *Computers & Education*, 130, 81–93.

Galmiche, D. (2018). The role of shame in language learning. *Journal of Languages, Texts, and Society*, 2, 99–129. https://www.nottingham.ac.uk/research/groups/ languagestexts-society/documents/lts-journal/issue-2/issue-2compressed.pdf#page=105

Gill, S. S., Xu, M., Patros, P., Wu, H., Kaur, R., Kaur, K., & Buyya, R. (2024). Transformative effects of ChatGPT on modern education: emerging era of AI chatbots. *Internet of Things and Cyber-Physical Systems*, 4, 19–23. https://www.sciencedirect.com/science/article/pii/S2667345223000354

Granger, S. (2014). The computer learner corpus: a versatile new source of data for SLA research. *Learner English on Computer*, pp. 3–18. Routledge, London. https://www.taylorfrancis.com/chapters/edit/10.4324/9781315841342-1/computer-learner-corpus-versatile-new-source-data-sla-research-sylviane-granger

Gustine, G. G. (2018). A survey on critical literacy as a pedagogical approach to teaching English in Indonesia. *Indonesian Journal of Applied Linguistics*, 7(3), 531–537. https://pdfs.semanticscholar.org/03b2/e28683428bffb069d94f676f732e4ff06969.pdf

Hager, G. D., Drobnis, A., Fang, F., Ghani, R., Greenwald, A., Lyons, T., & Tambe, M. (2019). Artificial intelligence for social good. *arXiv preprint arXiv:1901.05406*.

Hannafin, M. J., & Land, S. M. (1997). The foundations and assumptions of technology-enhanced student-centered learning environments. *Instructional Science*, 25(3), 167–202. https://link.springer.com/article/10.1023/A:1002997414652

Hariri, W. (2023). Unlocking the potential of ChatGPT: a comprehensive exploration of its applications, advantages, limitations, and future directions in natural language processing. *arXiv preprint arXiv:2304.02017*.

Horwitz, E. K., Tallon, M., & Luo, H. (2010). Foreign language anxiety. *Anxiety in Schools: The Causes, Consequences, and Solutions for Academic Anxieties*, 2, 96–115. https://core.ac.uk/download/pdf/328025363.pdf

Javaid, M., Haleem, A., & Singh, R. P. (2023). ChatGPT for healthcare services: an emerging stage for an innovative perspective. *Bench Council Transactions on Benchmarks, Standards and Evaluations*, 3(1), 100105. https://www.sciencedirect.com/science/article/pii/S2772485923000224

Juhn, Y. & Liu, H. (2020). Artificial intelligence approaches using natural language processing to advance EHR-based clinical research. *Journal of Allergy and Clinical Immunology*, 145(2), 463–469. https://www.sciencedirect.com/science/article/pii/S0091674919326041

Keerthiwansha, N. B. S. (2018). Artificial intelligence education (AIEd) in English as a second language (ESL) classroom in Sri Lanka. *Artificial Intelligence*, 6(1), 31–36. https://wairco.org/IJCCIT/August2018Paper4SL.pdf

Khang, A. (2023). *Advanced Technologies and AI-Equipped IoT Applications in High-Tech Agriculture* (1 ed.). IGI Global Press, Hershey, PA. https://doi.org/10.4018/978-1-6684-9231-4

Khang, A., Kali, C. R., Surabhika, P., Pokkuluri Kiran, S., & Santosh Kumar, P. (2023). Revolutionizing agriculture: exploring advanced technologies for plant protection in the agriculture sector. *Handbook of Research on AI-Equipped IoT Applications in High-Tech Agriculture* (pp: 1–22). IGI Global Press, Hershey, PA. https://doi.org/10.4018/978-1-6684-9231-4.ch001

Khang, A., Kali, C. R., Suresh Kumar, S., Amaresh, K., Sudhansu Ranjan, D., & Manas Ranjan, P. (2023). Enabling the future of manufacturing: integration of robotics and IoT to smart factory infrastructure in industry 4.0. *AI-Based Technologies and Applications in the Era of the Metaverse* (1st ed., pp. 25–50). IGI Global Press, Hershey, PA. https://doi.org/10.4018/978-1-6684-8851-5.ch002

Khang, A., Muthmainnah, M., Seraj, P. M. I., Yakin, A. A., Obaid, A. J., & Panda, M. R. (2023). AI-aided teaching model for the education 5.0 ecosystem. *AI-Based Technologies and Applications in the Era of the Metaverse* (1st ed., pp. 83–104). IGI Global Press, Hershey, PA. https://doi.org/10.4018/978-1-6684-8851-5.ch004

Khang, A., Shah, V., & Rani, S. (2023). *AI-Based Technologies and Applications in the Era of the Metaverse* (1 ed.). IGI Global Press, Hershey, PA. https://doi.org/10.4018/978-1-6684-8851-5

Khang, A., Vugar, A., Vladimir, H., & Shah, V. (2024). Advanced IoT technologies and applications in the industry 4.0 digital economy (1 ed.). CRC Press, Boca Raton, FL. https://doi.org/10.1201/9781003434269

Kim, N. Y. (2019). A study on the use of artificial intelligence chatbots for improving English grammar skills. *Journal of Digital Convergence*, 17(8), 37–46. https://search.ebscohost.com/login.aspx?direct=true&profile=ehost&scope=site&authtype=crawler&jrnl=27136434&AN=138623173&h=YtEExWMlT7pkP0xHJe8S0qASiJnXoj8jv71YjSIeFEJ3W8MaxbjXbc3Ua2EtLZ1xt%2FL0pvYXYIXrIjEpOIwijg%3D%3D&crl=c

Liu, Y. & Liu, M. (2019). Research on college English teaching under the background of artificial intelligence. *Journal of Physics: Conference Series*, 1345(4), 042064. https://iopscience.iop.org/article/10.1088/1742-6596/1345/4/042064/meta

Lowenberg, P. H. (1991). English as an additional language in Indonesia. *World Englishes*, 10(2), 127–138. https://onlinelibrary.wiley.com/doi/abs/10.1111/j.1467-971X.1991.tb00146.x

Marcus, G. & Davis, E. (2019). Rebooting AI: building artificial intelligence we can trust. *Vintage*. Knopf Doubleday Publishing Group, New York. https://www.google.com/books?hl=en&lr=&id=OmeEDwAAQBAJ&oi=fnd&pg=PA3&dq=Rebooting+AI:+Building+artificial+intelligence+we+can+trust.+Vintage.&ots=Lx1PcRZRbT&sig=RPw5JRLyhwMQh0ZCy8o6KDajRaQ

Menabò, L., Sansavini, A., Brighi, A., Skrzypiec, G., & Guarini, A. (2021). Promoting the integration of technology in teaching: an analysis of the factors that increase the intention to use technologies among Italian teachers. *Journal of Computer Assisted Learning*, 37(6), 1566–1577.

Murphy, R. F. (2019). Artificial intelligence applications to support K-12 teachers and teaching. *Rand Corporation*, 10, 20. https://onlinelibrary.wiley.com/doi/abs/10.1111/jcal.12554

Muthmainnah, M., Khang, A., Seraj, P. M. I., Yakin, A. A., Oteir, I., & Alotaibi, A. N. (2023). An innovative teaching model - the potential of metaverse for English learning. *AI-Based Technologies and Applications in the Era of the Metaverse* (1st ed., pp. 105–126). IGI Global Press, Hershey, PA. https://doi.org/10.4018/978-1-6684-8851-5.ch005

Muthmainnah, M., Seraj, P. M. I., & Oteir, I. (2022). Playing with AI to investigate human-computer interaction technology and improving critical thinking skills to pursue 21 st century age. *Education Research International*, 2022, 1–17. https://www.hindawi.com/journals/edri/2022/6468995/

Nagro, S. A. (2021). The role of artificial intelligence techniqies in improving the behavior and practices of faculty members when switching to E-learning in light of the covid-19 crisis. *International Journal of Education and Practice*, 9(4), 687–714. https://eric.ed.gov/?id=EJ1329070

Obari, H. & Lambacher, S. (2019). Improving the English skills of native Japanese using artificial intelligence in a blended learning program. *CALL and complexity-short papers from EUROCALL*, pp. 327–333. Research-Publishing, The Netherlands. https://www.google.com/books?hl=en&lr=&id=EHnCDwAAQBAJ&oi=fnd&pg=PA327&dq=Improving+the+English+skills+of+native+Japanese+using+artificial+intelligence+in+a+blended+learning+program&ots=pvYpbBVSoa&sig=KWkKWXFt1hoWWQeWlxg7BSfaHrw

Papadakis, N. M., Aletta, F., Kang, J., Oberman, T., Mitchell, A., & Stavroulakis, G. E. (2022). Translation and cross-cultural adaptation methodology for soundscape attributes-A study with independent translation groups from English to Greek. *Applied Acoustics*, 200, 109031. https://www.sciencedirect.com/science/article/pii/S0003682X22004054

Patra, I., Alazemi, A., Al-Jamal, D., & Gheisari, A. (2022). The effectiveness of teachers' written and verbal corrective feedback (CF) during formative assessment (FA) on male language learners' academic anxiety (AA), academic performance (AP), and attitude toward learning (ATL). *Language Testing in Asia*, 12(1), 1–21. https://languagetestingasia.springeropen.com/articles/10.1186/s40468-022-00169-2

Pedro, F., Subosa, M., Rivas, A., & Valverde, P. (2019). Artificial intelligence in education: challenges and opportunities for sustainable development. https://repositorio.minedu.gob.pe/handle/20.500.12799/6533

Popkova, E. G. & Gulzat, K. (2019, May). Technological revolution in the 21 st century: digital society vs. artificial intelligence. Institute of Scientific Communications Conference, pp. 339–345. Springer, Cham. https://link.springer.com/chapter/10.1007/978-3-030-32015-7_38

Prentzas, J. (2013). Artificial intelligence methods in early childhood education. *Artificial Intelligence, Evolutionary Computing and Metaheuristics*, pp. 169–199). Springer, Berlin, Heidelberg. https://link.springer.com/chapter/10.1007/978-3-642-29694-9_8

Ramu, M. M., Shaik, N., Arulprakash, P., Jha, S. K., & Nagesh, M. P. (2022). Study on potential AI applications in childhood education. *International Journal of Early Childhood*, 14(03), 2022. https://www.researchgate.net/profile/Ramu_Modapothula/publication/361605765_Study_on_Potential_AI_Applications_in_Childhood_Education/links/62bc0fb360e77b7db83cac93/Study-on-Potential-AI-Applications-in-Childhood-Education.pdf

Rana, G., Khang, A., Sharma, R., Goel, A. K., & Dubey A. K. (2021). *Reinventing Manufacturing and Business Processes through Artificial Intelligence*. CRC Press, Boca Raton, FL. https://doi.org/10.1201/9781003145011

Robaldo, L., Villata, S., Wyner, A., & Grabmair, M. (2019). Introduction for artificial intelligence and law: special issue "natural language processing for legal texts". *Artificial Intelligence and Law*, 27(2), 113–115. https://link.springer.com/article/10.1007/s10506-019-09251-2

Roll, I. & Wylie, R. (2016). Evolution and revolution in artificial intelligence in education. *International Journal of Artificial Intelligence in Education*, 26(2), 582–599. https://link.springer.com/article/10.1007/s40593-016-0110-3

Salas-Pilco, S. Z., Xiao, K., & Hu, X. (2022). Artificial intelligence and learning analytics in teacher education: a systematic review. *Education Sciences*, 12(8), 569. https://www.mdpi.com/2227-7102/12/8/569

Seldon, A., Abidoye, O., & Metcalf, T. (2020). *The Fourth Education Revolution Reconsidered: Will Artificial Intelligence Enrich or Diminish Humanity?*. Legend Press Ltd, London. https://www.google.com/books?hl=en&lr=&id=RIFJEAAAQBAJ&oi=fnd&pg=PT11&dq=The+Fourth+Education+Revolution+Reconsidered:+Will+Artificial+Intelligence+Enrich+or+Diminish+Humanity%3F.+Legend+Press&ots=BMhxPcbnMz&sig=VpVEpN3-bGaxu9Gih9jIC_O-kzM

Sharma, K., Giannakos, M., & Dillenbourg, P. (2020). Eye-tracking and artificial intelligence to enhance motivation and learning. *Smart Learning Environments*, 7(1), 1–19. https://slejournal.springeropen.com/articles/10.1186/s40561-020-00122-x

Soliman, M. W. M. (2016). A proposed perspective for designing a 3D virtual learning environment based on artificial intelligence for teaching English language for medical students at the University of Bisha. *Journal of Research in Curriculum Instruction and Educational Technology*, 2(1), 101–128. https://journals.ekb.eg/article_24535.html

Sołtysik, M., Gawłowska, M., Sniezynski, B., & Gunia, A. (Eds.). (2023). *Artificial Intelligence, Management and Trust*. Taylor & Francis, New York. https://www.tandfonline.com/doi/abs/10.1080/12460125.2020.1819094

Spada, N. & Lightbown, P. M. (2022). In IT together: teachers, researchers, and classroom SLA. *The Modern Language Journal*, 106(3), 635–650. https://onlinelibrary.wiley.com/doi/abs/10.1111/modl.12792

Stokes-Beverley, C. & Simoy, I. (2016). Advancing educational technology in teacher preparation: policy brief. Office of Educational Technology, US Department of Education. https://eric.ed.gov/?id=ED571881

Stone, P., Brooks, R., Brynjolfsson, E., Calo, R., Etzioni, O., Hager, G., & Teller, A. (2022). Artificial intelligence and life in 2030: the one hundred year study on artificial intelligence. *arXiv preprint arXiv:2211.06318*.

Sun, X. (2021). 5G joint artificial intelligence technology in the innovation and reform of university english education. *Wireless Communications and Mobile Computing*, 2021, 4892064. https://www.hindawi.com/journals/wcmc/2021/4892064/

Sun, Z., Anbarasan, M., & Praveen Kumar, D. J. C. I. (2021). Design of online intelligent English teaching platform based on artificial intelligence techniques. *Computational Intelligence*, 37(3), 1166–1180. https://onlinelibrary.wiley.com/doi/abs/10.1111/coin.12351

Venugopal, V., Bishnoi, S., Singh, S., Zaki, M., Grover, H. S., Bauchy, M., & Krishnan, N. A. (2021). Artificial intelligence and machine learning in glass science and technology: 21 challenges for the 21st century. *International journal of applied glass science*, 12(3), 277–292. https://ceramics.onlinelibrary.wiley.com/doi/abs/10.1111/ijag.15881

Vincent-Lancrin, S. & van der Vlies, R. (2020). Trustworthy artificial intelligence (AI) in education: promises and challenges. https://www.oecd-ilibrary.org/content/paper/a6c90fa9-en

Wogu, I. A. P., Misra, S., Olu-Owolabi, E. F., Assibong, P. A., Udoh, O. D., Ogiri, S. O., & Damasevicius, R. (2018). Artificial intelligence, artificial teachers and the fate of learners in the 21st century education sector: Implications for theory and practice. *International Journal of Pure and Applied Mathematics*, 119(16), 2245–2259.

https://www.researchgate.net/profile/Sanjay-Misra-6/publication/326679815_
ARTIFICIAL_INTELLIGENCE_ARTIFICIAL_   teachers_and_the_fate_of_learners_
in_the_21st_century_education_sector_impli-cations_for_theory_and_practice/
links/5b5e4855aca272a2d6745be2/artificial-intelligence-artificial-teachers-and-the-
fate-of-learners-in-the-21st-century-education-sector-impli-cations-for-theory-
and-practice.pdf

Xin, D. (2021). Application value of multimedia artificial intelligence technology in English
teaching practice. *Mobile Information Systems*, 2021(12), 1–11. https://downloads.
hindawi.com/journals/misy/2021/3754897.pdf

Zafar, S. & Meenakshi, K. (2012). A study on the relationship between extroversion-
introversion and risk-taking in the context of second language acquisition. *International
Journal of Research studies in language learning*, 1(1), 33–40. https://www.academia.
edu/download/37014762/A_study_on_the_relationship_between.pdf

Zhao, Y. & Ellis, R. (2022). The relative effects of implicit and explicit corrective feedback
on the acquisition of 3rd person-s by Chinese university students: a classroom-based
study. *Language Teaching Research*, 26(3), 361–381. https://journals.sagepub.com/doi/
abs/10.1177/1362168820903343

15 The Role of Artificial Intelligence in Developing an Effective Training Assessment Tool

Swati Bhatia, Veenu Arora, and Shweta Batra

15.1 INTRODUCTION

It is essential to evaluate the effectiveness of training assessment tools in achieving intended learning outcomes. There is a need for a quantitative strategy to appropriately measure the impact of the emergence of artificial intelligence (AI)-based training assessment systems. The assessment of important metrics and the discovery of causal relationships between the application of AI technologies and better learning outcomes are made possible by quantitative evaluation. This chapter concentrates on the quantitative evaluation of an AI-based training assessment tool's effectiveness.

Assessment tools based on AI have the potential to improve the effectiveness, precision, and individualization of assessments. However, it is crucial to evaluate their impact to determine whether they effectively contribute to enhanced learning outcomes. Quantitative evaluation provides empirical proof of the impact, enabling organizations to make data-driven decisions regarding the implementation and enhancement of AI-based training assessment tools (Khang & Rani et al., 2023).

The purpose of this chapter is to provide a framework for conducting a quantitative evaluation of the impact of an AI-based training assessment tool. It outlines the steps involved in the evaluation process, including defining evaluation goals, selecting appropriate metrics, collecting and analyzing data, and interpreting the results. The chapter also addresses the challenges and considerations specific to evaluating AI-based training assessment tools quantitatively and discusses the potential benefits and implications of using a quantitative approach (Khang & Hajimahmud et al., 2023).

This chapter's objective is to provide a framework for quantitatively assessing the impact of an AI-based training assessment tool. It describes the stages involved in the evaluation process, such as defining evaluation objectives, selecting suitable metrics, collecting and analyzing data, and interpreting the results. The chapter also discusses the challenges and considerations specific to quantitatively evaluating

DOI: 10.1201/9781003440901-15

AI-based training assessment tools, as well as the potential benefits and impacts of using a quantitative approach (Snehal et al., 2023).

15.2 AI APPLICATIONS IN TRAINING ASSESSMENT

15.2.1 Autoscoring

AI-assisted scoring is a common training assessment tool. Human assessors spent a lot of time assessing multiple-choice tests and essay evaluations. Machine learning algorithms in AI-based systems analyze replies and assign scores based on established criteria. AI systems can better score open-ended questions and essays using natural language processing (NLP) approaches. Automated scoring improves efficiency, reduces subjectivity, and speeds up feedback. It lets organizations test massive volumes of assessments rapidly and consistently, offering learners immediate results and achievable feedback.

15.2.2 Personal Feedback

Personalized feedback improves learning and engages students. AI may analyze learner replies and deliver personalized feedback based on performance, knowledge gaps, and learning objectives. AI systems can deliver personalized instruction, reinforcement, and improvement suggestions by using machine learning algorithms and natural language understanding. AI-powered feedback considers the learner's progress, learning style, and preferences. To aid learning, it can provide videos and interactive simulations. Personalized feedback improves learning and lets students track their progress.

15.2.3 Customizable and Adaptive Assessments

AI systems alter adaptive tests' difficulty and content based on the learner's performance. AI systems can continuously analyze the learner's replies to adjust the examination to match their knowledge level and challenge level. Adaptive tests boost engagement, reduce assessment fatigue, and more accurately assess learners. They reduce boredom and frustration by assessing students at their best. Adaptive evaluations also reveal learners' strengths and shortcomings, enabling targeted interventions and personalized learning pathways (Babasaheb et al., 2023a).

15.2.4 Predictive Analytics

Predictive analytics uses historical and real-time data to create predictions and decisions. Predictive analytics can identify training program quitters and underachievers. AI systems can detect challenges and enable early intervention by analyzing learner demographics, performance history, and engagement patterns. Predictive analytics can also optimize training interventions by finding the best instructional tactics, content kinds, and delivery formats for particular learner profiles. Organizations may optimize training programs by understanding what makes them successful (Babasaheb et al., 2023b).

15.3 ESTABLISHING OBJECTIVES

15.3.1 EVALUATION PURPOSES

Objectives for quantitative evaluations must be explicit. It is possible to evaluate the impact of the AI-based training assessment instrument on learner performance, engagement, satisfaction, and other outcomes. The evaluation procedure is guided by specific evaluation objectives.

15.3.2 EVALUATE QUESTIONS

Evaluation questions must be created. Evaluation queries should correspond to objectives and quantify specific impacts. Evaluation queries could focus on the impact of AI technology on learner performance, motivation, and contentment.

15.3.3 HYPOTHESIS

Evaluation queries can direct data analysis in the direction of hypotheses. Variable relationships or distinctions must be stated in hypotheses. For instance, the AI-based assessment tool for training may enhance performance outcomes.

15.4 SELECTING APPROPRIATE ASSESSMENT METRICS

15.4.1 PERFORMANCE METRICS

The impact of the AI-based training assessment instrument on learner performance is measured by performance metrics. These metrics could include assessment scores, completion rates, assessment completion times, and other pertinent performance indicators. The selection of appropriate performance metrics depends on the nature of the assessments and the learning objectives.

15.4.2 ENGAGEMENT METRICS

Engagement metrics assess the level of learner engagement with the AI-based assessment instrument for training. These metrics could include metrics such as the amount of time spent utilizing the tool, the frequency with which the tool is utilized, or interaction patterns within the tool. Metrics for engagement reveal the extent to which learners are actively utilizing and benefiting from the AI tool.

15.4.3 CONTENTMENT METRICS

Learners' perceptions and content with the AI-based training assessment tool are captured by satisfaction metrics. These metrics could include ratings or responses on a Likert scale for aspects such as the tool's usability, utility, relevance, and overall satisfaction. The subjective experiences and perceptions of the tool's impact can be evaluated with the aid of satisfaction metrics.

15.5 TRADITIONAL TRAINING ASSESSMENT MODELS

Traditional training assessment models serve as a basis for evaluating learning outcomes. These models incorporate a variety of evaluation strategies, including pre- and post-tests, quizzes, examinations, observations, interviews, and performance evaluations. The most prevalent training evaluation models include the following:

15.5.1 KIRKPATRICK'S FOUR EVALUATION LEVELS

The four levels of Kirkpatrick's model are reaction, learning, behavior, and results. It evaluates learners' responses to training, knowledge acquisition, behavior modification, and the impact of training on organizational results.

15.5.2 BLOOM'S TAXONOMY

It categorizes learning objectives into cognitive domains, such as knowledge, comprehension, application, analysis, synthesis, and evaluation. It directs the development of assessments that correspond to the intended learning outcomes, as shown in Figure 15.1.

15.5.3 ADDIE MODEL

The Analysis, Design, Development, Implementation, and Evaluation (ADDIE) model is a systematic instructional design process that emphasizes assessment. It ensures that assessments are aligned with training goals and integrated throughout the training process.

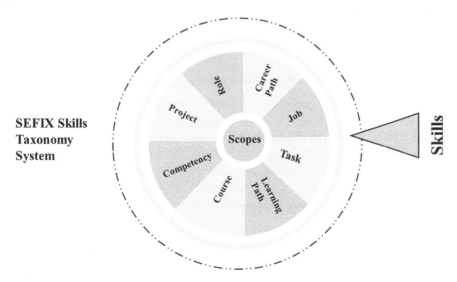

FIGURE 15.1 Skills taxonomy (Khang, 2021).

15.6 MODERN TOOLS: ARTIFICIAL INTELLIGENCE INTEGRATION IN TRAINING ASSESSMENT

AI offers numerous opportunities to improve training assessment models. By incorporating AI techniques and tools, training evaluations can become more adaptive, individualized, productive, and insightful. The following are methods for incorporating AI into training assessment models.

15.6.1 ADAPTIVE ASSESSMENTS

AI enables the creation of adaptive assessments that modify the level of difficulty and content based on the performance and abilities of the learner. In real time, machine learning algorithms analyze learner responses and adapt the examination accordingly. This ensures that students are appropriately challenged and receive assessments based on their level of knowledge.

15.6.2 NATURAL LANGUAGE PROCESSING (NLP) FOR TEXT-BASED EVALUATIONS

In text-based assessments, NLP algorithms can be implemented to analyze written responses. These algorithms evaluate the content of written assignments, essays, and open-ended questions and provide automated scoring and feedback. NLP improves the efficacy and consistency of scoring while providing learners with personalized feedback.

15.6.3 AUTOMATIC SCORING AND RESPONSE

AI-based assessment tools can automate scoring processes, eliminating the need for manual grading. Large datasets and human input are used to train machine learning algorithms to generate reliable and consistent scoring. Automated feedback systems can provide learners with immediate and individualized feedback, fostering self-directed learning and continuous improvement.

15.6.4 ANALYSIS OF DATA AND PREDICTIVE INSIGHTS

Data analytics techniques propelled by AI enable the analysis of large volumes of learner data to extract valuable insights. Algorithms for machine learning identify patterns, trends, and correlations in the performance, engagement, and outcomes of learners. Predictive analytics can predict learners' future performance and identify enhancement opportunities, enabling proactive interventions (Khang, 2023).

15.6.5 INTELLIGENT RECOMMENDATION SYSTEMS

Intelligent recommendation systems can prescribe personalized learning resources, activities, or assessments based on the preferences, learning styles, and performance data of the learners. These systems analyze learner data with machine learning algorithms and generate targeted recommendations. Personalized recommendations increase learner engagement and facilitate individualized learning paths.

15.6.6   ASSESSMENTS BASED ON VIRTUAL REALITY (VR) AND SIMULATIONS

Virtual reality (VR) and simulation-based evaluations provide realistic and immersive learning environments. AI can be incorporated into these evaluations to provide intelligent feedback and adapt simulations based on the interactions and performance of the learners. Algorithms based on AI can analyze the decisions and actions of students within simulations and provide individualized advice and insights.

15.6.7   CHATBOTS AND VIRTUAL ASSISTANTS

Chatbots and virtual assistants give learners immediate access to individualized support and direction. These AI-powered tools interact with learners, answer queries, and provide assistance by analyzing natural language. Chatbots can provide learners with explanations, additional resources, and directions to pertinent content. Virtual assistants can be integrated into training platforms or learning environments to increase learner engagement and provide real-time support by simulating human-like interactions.

15.7   ADVANTAGES OF AI IN TRAINING ASSESSMENT

15.7.1   PRODUCTIVITY AND ACCURACY ENHANCEMENTS

By automating time-consuming tasks such as grading and generating feedback, AI-powered training assessment tools significantly increase productivity. Utilizing machine learning algorithms, these tools are able to process and analyze immense quantities of assessments in an efficient and consistent manner. This allows organizations to evaluate more learners in less time, enhancing scalability and reducing administrative burdens.

In addition, AI systems can enhance the accuracy of evaluation results by eliminating human errors and inconsistencies. They can provide objective and unbiased evaluations, ensuring that the evaluation process is fair and equitable. Large datasets and human input can be used to train automated scoring algorithms, resulting in consistent and reliable assessments.

15.7.2   IMMEDIATE RESPONSE AND CUSTOMIZATION

AI enables the instantaneous transmission of feedback to learners fostering a timely understanding of their strengths and development areas. Learners are no longer required to wait for manual grading or assessments, allowing them to promptly address their deficiencies and strengthen their strengths. Immediate feedback enhances the learning experience by boosting engagement and motivation and enabling learners to modify their learning strategies in real time. In addition, AI-powered assessment tools for training facilitate individualization by adapting feedback and evaluations to the specific needs and preferences of each learner. Individualized feedback and adaptive assessments provide learners with varying levels of knowledge and learning styles with targeted instruction and challenges. Individualization not only improves learning outcomes but also fosters a sense of ownership and autonomy among learners.

15.7.3 ENHANCEMENT OF DECISION-MAKING

AI-based training assessment tools generate actionable insights and analytics that facilitate data-driven decision-making. By analyzing immense quantities of learner data, such as performance trends, knowledge gaps, and engagement patterns, organizations can identify opportunities for improvement and make well-informed training intervention decisions. The capabilities of predictive analytics enable organizations to address impending challenges and hazards in a proactive manner. If they intervene early, they can provide additional support to students who are at risk of underperforming or falling out. In addition, AI-generated insights can help organizations optimize their training programs by identifying the most effective instructional strategies, content categories, and delivery formats for different learner profiles.

15.7.4 SCALABILITY AND FINANCIAL SUSTAINABILITY

AI-powered training assessment tools offer scalability and cost-effectiveness by automating manual processes and reducing the need for human evaluators. Organizations are able to assess a large number of students concurrently without compromising assessment quality or timeliness. This scalability is especially beneficial when evaluating a geographically dispersed or rapidly expanding workforce. Moreover, AI-based assessment tools can help organizations save money on costs associated with human evaluation, such as the hiring and training of assessors, as well as the time and resources required for manual assessment processes. By streamlining and automating assessment workflows, organizations can allocate resources more efficiently and focus on other essential aspects of learning and development programs.

15.8 DIFFICULTIES AND ETHICAL FACTORS

15.8.1 DATA PRIVACY AND SECURITY

Data privacy and security AI-based training assessment tools rely on vast quantities of data, such as learner responses, performance history, and engagement patterns. The protection of this information's confidentiality and security is of the utmost importance. Organizations must abide by applicable data protection laws and regulations, such as the General Data Protection Regulation (GDPR), and implement suitable safeguards to secure learner data from unauthorized access, breaches, and misconduct. The collection and utilization of data must also be conducted openly. Organizations should disclose to learners the categories of data collected, the purposes for which it will be used, and any potential consequences. Establishing stringent data governance frameworks and protocols can aid businesses in maintaining confidence in AI-powered assessment systems (Shashi et al., 2023).

15.8.2 PREJUDICE AND JUSTICE

AI systems are susceptible to both explicit and implicit biases prevalent in the training data. Unfair assessments and unequal opportunities for learners can result from biases. It is essential to diligently curate training datasets to eliminate bias and ensure fairness and equity in assessments. Organizations should regularly evaluate AI models for biases and implement strategies to mitigate them. This includes evaluating the impact of variables such as gender, race, and socioeconomic status on assessment outcomes. It is crucial to guarantee diversity and representation in training data in order to create objective and fair AI models.

15.8.3 TRANSPARENCY AND EXPLANATION

Frequently, AI algorithms operate as black boxes, making it challenging to comprehend their decision-making processes. The absence of transparency and explainability can hinder the adoption and confidence in AI-based assessment tools. Organizations should strive to make AI systems transparent and explicable, so learners can understand how their assessments are scored. Through the development of AI models that provide explanations or justifications for their decisions, learners and institutions can gain insight into the assessment procedure. Transparency and explainability can be enhanced using techniques such as interpretable machine learning and rule-based models.

15.8.4 HUMAN OBSERVATION AND INTERVENING

It is essential to establish a balance between automation and human input, despite the numerous benefits of AI-powered evaluation tools. Human oversight and intervention are required to address limitations, monitor system performance, and ensure the ethical use of AI in training assessment. Organizations should develop evaluation and validation procedures for AI-generated outcomes. Evaluation of complex or subjective tasks requiring human judgment should be performed by specialists. Moreover, there should be mechanisms in place to manage situations in where learners contest assessment results or request human intervention (Pallavi et al., 2023).

15.9 CASE STUDIES AND ILLUSTRATIONS

15.9.1 EVALUATION INSTRUMENTS BASED ON AI IN CORPORATE TRAINING

Numerous organizations have adopted AI-based evaluation tools to assess and enhance corporate training programs. These tools use algorithms for machine learning to automate scoring, generate analytics, and provide personalized feedback. For instance, a multinational technology company uses an AI-powered assessment platform that evaluates coding exercises automatically and provides learners with real-time feedback. This has considerably reduced the time and effort required for manual grading, allowing for faster and more accurate evaluations.

15.9.2 EDUCATIONAL ASSESSMENTS ENABLED BY AI

AI has revolutionized the assessment process in the field of education. Individualized examinations tailored to each learner's learning needs are generated by AI-enabled assessment platforms. An AI-based assessment tool for mathematics, for example,

Data Engineer Job Title   Assessment Results

John Smith - Assessment Results	Evaluated by Employee	Expected by Employee	Evaluated by Manager	
No	Description	Senior Data Engineer	Principal Data Engineer	[Mid-level] Data Engineer	Senior Data Engineer
1	Section I: Description for Roles and Responsibilities	30.0	30.00	22.50	22.50
2	Section II: Description for Professional Skills	26.25	35.00	26.25	26.25
3	Section III: Description for Soft Skills	15.0	15.00	15.0	15.0
5	Section V: Description for Qualification	10.0	10.00	7.50	7.50
4	Section IV: Description for Contributions	6.67	10.00	3.33	3.33
	Total Scores	87.92	100.0	74.58	74.58

FIGURE 15.2   Assessment results of the data analyst job title.

analyzes responses from learners and adjusts the level of difficulty of questions based on their performance, thereby ensuring a challenging yet appropriate assessment experience. These tools offer immediate feedback, track progress, and facilitate targeted interventions to improve learning outcomes (Ashwini & Khang, 2023).

15.9.3   AI in Skill-based Assessments

AI has also been integrated into competency-based assessments, such as language proficiency or job-specific competencies. Using natural language comprehension and automated scoring algorithms, language assessment tools propelled by AI evaluate the writing and speaking skills of students. In addition to providing accurate and consistent assessments, these tools also provide individualized feedback to aid language learners in enhancing their abilities. Likewise, AI-based simulations and VR assessments enable realistic and immersive evaluations of practical skills in fields such as healthcare, aviation, and engineering, as shown in Figure 15.2.

15.10   IMPLEMENTATION CONSIDERATIONS

15.10.1   Compatibility with Existing Learning Management Systems

Existing learning management systems (LMS) should be incorporated with AI-powered assessment tools. Effortless integration enables a unified user experience, streamlined data administration, and efficient learner progress monitoring. Compatibility with industry standards, such as the Experience API (xAPI), assures interoperability and data exchange between different learning platforms.

15.10.2   Collection and Analysis of Data

For organizations to maximize the value of AI in training evaluation, they must collect and analyze germane data. This includes responses from learners, performance metrics, engagement data, and contextual data. Data collection must adhere to privacy regulations and ethical principles, ensuring informed consent and open data practices. In addition, businesses should implement robust data analytics processes to extract actionable insights and inform their decision-making.

15.10.3   USER ACCEPTANCE AND TRAINING

Adoption and user acceptance are essential to the successful implementation of AI-powered assessment tools. Organizations must provide learners and instructors with adequate training and support to ensure they are conversant with the tools and understand their benefits. Clear communication regarding the purpose, functionality, and expected outcomes of AI-based assessments facilitates the development of stakeholder confidence and acceptance.

15.10.4   CONTINUOUS IMPROVEMENTS AND UPGRADES

To address emergent challenges, the performance of AI-based assessment tools should be continually monitored and enhanced. Organizations should collect user feedback, evaluate system performance, and implement system updates and enhancements based on learner needs and evolving assessment requirements. Continuous evaluation and validation ensure that AI models remain accurate, unbiased, and aligned with organizational goals (Khang & Muthmainnah et al., 2023).

15.11   FUTURE PROSPECTS AND OPPORTUNITIES

15.11.1   INNOVATIONS IN NATURAL LANGUAGE PROCESSING

Advances in NLP will enable AI systems to comprehend and evaluate complex language-based assessments with greater precision. Improved language comprehension capabilities, including sentiment analysis, context comprehension, and semantic analysis, will enhance the accuracy and sophistication of automated scoring and feedback generation (Khang & Shah et al., 2023).

15.11.2   EMOTIONAL IDENTIFICATION AND SENTIMENT ANALYSIS

Emotion recognition and sentiment analysis capabilities enable AI systems to assess the emotional states of learners during examinations. By analyzing the facial expressions, tone of voice, and written responses of learners, AI can determine their emotional states, such as frustration, engagement, and confusion. This information can be used to tailor interventions, modify assessments, and enhance the overall learning experience.

15.11.3   ASSESSMENTS IN VIRTUAL REALITY AND IMMERSION

VR and immersive technologies offer exciting possibilities for evaluating practical skills and complex scenarios. VR assessments, powered by AI, can simulate authentic environments, allowing students to demonstrate their abilities in a controlled yet authentic setting. These evaluations can capture copious amounts of information on the actions, decisions, and problem-solving strategies of the learners, allowing for exhaustive and objective evaluations (Khang & Kali et al., 2023).

15.11.4 ETHICAL CONSIDERATIONS AND REGULATIONS FOR AI-BASED EVALUATION TOOLS

The ethical implications of AI-based assessment tools will continue to be a focal point of research and discourse. In order to ensure impartiality, transparency, and accountability in the application of AI, ethical guidelines and regulations will be required. Organizations and policymakers must address concerns regarding data privacy, bias, explainability, and human oversight in order to cultivate trust and the responsible use of AI in training assessment (Muthmainnah et al., 2023).

15.12 CONCLUSION

This chapter examined the function of AI in the development of effective training assessment instruments. AI provides numerous applications, such as automated scoring, personalized feedback, adaptive assessments, and predictive analytics. Assessment tools enabled by AI offer numerous advantages, such as increased efficiency and precision, immediate feedback and individualization, enhanced decision-making, scalability, and cost-effectiveness.

The incorporation of AI in training evaluation has the potential to revolutionize how organizations evaluate the efficacy of their training programs. Using assessment tools based on AI, organizations are able to evaluate large numbers of learners quickly and consistently, providing immediate feedback and individualized support. Capabilities in predictive analytics enable organizations to optimize their training interventions and make decisions based on data (Shah et al., 2023).

AI presents numerous opportunities, but it also raises obstacles and ethical concerns that must be addressed. Ensuring data privacy and security, mitigating biases, and fostering transparency and human oversight are essential for the responsible application of AI in training assessment. The future of training assessment contains promising prospects, including advancements in NLP, emotion recognition, immersive assessments, and ethical frameworks, thanks to advances in AI and ongoing research. AI has the potential to revolutionize training evaluation, augmenting the learning experience and empowering organizations to make data-driven decisions for continuous improvement. By utilizing the power of AI, organizations are able to develop effective and individualized training assessment tools that contribute to individual and organizational success (Khang & Misra et al., 2023).

REFERENCES

Ashwini, Y. S. & Khang, A. (2023). Challenges faced by marketers in developing and managing contents in workforce development system. *Designing Workforce Management Systems for Industry 4.0: Data-Centric and AI-Enabled Approaches* (1st ed., pp. 332–359). CRC Press, Boca Raton, FL. https://doi.org/10.1201/9781003357070-18

Babasaheb, J., Sphurti, B., & Khang, A. (2023a). Industry revolution 4.0: workforce competency models and designs. *Designing Workforce Management Systems for Industry 4.0: Data-Centric and AI-Enabled Approaches* (1st ed., pp. 14–31). CRC Press, Boca Raton, FL. https://doi.org/10.1201/9781003357070-2

Babasaheb, J., Sphurti, B., & Khang, A. (2023b). Design of competency models in the human capital management system. *Designing Workforce Management Systems for Industry 4.0: Data-Centric and AI-Enabled Approaches* (1st ed., pp. 32–50). CRC Press, Boca Raton, FL. https://doi.org/10.1201/9781003357070-3

Khang, A. (2021). "Material4Studies," Material of Computer Science, Artificial Intelligence, Data Science, IoT, Blockchain, Cloud, Metaverse, Cybersecurity for Studies. Retrieved from https://www.researchgate.net/publication/370156102_Material4Studies

Khang, A. (2023). *Advanced Technologies and AI-Equipped IoT Applications in High-Tech Agriculture* (1 ed.). IGI Global Press, Hershey, PA. https://doi.org/10.4018/978-1-6684-9231-4

Khang, A., Hajimahmud, V. A., Gupta, S. K., Babasaheb, J., & Morris, G. (2023). *AI-Centric Modelling and Analytics: Concepts, Designs, Technologies, and Applications* (1 ed.). CRC Press, Boca Raton, FL. https://doi.org/10.1201/9781003400110

Khang, A., Kali, C. R., Suresh Kumar, S., Amaresh, K., Sudhansu Ranjan, D., & Manas Ranjan, P. (2023). Enabling the future of manufacturing: integration of robotics and IoT to smart factory infrastructure in industry 4.0. *AI-Based Technologies and Applications in the Era of the Metaverse* (1st ed., pp. 25–50). IGI Global Press, Hershey, PA. https://doi.org/10.4018/978-1-6684-8851-5.ch002

Khang, A., Misra, A., Gupta, S. K., & Shah, V. (2023). *AI-Aided IoT Technologies and Applications in the Smart Business and Production*. CRC Press, Boca Raton, FL. https://doi.org/10.1201/9781003392224

Khang, A., Muthmainnah, M., Seraj, P. M. I., Yakin, A. A., Obaid, A. J., & Panda, M. R. (2023). AI-aided teaching model for the education 5.0 ecosystem. *AI-Based Technologies and Applications in the Era of the Metaverse* (1st ed., pp. 83–104). IGI Global Press, Hershey, PA. https://doi.org/10.4018/978-1-6684-8851-5.ch004

Khang, A., Rani, S., Gujrati, R., Uygun, H., & Gupta, S. K. (2023). *Designing Workforce Management Systems for Industry 4.0: Data-Centric and AI-Enabled Approaches*. CRC Press, Boca Raton, FL. https://doi.org/10.1201/9781003357070

Khang, A., Shah, V., & Rani, S. (2023). *AI-Based Technologies and Applications in the Era of the Metaverse* (1 ed.). IGI Global Press, Hershey, PA. https://doi.org/10.4018/978-1-6684-8851-5

Muthmainnah, M., Khang, A., Seraj, P. M. I., Yakin, A. A., Oteir, I., & Alotaibi, A. N. (2023). An innovative teaching model - the potential of metaverse for English learning. *AI-Based Technologies and Applications in the Era of the Metaverse* (1st ed., pp. 105–126). IGI Global Press, Hershey, PA. https://doi.org/10.4018/978-1-6684-8851-5.ch005

Pallavi, J., Vandana, T., Ravisankar, M., & Khang, A. (2023). Data-driven AI models in the workforce development planning. *Designing Workforce Management Systems for Industry 4.0: Data-Centric and AI-Enabled Approaches* (1st ed., pp. 179–198). CRC Press, Boca Raton, FL. https://doi.org/10.1201/9781003357070-10

Shah, V. & Khang, A. (2023). Metaverse-enabling IoT technology for a futuristic healthcare system. *AI-Based Technologies and Applications in the Era of the Metaverse* (1st ed., pp. 165–173). IGI Global Press, Hershey, PA. https://doi.org/10.4018/978-1-6684-8851-5.ch008

Shashi, K. G., Khang, A., Parin, S., Chandra Kumar, D., Anchal, P. (2023). Data mining processes and decision-making models in personnel management system. *Designing Workforce Management Systems for Industry 4.0: Data-Centric and AI-Enabled Approaches* (1st ed., pp. 89–112). CRC Press, Boca Raton, FL. https://doi.org/10.1201/9781003357070-6

Snehal, M., Babasaheb, J., & Khang A. (2023). Workforce management system: concepts, definitions, principles, and implementation. *Designing Workforce Management Systems for Industry 4.0: Data-Centric and AI-Enabled Approaches* (1st ed., pp. 1–13). CRC Press, Boca Raton, FL. https://doi.org/10.1201/9781003357070-1

16 Revolutionizing Recruitment
Skill-Based Interview Models in the Artificial Intelligence-Driven Economy

Sharmila Devi R. and Swamy Perumandla

16.1 INTRODUCTION

The skill-based interview model is a methodology employed to evaluate a candidate's aptitude, proficiencies, and competencies, as opposed to solely considering their educational background and professional history. This approach prioritizes the evaluation of the candidate's current capabilities rather than placing emphasis on their previous track record or credentials. The advent of an AI-centric economy has brought about significant changes in the labor market, resulting in a paradigm shift in employment dynamics. This transformation has led to the restructuring of conventional job roles while simultaneously giving rise to fresh and innovative positions. The imperative for individuals possessing the requisite expertise to effectively utilize the capabilities of artificial intelligence (AI) is of utmost importance in industries such as healthcare, banking, manufacturing, and entertainment.

The conventional standards used to assess job applicants have proven inadequate due to their heavy reliance on employment backgrounds and academic accomplishments (Hitharth & Dhanya, 2022). In response to this issue, organizations are increasingly adopting skill-based interview methods. The primary objective of these interviews is to assess the talents and competencies of candidates, with a particular emphasis on fundamental qualities such as technical proficiency, flexibility, critical thinking, and collaborative abilities. This methodology assists enterprises in cultivating a workforce that excels in the AI-driven economy and effectively addresses the disparity between AI-related job openings and prospective individuals.

16.2 THE EVOLUTION OF EMPLOYMENT IN THE AI-ORIENTED ECONOMY

The economy that is focused on AI is causing significant changes in several industries through the implementation of AI, automation, and rapid technical progress. This transition necessitates a reconceptualization of conventional roles and the

DOI: 10.1201/9781003440901-16

establishment of novel ones. The advent of distinct and interdisciplinary positions is a defining characteristic of this paradigm shift, requiring a fusion of proficiencies across several fields of study. The dynamic interaction between a variety of skill sets cultivates a climate conducive to the generation of novel ideas and the resolution of complex issues (Sharma et al., 2022).

The need for professionals who possess adaptability and a commitment to continual learning is significant, as they are required to adopt a growth mindset that enables them to effectively navigate changes, develop, and acquire new abilities consistently (Khamesra, 2023). Professionals that possess adaptability comprehend that their expertise is dynamic and may be augmented and shaped. It is imperative for organizations to cultivate a culture that promotes the ongoing pursuit of education and the enhancement of abilities.

16.3 CHALLENGES WITH TRADITIONAL INTERVIEW APPROACHES

In the rapidly evolving landscape of the AI-oriented economy, traditional interview approaches find themselves facing significant challenges. In this, we delve into the core challenges that traditional interviews encounter in the face of AI-oriented roles.

16.3.1 RELIANCE ON RESUMES AND PAST EXPERIENCES

Conventional interviews have often relied heavily on resumes and candidates' past experiences to make hiring decisions. However, in an economy that is driven by the adoption of advanced technologies and the emergence of AI-driven solutions, past experiences alone may not be a reliable indicator of a candidate's future performance. Resumes might emphasize accomplishments that were relevant in a different context but fail to capture the breadth of skills required in AI-oriented roles. For example, a candidate with a strong background in traditional software development might have limited exposure to machine learning algorithms, even if the role they are applying for demands a deep understanding of such concepts. Reliance on past experiences alone risks overlooking potential candidates who possess the aptitude and capacity to quickly learn and excel in new domains (Roulston, 2011).

16.3.2 INADEQUACY IN EVALUATING ESSENTIAL SKILLS

AI-oriented roles require a unique blend of technical prowess, creative problem-solving, adaptability, and the ability to collaborate across disciplines. Traditional interviews, often focused on standard questions and behavioral assessments, may struggle to accurately gauge these multifaceted skills. For instance, asking a candidate to recite facts or definitions might not effectively reveal their capacity to think critically and apply their knowledge to real-world scenarios.

Technical proficiency, a cornerstone of AI-driven positions, is particularly challenging to assess through traditional interviews. Candidates might easily manipulate their responses to appear more knowledgeable than they truly are, leaving interviewers with an inaccurate assessment of their skill levels. This inadequacy to assess critical skills could lead to hiring individuals ill-suited for the intricacies of AI-oriented roles (Verma et al., 2022).

16.3.3 GROWING DISCREPANCY BETWEEN JOB DESCRIPTIONS AND ROLE REQUIREMENTS

As AI technologies continue to reshape industries and create new opportunities, the descriptions of job roles can quickly become outdated. The rapid pace of technological advancement can lead to a significant lag between the creation of job descriptions and the actual requirements of the role (Kashani & Ivry, 2021). Consequently, traditional interviews that rely on these descriptions might not accurately reflect the evolving demands of AI-oriented positions (Khang & Hajimahmud et al., 2023).

This discrepancy can result in a misalignment between the skills that candidates possess and those that are truly needed for success in AI-driven roles. Organizations might miss out on candidates who exhibit potential but don't fit the mold of the traditional job description. Conversely, they might inadvertently hire candidates who appear to meet the description but lack the skills necessary to thrive in the dynamic landscape of the AI-oriented economy.

16.4 UNDERSTANDING SKILLS-BASED INTERVIEW MODELS

Skill-based interviews, as the name suggests, prioritize the assessment of candidates' skills and competencies over their academic credentials or work history. These interviews focus on understanding how well a candidate can apply their knowledge and skills to practical situations and scenarios that mirror the challenges they might encounter in their prospective roles. By gauging candidates' practical proficiency, skill-based interviews aim to provide a more accurate representation of their potential performance in the real world. At the heart of skill-based interviews are several core principles that distinguish them from traditional methods. These principles include a heightened emphasis on evaluating technical expertise, problem-solving capabilities, adaptability, and collaboration skills. By structuring interview questions and assessments around these principles, organizations can identify candidates who possess the multifaceted abilities required in the AI-oriented economy (Khang & Misra et al., 2023).

16.4.1 EMPHASIS ON ASSESSING TECHNICAL PROFICIENCY, PROBLEM-SOLVING, ADAPTABILITY, AND COLLABORATION

- **Technical Proficiency**: In AI-oriented roles, technical skills play a pivotal role. skill-based interviews delve into a candidate's technical acumen, encompassing their ability to code; navigate machine learning frameworks; and troubleshoot technical challenges. Practical coding exercises, whiteboard sessions, and simulation-based scenarios are commonly employed to evaluate a candidate's hands-on skills (Greiff et al., 2013).
- **Problem-Solving and Critical Thinking**: AI-driven positions often require individuals to tackle complex challenges with innovative solutions.

Skill-based interviews design scenarios that mirror real-world problems, evaluating candidates' analytical thinking, creative problem-solving, and their capacity to approach multifaceted challenges systematically (Belani et al., 2019).

- **Adaptability and Learning Agility**: The pace of technological change in the AI-oriented economy demands candidates who are open to learning new tools, languages, and methodologies. Interview questions are structured to assess a candidate's willingness to adapt to new technologies and their history of successfully embracing change in their previous roles.
- **Communication and Collaboration**: Soft skills remain pivotal in the context of cross-disciplinary collaborations in AI-oriented roles. Candidates are evaluated on their ability to effectively communicate complex technical concepts to non-technical stakeholders. The capacity to collaborate harmoniously in cross-functional teams is also a critical component of these interviews (Munir, 2022).

16.4.2 SHIFT FROM CREDENTIALS TO COMPETENCY-BASED EVALUATIONS

A defining aspect of skill-based interview models is the shift from evaluating candidates based solely on their credentials—such as degrees and work experience—to assessing their competencies. While credentials may provide some insight into a candidate's background, they often fail to capture the full scope of a candidate's abilities, particularly in the context of rapidly evolving AI-oriented roles.

Competency-based evaluations focus on what candidates can do rather than what they have done. This shift acknowledges that the ability to succeed in AI-driven roles transcends traditional metrics. By assessing candidates' practical skills and evaluating their potential to contribute meaningfully in a rapidly changing landscape, skill-based interview models align more closely with the multifaceted nature of AI-oriented positions (Zhao et al., 2011).

16.5 COMPONENTS OF SKILLS-BASED INTERVIEW MODELS

Skill-based interview models are designed to comprehensively assess candidates' abilities and aptitudes, focusing on the specific competencies required for success in AI-oriented roles (Faruqe et al., 2021). In this chapter, we dissect the key components of skill-based interviews and elaborate on the crucial role each component plays in identifying candidates who are well-equipped to excel in the dynamic landscape of the AI-oriented economy (Khang & Rani et al., 2023).

16.5.1 TECHNICAL PROFICIENCY ASSESSMENT

The field of AI thrives on technical innovation and proficiency. AI-related positions demand candidates who possess a deep understanding of programming languages, machine learning frameworks, data manipulation, and problem-solving. Technical proficiency assessment involves evaluating a candidate's ability to translate their theoretical knowledge into practical solutions. This component highlights:

- **Importance of Technical Skills**: In AI-oriented positions, technical skills form the foundation of a candidate's ability to implement and execute complex algorithms, design machine learning models, and analyze data effectively. These skills directly impact an organization's ability to leverage AI-driven solutions.
- **Methods for Evaluation**: Skill-based interviews employ various methods to assess technical proficiency, including coding exercises, algorithm design challenges, and simulated real-world projects. These methods provide interviewers with insights into a candidate's coding skills, their familiarity with popular AI and machine learning frameworks, and their capacity to troubleshoot technical issues.

16.5.2 Problem-Solving and Critical Thinking

The AI-oriented economy thrives on innovation and solving complex problems. AI professionals are expected to devise novel solutions that leverage data insights and machine learning techniques. Problem-solving and critical thinking assessments aim to uncover a candidate's ability to dissect intricate challenges and develop creative solutions. This component emphasizes:

- **Significance of Analytical Skills**: In AI-driven roles, analytical skills are essential for processing large volumes of data, identifying patterns, and making data-driven decisions. Candidates who excel in problem-solving demonstrate their capacity to tackle complex issues strategically.
- **Designing Real-World Scenarios**: Skill-based interviews create scenarios that mirror the challenges candidates might face in their roles. These scenarios assess a candidate's ability to approach problems methodically, considering multiple dimensions, and arriving at innovative solutions.

16.5.3 Adaptability and Learning Agility

The AI-oriented landscape is marked by rapid technological advancements and constant change. Adaptability is a hallmark of successful professionals in this field, given the frequent shifts in tools, methodologies, and paradigms. This component focuses on evaluating candidates' willingness and ability to adapt. It includes:

- **Adapting to Rapid Change**: AI-oriented roles demand professionals who can swiftly embrace new technologies, tools, and frameworks as they emerge. Adaptability is crucial in ensuring that organizations remain at the forefront of technological innovation.
- **Gauging Learning Willingness**: Crafting interview questions that delve into a candidate's history of learning new skills or technologies demonstrates their willingness to learn and their capacity to stay up-to-date with the evolving AI landscape.

16.5.4 COMMUNICATION AND COLLABORATION

While technical skills are pivotal, they must be complemented by effective communication and collaboration abilities. AI-driven roles often involve interdisciplinary collaboration, requiring professionals to convey technical insights to non-technical stakeholders. This component underscores:

- **Role of Soft Skills**: Collaboration is vital in AI projects that involve engineers, designers, marketers, and domain experts. Soft skills, including communication and teamwork, enable professionals to work cohesively across disciplines and bridge knowledge gaps.
- **Assessing Technical Communication**: Skill-based interviews evaluate a candidate's ability to explain technical concepts in clear, concise terms, ensuring that their contributions can be understood by team members from diverse backgrounds. By addressing each component, organizations can ensure that they are building a workforce that is poised to excel in the rapidly evolving AI-oriented economy.

16.6 COMPONENTS OF IMPLEMENTING SKILLS-BASED INTERVIEWS IN THE AI-ORIENTED ECONOMY

16.6.1 SKILL-BASED INTERVIEW MODELS

This breakdown outlines the essential components of implementing skill-based interviews in the AI-oriented economy as Figure 16.1.

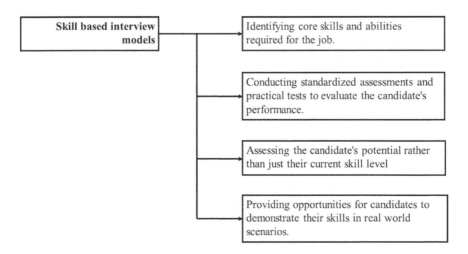

FIGURE 16.1 Components of Skill-based interview models.

16.6.2 STRUCTURED INTERVIEW FRAMEWORKS

- Developing a consistent and unbiased interview process ensures fair evaluations across candidates.
- Emphasizing interviewer training and adherence to standardized questions minimizes bias and subjectivity.

16.6.3 PRACTICAL ASSESSMENTS

- Incorporating hands-on tasks and real-world assessments allows candidates to showcase their skills in practical scenarios.
- Examples of practical assessments include coding challenges, data analysis tasks, and designing machine learning models based on provided datasets.

16.6.4 BEHAVIORAL QUESTIONS

- Behavioral questions uncover candidates' past experiences, decisions, and problem-solving approaches.
- These questions provide insights into how candidates have handled challenges in the past and assess their alignment with the organization's values and goals.

16.7 BENEFITS AND CHALLENGES OF SKILL-BASED INTERVIEW MODELS

This section explores the benefits and complexities of skill-based interview models in the AI-driven economy, highlighting their focus on assessing individuals applying for AI-related professions.

16.7.1 ADVANTAGES OF ACCURATE SKILLS ASSESSMENT FOR AI-ORIENTED POSITIONS

- **Precise Evaluation**: Skill-based interviews directly assess the skills and competencies required for success in AI-oriented roles, providing a more accurate representation of a candidate's potential performance.
- **Matching Role Requirements**: By focusing on skills directly relevant to the job, organizations can identify candidates whose capabilities align closely with the demands of AI-driven positions.
- **Predictive Performance**: Skill-based assessments offer insights into a candidate's problem-solving abilities, adaptability, and technical expertise, providing indicators of their future performance in dynamic and rapidly evolving AI contexts.
- **Improved Hiring Outcomes**: Accurate skills assessment leads to better matches between candidates and roles, reducing turnover rates and increasing overall organizational success.

16.7.2 Reducing Bias and Subjectivity in Hiring Decisions

- **Objective Evaluation**: Skill-based interviews emphasize standardized assessments, minimizing the influence of interviewer bias and personal preferences.
- **Meritocracy**: Candidates are evaluated based on their demonstrated skills, irrespective of their background, education, or personal connections.
- **Diversity and Inclusion**: By focusing on skills, organizations can identify candidates from diverse backgrounds who possess the necessary competencies, fostering a more inclusive workforce.
- **Fairness**: Skill-based models help ensure that candidates are evaluated consistently, promoting fairness and equity in the hiring process.

16.7.3 Challenges Related to Standardization, Role-Specific Assessment Criteria, and Scalability

- **Standardization**: Designing and maintaining standardized interview questions and assessments across diverse roles can be challenging, requiring ongoing efforts to refine the process.
- **Role-Specific Criteria**: Skills and competencies vary among different AI-oriented roles. Developing assessment criteria tailored to specific positions can be complex, particularly for niche roles.
- **Scalability**: Implementing skill-based interviews across a large number of candidates can be resource-intensive, necessitating efficient processes and well-trained interviewers.
- **Subjective Evaluation**: Despite efforts to minimize bias, subjective elements may still influence interviewers' evaluations, potentially leading to inconsistent outcomes.
- **Resource Allocation**: Building and executing skill-based interviews demand investments in time, training, and assessment development.

16.8 PROSPECTS FOR ADVANCEMENT IN SKILL-BASED INTERVIEW MODELS WITHIN THE AI-DRIVEN ECONOMY

16.8.1 Incorporating AI in Assessment

Enhancing Precision and Efficiency: The integration of AI into skill-based interview models offers a promising avenue for growth in the context of the AI-driven economy. As AI continues to reshape industries and redefine job roles, the utilization of AI algorithms to evaluate candidate responses emerges as a significant achievement. This advancement not only holds the potential for heightened efficiency but also provides a pathway to more accurate evaluations. By leveraging AI algorithms, organizations can align candidates' capabilities more precisely with the demands of AI-driven positions, resulting in a more competent and suitable workforce (Khang, 2024).

16.8.2 PERSONALIZED ASSESSMENTS

Tailoring Evaluation to AI-Oriented Roles: Looking ahead, a compelling trend within skill-based interview models is the move toward personalized assessments tailored to the unique demands of AI-oriented roles. With the burgeoning need for professionals adept in AI technologies, the adaptation of interview methodologies becomes crucial. The envisioned shift toward customized assessments signifies a strategic evolution, enabling organizations to gauge candidates' skills within the specific context of their prospective roles. This tailored approach promises to yield a workforce that not only possesses the required competencies but is also closely aligned with the intricacies of AI-driven responsibilities.

16.8.3 HARNESSING VR SIMULATIONS AND DATA-DRIVEN INSIGHTS

Elevating Immersive Assessment: The application of virtual reality (VR) simulations holds significant potential as an immersive assessment tool. By immersing candidates in virtual environments and presenting realistic scenarios, this approach provides a dynamic platform for showcasing problem-solving abilities and capacity in real-world situations. Additionally, the integration of data-driven insights into skill-based interview models stands as a critical advancement. Through the analysis of performance data and trends, organizations can extract valuable insights that contribute to the continuous refinement of evaluation methodologies. This data-centric approach enhances precision and effectiveness in assessing candidates' skills.

16.8.4 CONTINUOUS LEARNING EVALUATION

Adapting to the AI Landscape: The concept of continuous learning evaluation emerges as a pivotal element in the future of skill-based interview models. As the AI landscape evolves rapidly, the adaptability of the workforce is paramount. Continuous assessments that evaluate employees' abilities and their agility in acquiring new skills ensure that organizations maintain a talent pool aligned with the latest advancements. In this way, the future workforce remains up-to-date and equipped to tackle the challenges of an ever-changing AI landscape (Khang & Muthmainnah et al., 2023).

16.8.5 ETHICAL CONSIDERATIONS

Assessing AI Ethics and Problem-Solving: As the ethical implications of AI gain prominence, skill-based interview models are poised to incorporate ethical evaluation components. With the growing significance of ethical decision-making in AI-driven roles, assessing candidates' comprehension of AI ethics and their adeptness in addressing ethical quandaries becomes essential. Integrating ethical considerations into assessment methodologies ensures that candidates not only possess technical skills but also possess the ethical sensibility required to navigate the ethical complexities of the AI-oriented economy.

16.8.6 Trends in Skill-Based Interview Models in the AI-Driven Economy

- **Adaptive Simulations**: Skill-based interview models are increasingly incorporating adaptive simulations powered by AI. These simulations dynamically adjust the complexity of challenges based on candidate responses, providing a more accurate assessment of problem-solving skills and adaptability.
- **Multidimensional Evaluation**: Beyond technical expertise, skill-based interviews now emphasize assessing a wide range of competencies, including communication, collaboration, critical thinking, and emotional intelligence, aligning with the holistic demands of the AI-driven workplace.
- **Algorithmic Fairness**: Addressing algorithmic bias and fairness is a growing concern. Skill-based interview models are incorporating measures to ensure equitable evaluations, such as auditability of AI algorithms and continuous monitoring for bias.
- **Remote Assessment**: The rise of remote work has prompted the development of skill-based interview models that can be conducted virtually. AI-driven platforms offer real-time collaboration tools, enabling candidates to engage in team-based simulations regardless of their geographical location.
- **Personalized Feedback**: AI-powered assessment tools provide candidates with personalized feedback on their performance, helping them understand their strengths and areas for improvement. This feedback-centric approach enhances the learning process and contributes to ongoing skill development.
- **Industry-Specific Scenarios**: Skill-based interview scenarios are becoming more tailored to specific industries and job roles within the AI-driven economy. This trend ensures that candidates are assessed based on their alignment with the unique challenges of their intended roles.
- **Real-Time Analytics**: Skill-based interview platforms are integrating real-time analytics to provide recruiters with insights into candidates' performance. These analytics enable data-driven decision-making and streamline the candidate selection process.
- **Integration with Learning Platforms**: To encourage continuous learning, some skill-based interview models are being integrated with online learning platforms. This integration allows candidates to access relevant resources and materials to enhance their skills before the interview.
- **Augmented Reality (AR) and Virtual Reality (VR)**: AR and VR technologies are being utilized to create immersive skill-based interview experiences. Candidates can engage in realistic simulations that closely mimic the AI-driven work environment.
- **Cognitive and Behavioral Assessment**: AI algorithms are being developed to analyze candidates' cognitive and behavioral responses during skill-based interviews. This approach provides deeper insights into candidates' decision-making processes and behavioral tendencies (Agrawal et al., 2020).

- **Hybrid Models**: Combining traditional interview methods with skill-based assessments is gaining traction. Hybrid models offer a balanced evaluation that considers both candidates' soft skills and their ability to navigate AI-driven complexities (Ramalingam et al., 2021).
- **Continuous Assessment**: Rather than a single-point evaluation, skill-based interview models are evolving to allow continuous assessment over time. This approach accommodates the dynamic nature of the AI-driven economy and candidates' ongoing skill enhancement.

In the dynamic landscape of the AI-driven economy, skill-based interview models are undergoing rapid evolution to meet the multifaceted demands of modern workplaces. These trends highlight the integration of AI technologies, personalized assessment, ethical considerations, and adaptability in creating effective evaluation methods for a skilled workforce (Alekseeva et al., 2021).

16.9   CONCLUSION

Skill-based interview models play a vital role in the AI-oriented economy by aiding firms in the identification and selection of suitable individuals for AI-oriented positions. These models assess the technical proficiency, problem-solving abilities, flexibility, and collaboration skills of applicants, all of which are crucial for achieving success in the AI-driven environment (Khang & Shah et al., 2023).

The utilization of AI-based assessments offers a more precise depiction of a candidate's prospective performance, ensuring that recruiting decisions are in line with the demands of the AI-driven economy. These models additionally play a part in promoting the expansion and prosperity of the AI-focused economy by cultivating a workforce that possesses the necessary skills to effectively traverse the intricacies of AI-driven positions. The use of skill-based evaluations is increasingly recognized as a strategic imperative and a key requirement for achieving sustainable growth and success, given the ongoing transformation of sectors and the redefinition of employment by AI (Khang & Kali et al., 2023).

REFERENCES

Alekseeva, L., Azar, J., Giné, M., Samila, S., & Taska, B. (2021). The demand for AI skills in the labor market. *Labour Economics*, 71, 102002. https://doi.org/10.1016/j.labeco.2021.102002

Belani, H., Vukovic, M., & Car, Z. (2019). Requirements engineering challenges in building AI-based complex systems. *2019 IEEE 27th International Requirements Engineering Conference Workshops (REW)*. IEEE, Jeju, Korea. https://ieeexplore.ieee.org/abstract/document/8933653/

Faruqe, F., Watkins, R., & Medsker, L. (2021). Competency model approach to AI Literacy: Research-based path from initial framework to model. *arXiv [cs.AI]*. https://arxiv.org/abs/2108.05809

Greiff, S., Holt, D. V., & Funke, J. (2013). Perspectives on problem solving in educational assessment: analytical, interactive, and collaborative problem solving. *The Journal of Problem Solving*, 5(2), 71–91. https://doi.org/10.7771/1932-6246.1153

Hitharth Sai, K. B. & Dhanya, N. M. (2022). Predicting employability of candidates: comparative study of different machine learning models. *Proceedings of Emerging Trends and Technologies on Intelligent Systems*, pp. 179–190. Springer, Singapore. https://link.springer.com/chapter/10.1007/978-981-16-3097-2_15

Kashani, S. & Ivry, A. (2021). Deep learning interviews: hundreds of fully solved job interview questions from a wide range of key topics in AI. *arXiv [cs.LG]*. https://arxiv.org/abs/2201.00650

Khamesra, M. (2023). The impact of artificial intelligence on labour economics. *Times of India*, 5, 1. https://timesofindia.indiatimes.com/blogs/voices/the-impact-of-artificial-intelligence-on-labour-economics/

Khang, A., Hajimahmud, V. A., Gupta, S. K., Babasaheb, J., & Morris, G. (2023). *AI-Centric Modelling and Analytics: Concepts, Designs, Technologies, and Applications* (1 ed.). CRC Press, Boca Raton, FL. https://doi.org/10.1201/9781003400110

Khang, A., Kali, C. R., Suresh Kumar, S., Amaresh, K., Sudhansu Ranjan, D., & Manas Ranjan, P. (2023). Enabling the future of manufacturing: integration of robotics and IoT to smart factory infrastructure in industry 4.0. *AI-Based Technologies and Applications in the Era of the Metaverse* (1st ed., pp. 25–50). IGI Global Press, Hershey, PA. https://doi.org/10.4018/978-1-6684-8851-5.ch002

Khang, A., Misra, A., Gupta, S. K., & Shah, V. (2023). *AI-Aided IoT Technologies and Applications in the Smart Business and Production*. CRC Press, Boca Raton, FL. https://doi.org/10.1201/9781003392224

Khang, A., Muthmainnah, M., Seraj, P. M. I., Yakin, A. A., Obaid, A. J., & Panda, M. R. (2023). AI-aided teaching model for the education 5.0 ecosystem. *AI-Based Technologies and Applications in the Era of the Metaverse* (1st ed., pp. 83–104). IGI Global Press, Hershey, PA. https://doi.org/10.4018/978-1-6684-8851-5.ch004

Khang, A., Rani, S., Gujrati, R., Uygun, H., & Gupta, S. K. (2023). *Designing Workforce Management Systems for Industry 4.0: Data-Centric and AI-Enabled Approaches*. CRC Press, Boca Raton, FL. https://doi.org/10.1201/9781003357070

Khang, A., Shah, V., & Rani, S. (2023). *AI-Based Technologies and Applications in the Era of the Metaverse* (1 ed.). IGI Global Press, Hershey, PA. https://doi.org/10.4018/978-1-6684-8851-5

Khang, A. (2024). *AI-Oriented Competency Framework for Talent Management in the Digital Economy: Models, Technologies, Applications, and Implementation*. CRC Press, Boca Raton, FL. https://doi.org/10.1201/9781003440901

Munir, F. (2022). More than technical experts: engineering professionals' perspectives on the role of soft skills in their practice. *Industry and Higher Education*, 36(3), 294–305. https://doi.org/10.1177/09504222211034725

Ramalingam, H., Usmani, R. S. A., Hashem, I. A. T., & Pillai, T. R. (2021). A hybrid model to profile and evaluate soft skills of computing graduates for employment. *International Journal of Advanced Computer Science and Applications: IJACSA*, 12(7), 2021. https://doi.org/10.14569/ijacsa.2021.0120761

Roulston, K. (2011). Working through challenges in doing interview research. *International Journal of Qualitative Methods*, 10(4), 348–366. https://doi.org/10.1177/160940691101000404

Sharma, M., Luthra, S., Joshi, S., & Kumar, A. (2022). Implementing challenges of artificial intelligence: evidence from public manufacturing sector of an emerging economy. *Government Information Quarterly*, 39(4), 101624. https://doi.org/10.1016/j.giq.2021.101624

Verma, A., Lamsal, K., & Verma, P. (2022). An investigation of skill requirements in artificial intelligence and machine learning job advertisements. *Industry and Higher Education*, 36(1), 63–73. https://doi.org/10.1177/0950422221990990

Zhao, S. & Du, J. (2011). The application of competency-based talent assessment systems in China. *Human Systems Management*, 30(1–2), 23–37. https://doi.org/10.3233/hsm-2011-0738

17 Design and Modeling of Artificial Intelligence-Oriented Competency Framework for Information Technology Sector

Alex Khang

17.1 INTRODUCTION

A structural approach with the resources of various competency models is visible in the real world for roles in the pipeline of career development as well as on feasibility, privacy, user safety, and social issues are studied and proposed with the help of AI technology. The competency framework is a managerial model that is created and sustained by human resource management and talent development teams. It is a shared, immersive resource where people can interact with each other and with digital objects and environments in real time.

The AI-oriented competency framework is often thought of as a digital extension of modern competency adding skills and knowledge related to sectors in the AI ecosystem, and it has the potential to transform the way we work, communicate, and collaborate together in the real world. A structural approach for a proposed framework supported by live cases of the workforce management and development strategy will enhance the use of emerging technologies, which will increase the chances for employees to adapt technologies in the AI ecosystem for not only IT companies but also all organizations in the world (Khang, 2024).

The chapter also aims to explore the different methods that can be used to assess whether an individual can meet the criteria required by the software development industry. In our study, the IT software company requires a framework to fill the gaps between competency components like mandatory roles and required responsibilities, core knowledge, key skills, certifications, contributions, and experiences. This includes starting with new hires, appointing full-time or part-time employees, joining in development projects, and until the product is delivered to the client, it is necessary to implement an appropriate employee evaluation process with the help of proper AI techniques and data analysis methods.

17.2   WHY AI-ORIENTED COMPETENCY FRAMEWORK

For the past years, AI has completely transformed most business activities like forecasting, production, inventory, transportation, and sales planning. Companies may boost their productivity, reduce costs, and increase profitability while bringing high quality, good service, and responsiveness to clients by utilizing AI. Therefore, careful hiring, training, rotating, and delivering manpower between various stakeholders are necessary to effectively deploy an AI-oriented competency framework in your enterprise (Khang & Santosh et al., 2023).

As more and more AI applications appear in real life, many companies face a lot of challenges in using AI technologies including forecasting sales, resource planning, manufacturing, transportation cost, and customer segmentation on the analysis of business patterns, balancing supply and demand, and partner relationship management. The goals of this chapter are to understand what the skills of an artificial intelligence (AI) ecosystem are and how to design a competency framework in a human capital management (HCM) system, to help a company manage and develop more appropriate human resources (Khang & Muthmainnah et al., 2023).

The AI-Oriented Competency Framework is really a straightforward idea, it implies using any skill set along with the actual experience of a particular job role, adding the skills of emerging technologies, and then merging them into the framework. This supplement includes bits of knowledge and skills from fundamental to the professional, from the design to implementation phase that are more or less related to any component in the AI ecosystem (Arpita et al., 2023). The fact is that applying the advancements of an AI-Oriented Competency Framework to personnel management yields many astonishing advantages. We have all seen these advantages and benefits with employees, managers, partners, customers, and business owners in the strategy to move your organization toward the global market (Khang & Shah et al., 2023).

17.3   RELATED WORK

All of us know that the competency framework is a part of the workforce management and development system, it consists of hard skills, soft skills, management skills, life skills, social skills, behaviors, tools, and emerging platforms like artificial intelligence is preferred. Therefore, the managers will try to outline the role and importance of the application of AI technology in the entire journey to retain and develop the high-tech workforce (Pallavi et al., 2023).

The timely and exact detection of outdated knowledge and skills should be fully enhanced into the learning planning because a personalized learning path plays an essential role in career development. Therefore, a person who has no plan or slow addition of emerging skills can reduce job adaptability, or reduced income significantly and even increase the likelihood of job loss. Therefore, this study on modern competency framework by AI-oriented competency ranking techniques has been actively implemented in the Information Technology (IT) industry (most scenarios in this chapter are based on the IT industry).

This chapter presents skills taxonomy, Traditional Competency Framework (TCF - competency framework without skills in an AI ecosystem), and AI-oriented Competency Framework (AIoCF), a comparative analysis of two competency models, job description, and job assessment in an IT organization. The chapter contains the sequential processes in designing three competency frameworks, the assessment tools used, and screen grabs of output, interface, and essential system prerequisites required to implement the system properly (Khang & Shashi et al., 2023).

17.4　SKILLS TAXONOMY

A Skills Taxonomy System (STS) is a core component of a competency framework that has a larger number of skills than the available skills in an organization. It has code, name, and related attributes belonging to a particular industry (SCEDEX, 2021).

17.4.1　INDUSTRY CLASSIFICATION

The skills in the STS are not only organized under industry, sub-industry, platform, application, skill group, skill category, skill set, and skill chain but also arranged by critical, common or rarely, career paths, learning paths, and recruitment requirements as a Skills Ecosystem for Industry 4 (SEFIX) model (AIoCF, 2021). An STS is designed with many industries and its sub-industries as shown in Figure 17.1

17.4.2　WORKING LAYER

Each industry or sub-industry layer usually contains skills for working layers of career paths, jobs, roles, projects, or even tasks as shown in Figure 17.2.

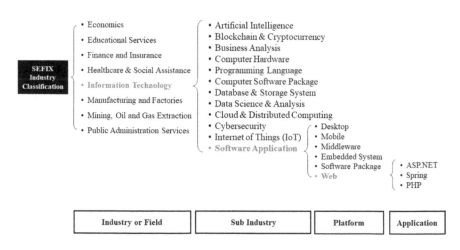

FIGURE 17.1　Industry and its sub-industry classification (Khang, 2021).

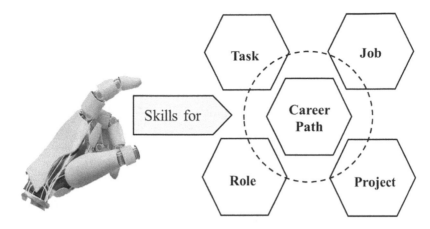

FIGURE 17.2   Skills for working layers in an industry or a sub-industry (Khang, 2021).

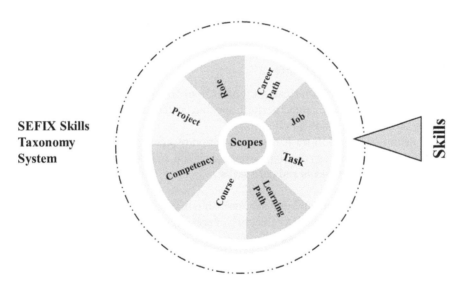

FIGURE 17.3   Skills taxonomy scopes (Khang, 2021).

17.4.3 Scope of Skills

The industry, sub-industry, platform, the application later, or even its components in STS contain optional skills or nice-to-have skills that act as additional competencies (Pooja & Khang, 2023). We convey STS using skills transforms and sharp integrating to offer a consistent bridge between the AIoCF System and STS, and different fusion rules have been launched for integrating updates which can be used for working and training scopes as shown in Figure 17.3.

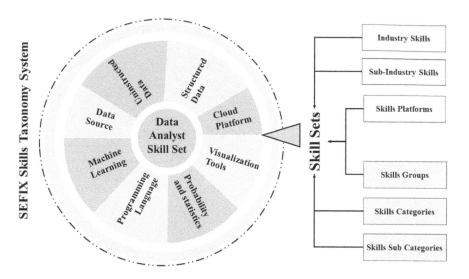

FIGURE 17.4   Skill set for data analyst role and data analytics project (Khang, 2021).

17.4.4   DEFINITION OF SKILL SET

Skill set refers to a particular category of an individual's skills, experiences, and abilities necessary to acquire and perform specific current or future activities such as working or learning as shown in Figure 17.4.

The definition of a skill set is the combination of knowledge, personal qualities, and abilities that people can develop through their life and work, it typically contains two types of skills: soft skills and hard skills.

- **Hard Skills**: are quantifiable and teachable things, they include the professional knowledge, specific engineering or non-engineering understanding, and abilities required for a work or a job. In the IT industry, hard skills are computer programming, database design, cloud platforms, big data analysis, etc.
  - Python/R Programming.
  - Database Architect Design.
  - Cloud Microservice Implementation.
  - Web API Development.
- **Soft Skills**: are societal, public, and personal skills, they are somewhat really difficult to quantify and define because they are related to people's personalities and ability to treat others in real life. In the era of AI and the digital economy, the soft skill set includes good communication, computer tools, listening, public speaking, critical thinking, emotion, empathy, self-control, and problem-solving abilities, among other skills. For example:
  - Communication skills.
  - Computer skills.

- Interpersonal skills.
- Leadership skills.
- Management skills.
- Active listening skills.
- Design Thinking,
- Problem-solving skills.
- Customer service skills.

17.4.5 Types of Skill Set

A skill set refers to a specific area of competence, knowledge, experience, and abilities required to do a corresponding work. In an organization, employees can typically use a collection of skills for a given work, and skills are added to a skill set collected from various fields with different levels (Ashwini & Khang, 2023). For instance, in the IT software sector, there are many skill sets created for managerial purposes such as recruitment-based skills, training-based skills, job-based skill sets, project-based skill sets, or role-based skill sets as shown in Figure 17.5.

17.4.6 Skill Chain Concept

A skill chain refers to a group of mandatory, dependent, relevant, optional, nice-to-have skills required to perform a respective task. Typically, a skill chain is an essential skill for someone working in a cross-platform environment or in an ecosystem that involves multiple technologies as shown in Figure 17.6.

- **Mandatory Skills**: are must-have knowledge, skills, and competences that a person needs for the job. These knowledge and skills are weighted more

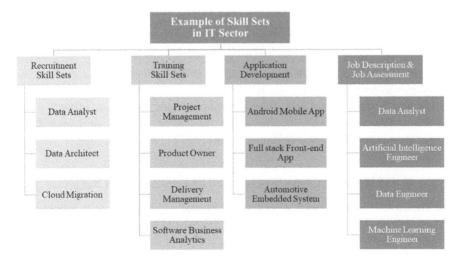

FIGURE 17.5 Example of skill sets in the information technology sector (Khang, 2021).

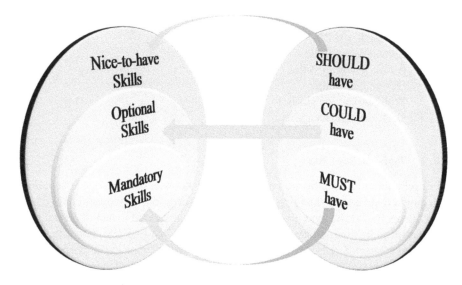

FIGURE 17.6   Example of a skill chain in the information technology sector (Khang, 2021).

heavily and will usually affect the size of your talent pool, which is indicated on the key knowledge and assessment page.
- **Optional Skills**: usually are knowledge, skills, and competences that may
be required or offered as additional options when working in an occupation
depending on the request of client, employer, interviewer, or on the working
context or on the culture.
- Nice-to-have skills are those extra skills not required to carry out the core
functions of the job better or for the product to be easily used and more
user-friendly.

17.4.7   AN EXAMPLE OF SKILL SET

In the new version of the SEFIX model, one key aspect of the AIoCF is its reliance
on role-based skill sets, which can include everything from knowledge, performance
capabilities to tools and even flight hours in a specialized field. These skill sets are
increasingly being updated and aligned on AI platforms, which can provide a clear
and transparent way to help managers plan for human resource development and
coordination more effectively as shown in Figure 17.7.

17.5   AI-ORIENTED COMPETENCY FRAMEWORK (AIOCF)

The AIoCF will provide users with an accurate and reliable competency framework
that can predict and identify various gaps in employee's skill sets. The results show
that both models performed quite well, but in the era of an AI-oriented economy, the
level of approach and assistance for processing situations in the work is 30% for TCF
and 70% for AIoCF as shown in Figure 17.8.

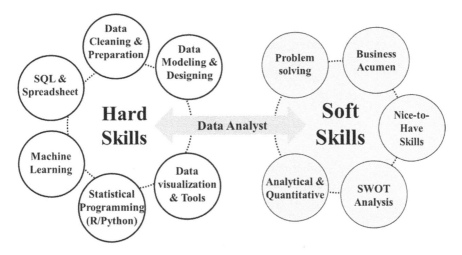

FIGURE 17.7 The list of skills for the data analyst role (Khang, 2021).

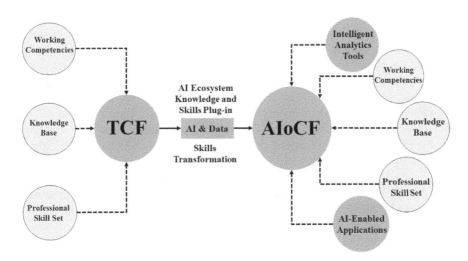

FIGURE 17.8 The list of skills of TCF and skills++ of AIoCF in the taxonomy system (Khang, 2021).

17.5.1 ARCHITECTURE OF AIoCF COMPETENCY MODEL

In order to create a new fused model to incorporate the best competency components from the traditional competency framework and components in the AI ecosystem, competency framework fusion aims to merge skills from several competency components (skills) collected from the same business domain or platform as shown in Figure 17.9.

FIGURE 17.9   The list of skill sets of TCF in the taxonomy system (Khang, 2021).

Professional skills – are hard skills and soft skills required for work or a job. In the IT industry, professional skills usually are computer programming, database design, cloud platforms, big data analysis, communication skills, foreign languages, design thinking, management skills, etc.

With the AIoCF model, the combined components of AI and Data fields to TCF are of higher importance than any of the source components. The key step in competency framework fusion based on AI ecosystem skills is that of knowledge combination, namely, the process of merging the knowledge in an appropriate way in order to obtain the best skill set in the fused competency framework as shown in Figure 17.10.

- **AI Skills**: are skills related to an AI ecosystem, they include professional knowledge, specific AI-engineering understanding, and the abilities required for AI-related jobs. In the IT industry, AI skills are AI science, mathematics, AI programming, machine learning (ML) model design, deep learning (DL) framework, AI tools, robotics, etc.
  - AI Programming languages.
  - ML and DL models.
  - ChatGPT applications.
  - Robotics.
- **Data Skills**: are skills related to the Data ecosystem, they include professional knowledge, specific data science, statistics, data engineering, data analytics, and data visualization understanding, and abilities required for data-related jobs. In the IT industry, hard skills are computer programming, database design, cloud platforms, big data analysis, etc.
  - Python/R Programming.
  - Data modeling and architecture.

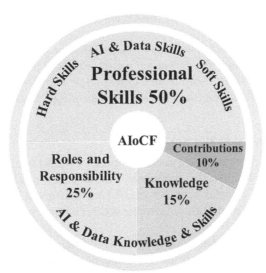

FIGURE 17.10 The list of skill sets of AIoCF in the taxonomy system (Khang, 2021).

- Data warehouse and data migration.
- Data visualization and tools.

The competency framework designed uses a type of matrix called a spreadsheet to make classifications of competency components based on 4 or greater than 4 categories which are;

- Role and Responsibilities
- Professional Skills
- Soft Skills
- Contributions

17.5.1.1 Responsibilities Component

Each role is responsible for defining one responsibility or many responsibilities within the organization, these are teams or committees tasked to do them. He and she can go about it in various ways, using several models, methodologies, tools, or software, and depending primarily on the types of the business or the operations of the organization, as well as the missions or goals, such as some of the common activities undertaken in the process of software development.

17.5.1.2 Roles Component

In the management model, a job role is not a responsibility, and vice versa, some people mistake a role as the job title, but there is more to it than just a designation. The whole management process in business generally and the Software engineering industry, in particular, is comprised of different roles. It is possible for two or more people to have one or the same role, depending on what they do; but there are roles that are solely focused on the project management side of things, while others are more on the technical side.

The most common roles you will find in an organization of Software Development and Services setup include the Business Leader, Project Manager, Data Analyst, Software Business Analyst, and Team Leader. These roles then come with corresponding responsibilities or the specific Key Results or KPI results that are expected from these roles.

17.5.1.3   Professional Skills Component

The Professional Skills component in the Competency Model is an ability and capacity acquired through deliberate, systematic, and sustained effort to smoothly and adaptively carry out complex activities or job functions involving technical skills, business skills, management skills, cognitive skills, and personal skills.

- **Specialized Engineering Skills**: Specialized Engineering skills may relate to a specific job, task, enterprise discipline, or area of particular knowledge and skills. Depending on their experience, interest, and further learning, individuals may possess one or several specialized skills.
- **Advanced Skills**: People with advanced skills or expert skills will be rewarded and the company shall seek to recruit, retain, and reward our employees to match the task in the short term as well as in the long term.
- **Subject Matter Expert and Proficient (SME)**: An SME is a person who has special skills or knowledge on a particular job or topic. SMEs are highly accessed by instructional designers to extract intelligence when developing courseware and learning programs.
- **Business Skills**: The grade of the business/domain skills is to give an employee a competitive advantage as they compete for positions in the best projects and eventually for the highly coveted jobs after.
- **Management Skills**: Management skills can be nearly anything that enables you to manage others effectively. While some skills will vary based on your industry, there are several that are universal across nearly every work environment such as Motivation, Problem Solving, Professionalism, Communication, Technical Skills, Innovation, and Negotiation.
- **Leadership Skills**: Leadership skills are skills that an employee uses when organizing other people to reach a shared goal such as Delegation, Effective Feedback, Conflict Resolution, Organization, and Team building.
- **Soft Skills**: Personal skills are categorized into two ways: soft skills and hard skills. Soft skills are attitude, personality, emotions, habits, natural languages, communication style, and social manners.

In contrast, professional skills are more specific and are often associated with a task or activity, most times job-related professional skills.

17.5.1.4   Knowledge and Qualification Component

In the Competency Model, AIoCF has defined the Knowledge and Qualification component as what refers to awareness of or familiarity with various technologies or platforms, business domains, procedures, or ways of creating things. AIoCF has divided the knowledge component into three objects:

- Specialized Knowledge and Methodology
- Education Background
- Professional Certifications

And based on three scopes:

- Personal Knowledge (Education and Qualifications)
- Procedural Knowledge (Methodologies and Knowledge)
- Propositional Knowledge (Professional Certifications)
- Technology (Specialized) Knowledge
  - In the context of the IT business enterprise or software engineering, Technology knowledge refers to how to create software products, software solutions, and software services, such as how to develop Enterprise Resource Planning software.
  - Methodology Knowledge: Methodology knowledge relates to academic programs, technology and business trends, resource references, and scientific facts of industry in the world such as an employee understands how Methodology can improve the performance and storage capacity of a huge database.
  - Qualification and Certifications: Because the qualifications and certifications prove that an employee has successfully completed some testing or assessment which will meet the skills of job requirements.

AIoCF adds the value of qualifications and certifications and provides a context for positioning them within the skills needed by industry and business.

17.5.1.5 Business Contribution Component

In the Competency Model, AIoCF has provided objective evidence of progress toward achieving a desired OKR or KPI result (by points or efforts) via working in scope or out of scope of projects. AIoCF has divided the Contribution component into two forms:

- Number of flight hours in the profession at the organization currently working
- Expertise contributions are individual efforts or time worked in specialized activities of one project or more projects which is in scope or out scope of their department.
  - Project Contributions are individual mandatory/voluntary efforts or time contributed to one project or more projects which is in scope or out scope of their department.
  - Personal contribution means employees giving something away while working whether it is knowledge, skills, or experiences.
  - Training/Coaching/Mentoring for subordinates or team members.

Depending on the business model, technology platform, culture, type of job, etc. each section contains one or more categories with or without AI and Data knowledge and skills as shown in Figure 17.11.

However, the AIoCF design also uses a type of matrix called an Excel sheet to make classifications of competency components based on 5 categories which are; (a) Role and Responsibilities (R&R) – (b) Professional Skills – (c) Soft Skills – (d) Qualification and – (e) Contributions as shown in Figure 17.12.

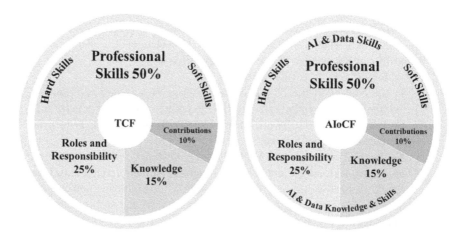

FIGURE 17.11   The comparison between TCF & AIoCF architecture (Khang, 2021).

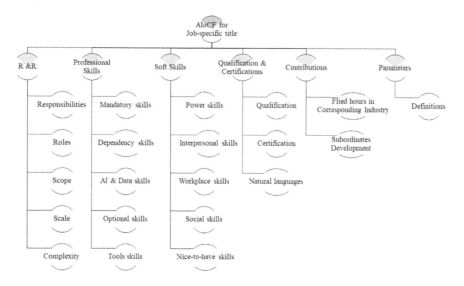

FIGURE 17.12   The components of AIoCF architecture.

Besides common competency components, there are some new components in AIoCF such as Dependency skills, Power skills, Social skills, and Workplace skills as follows.

- **Dependency Skills**: are specialized skills that require a good understanding of the product or service development lifecycle, as well as knowledge of the tools and technologies used to design, develop, test, deploy, and maintain a software product in various environments.
- **Power Skills**: are skills of adaptability, creativity, decision-making, and time management, which are person qualities and traits that machines simply can't duplicate. Also, note that these qualities will help set a person apart from others, helping managers advance their workforce development.
- **Social Skills**: are skills that humans use to communicate with others daily in a variety of ways including verbal, nonverbal, written, and visual. They are any competence facilitating interaction and communication with others where social rules and relations are created, communicated, and changed in verbal and nonverbal ways. The process of learning these skills is called socialization. Lack of such skills can cause social communication awkwardness.
- **Workplace Skills**: are the personal skills that ensure an employee does his or her job well, such as being adept at listening, teamwork, time management, solving problems, or mentoring others.

17.5.2 Patterns of AIoCF Model

It is a multiplatform (multirole) method that is well adapted to handle the various job roles. The merging of primary domain skills and child-domain skills enables the assessment of competency while maintaining the integrity of the competency framework. These child-domain skills can be properly combined to create a new competency framework called a child-domain competency framework, ensuring that the skills belong in the specific child-domain (Shashi & Khang, 2023).

Recent years have seen the study of multirole representation in a career path by some software developers as shown in the pattern of the Project Manager Role in Figure 17.13, who have found that multiplatform skills can be valuable in a variety of job positions, including managerial jobs as shown by the pattern of the Project Manager Role in Figure 17.14.

Similarly, a new pattern of the AIoCF competency framework for the Data analyst role is created using transform from the primary domain framework and adding some specific skills of specialization in data analytics as shown in Figure 17.15.

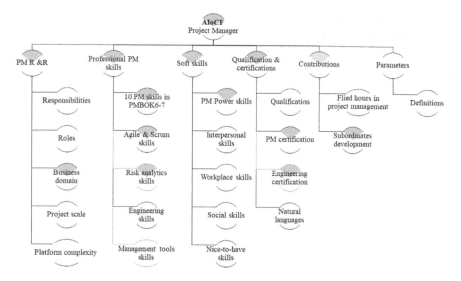

FIGURE 17.13   The components of AIoCF architecture of software engineer (Khang, 2021).

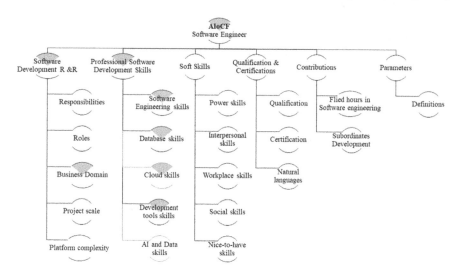

FIGURE 17.14   The components of AIoCF architecture of project managers (Khang, 2021).

Note that a competency framework can also be created using transformations from the primary domain framework but bring together some skills and factors such as scope, size, complexity, culture, and type of participants into the respective key skill. For example, incorporating the scope, scale, and complexity mode into a corresponding responsibility level is shown in Figure 17.16.

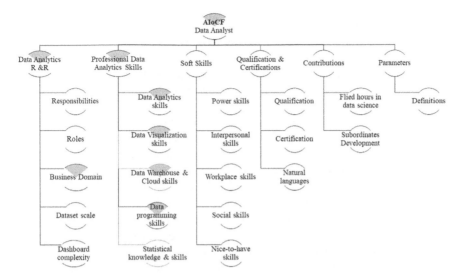

FIGURE 17.15   The components of AIoCF architecture of the data analyst (Khang, 2021).

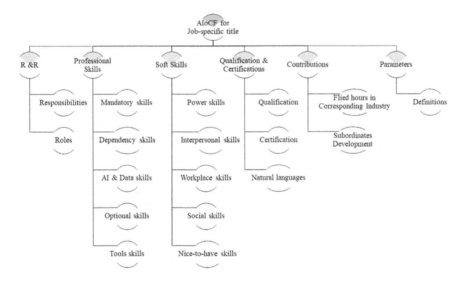

FIGURE 17.16   The components of other AIoCF architecture.

17.6   AIOCF METHODOLOGY

The attributes of each skill in the AIoCF reflect the extent to which an individual's performance is affected when participating in the competency assessment. They are the most common set of characteristics in many fields of business and technology that AIoCF uses as the foundation for effective management of labor resources in the era of the digital economy.

17.6.1 The Levels of Role and Responsibility

AIoCF has defined nine levels of responsibility (responsibility level), from level 0 to level 8 (Not Required – Follow – Assist – Apply – Create – Design – Initiate – Command – Orientate) to make them suitable for mapping the stages within a career path as shown in Figure 17.17a and b for implementation template.

- **Level 8 – Orientate**: Performs a top leadership role with overall responsibility and strategies for the whole business domain in the organization while supervising teams of professionals and the board of directors.
- **Level 7 – Command**: Performs a commander role with overall responsibility for projects based on multiple platforms and business domains while supervising teams of professionals and middle management.
- **Level 6 – Initiate**: Performs a senior management role with overall responsibility for projects while supervising teams of professionals.
- **Level 5 – Design**: Holds a senior position and may act as a manager or consultant. Independently conceives programs and defines problems to be studied or objectives to be achieved. Involved in long-term planning for the organization.
- **Level 4 – Create**: Works in a complex technological area without direct supervision. Receives broad task objectives, is responsible for significant technical decisions, and may train other professionals.
- **Level 3 – Apply**: Performs technical tasks with limited direct supervision and solves problems based on principles of applied science technology. Also make some decisions for which they assume responsibility.
- **Level 2 – Assist**: Performs technical procedures or solutions with occasional direct supervision.
- **Level 1 – Follow**: Receives on-the-job training in a variety of new environments while working under close supervision as they gain experience, typically working on small tasks within larger projects.

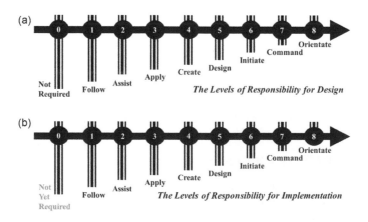

FIGURE 17.17   (a) Nine levels of role and responsibility in design mode. (b) Nine levels of role and responsibility in implementation mode.

- **Level 0 – Not Required**: Not required to perform any technical procedure.
- **Level 0 – Not Yet Required**: Not yet required to perform any technical procedure at this time as shown in Figure 17.17b.

Note that the levels of responsibility in some job roles are created based on a combination of many factors such as scope, scale, complexity, culture, and type of participants. For instance, include size and mode in the level 3 and level 2 as follows:

- **Level 3 – Apply**: Performs technical tasks (*of the large and highly complex programs*) with limited direct supervision and solves problems based on principles of applied science technology. Also make some decisions for which they assume responsibility.
- **Level 2 – Assist**: Performs technical procedures (*of the large and highly complex projects*) with occasional direct supervision.

In addition, within a working organization, an individual is to be found in a wide range of levels, depending on their characteristics, the levels of responsibility in AIoCF are identified by using the six characteristics (SEDEX, 2021) in order as

- **The Degree of Autonomy**: A degree or level of freedom and discretion allowed to an employee over his or her job. As a general rule, jobs with a high degree of autonomy engender a sense of responsibility and greater job satisfaction in the employee.
- **The Scope of Influence**: Influence is the level of involvement an employee has and impact is the ability of the stakeholder to bring out the desired change. Influence is the power to make other people agree with your opinions or many other employees have an influence on the levels of decision.
- **The Level of Complexity**: The level of complexity refers to the duties, the degree of autonomy, and the scope of responsibilities in the business unit or organization. In the Software industry, there are generally three levels of complexity: the entry and learning level, the independent level, and the team leader or specialist level.
- **The Volume of Knowledge**: Knowledge is a crucial component of the competence framework and it is authenticated by AIoCF. In the Software industry, to be competent and effective in any role an individual needs to get a mix of specific methodologies, technology knowledge, and the business domain knowledge.
- **The Experience in Business**: Experience demonstrates the ability to apply knowledge and skills in a practical environment at the business unit or organization. In real project(s), an individual has a skill at a particular level and has been practiced in a real-world situation.
- **The Degree of Achievements in Expertise Sharing**: Sharing experience/ knowledge at a workshop or seminar within the business unit or organization. An employee has activities in the professional community such as academic society and committees.

17.6.2 The Levels of Skills Proficiency

The AIoCF proficiency level has used seven levels, from level 0 to level 7 (Not Required – Fundamental Awareness – Limited Experience – Intermediate – Advanced – Expert - Master) of proficiency for applying the practiced and phrases skills to facilitate their use in ranking and competencies as shown in Figure 17.18a and b for implementation template.

- **Level 6 – Master (Mastering all Techniques and Solutions)**: In-depth knowledge and skills as an Insider, Scientist, Researcher, Producer, Creator, Inventor, and Core Handler. Acquire complete knowledge and skills in (a subject, technique, or art).
- **Level 4 – Expert (Recognized Authority)**: A prolonged or intense experience through practice and as a consultant in a particular field. Applies the competency in exceptionally difficult situations. You can provide guidance, troubleshoot, and answer questions related to this area of expertise and the field where the skill is used. Focus is strategic. You have demonstrated consistent excellence in applying this competency across multiple projects and/or organizations.
- **Level 4 – Advanced (Apply Theory)**: Has knowledge and experience in the application of this skill. Is a recognized specialist and advisor in this skill including user needs, generation of ideas, methods, tools, and leading or guiding others in best practice use. Be able to perform the actions associated with this skill without assistance. Applies the competency in considerably difficult situations. Are certainly recognized within your immediate organization as "a person to ask" when difficult questions arise regarding this skill. The focus is on broad organizational/professional issues.

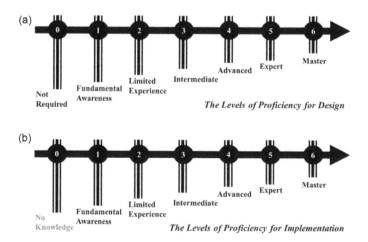

FIGURE 17.18 (a) The eight levels of proficiency in design mode. (b) The eight levels of proficiency in implementation mode.

- **Level 3 – Intermediate (Practitioner/Practical Application)**: Shares knowledge and experience of the skill with others, including tools and techniques, defining those most appropriate for the environment. Able to successfully complete tasks in this competency as requested. Applies competency in difficult situations. Help from an expert may be required from time to time, but you can usually perform the skill independently. The focus is on applying and enhancing knowledge or skills.
- **Level 2 – Limited Experience (Working)**: Applies knowledge and experience of the skill, including tools and techniques, adopting those most appropriate for the environment. Have the level of experience gained in a classroom and/or experimental scenarios or as a trainee on the job. Can apply the competency in the simplest situations. You are expected to need help when performing this skill. The focus is on developing through on-the-job experience. You understand and can discuss terminology, concepts, principles, and issues related to this competency.
- **Level 1 – Fundamental Awareness (Basic Knowledge)**: Have a common knowledge or an understanding of basic techniques and concepts. The focus is on learning and researching.
- **Level 0 – Not Required**: Not required to have any knowledge and skill.
- **Level 0 – No Knowledge**: Have no knowledge and experience as shown in Figure 17.18b.

17.7 CHALLENGES

In order to create an AIoCF job description for a specific job title, you can follow the step-by-step technique using the Excel format as shown in Table 17.1, and the example of a job description is shown in Figure 17.2.

Note that the child-domain competency framework is created using transform from the primary domain framework and retains all of the key skills and their levels (Table 17.2).

TABLE 17.1
AIoCF Job Description

Section 1: Role and Responsibilities			
No	Code	Type	Description
1	RR01	Mandatory	• Description for Responsibility 1 • Description for Responsibility 2
2	RR02	Optional	• Description for Role 1 • Description for Role 2
3	RR03	Optional	• Description for Scope 1 • Description for Scope 2
4	RR04	Optional	• Description for Scale 1 • Description for Scale 2
5	RR05	Optional	• Description for Complexity 1 • Description for Complexity 2

(Continued)

TABLE 17.1 (*Continued*)
AIoCF Job Description

Section 2: Professional [Business Domain Name] Skills

No	Code	Skill Category	Type	Description
1	PS01	Skill category 1	Mandatory	• Description for Skill 1 • Description for Skill 2
2	PS02	Skill category 2	Optional	• Description for Skill 1 • Description for Skill 2
3	PS03	Skill category 3	Mandatory	• Description for Skill 1 • Description for Skill 2

Section 3: Soft Skills

No	Code	Skill Category	Type	Description
1	SS01	Skill category 1	Mandatory	• Description for Skill 1 • Description for Skill 2
2	SS02	Skill category 2	Optional	• Description for Skill 1 • Description for Skill 2
3	SS03	Skill category 3	Optional	• Description for Skill 1 • Description for Skill 2

Section 4: Contributions (Optional)

No	Code	Type	Description
1	CW01	Optional	• Description for Contribution 1 • Description for Contribution 2
2	CW02	Optional	• Description for Contribution 1 • Description for Contribution 2

TABLE 17.2
Sample AIoCF Job Description of Data Engineer (DE)

Section 1: Role and Responsibilities
Implementing all the processes from data collection, cleaning, and preprocessing, to training data sets and deploying them to production, including but not limited to the following responsibilities:

No	Code	Type	Description
1	RR01	Mandatory	• Collaborating with data scientists and architects on projects • Data ingests, transforms, and integrates structured data • Delivers data to a scalable data warehouse platform • Interpreting trends and patterns • Data modeling, database architecture design • Analyzing and organizing raw data
2	RR02	Mandatory	• Building data systems and pipelines • Preparing data for prescriptive and predictive modeling • Building algorithms and prototypes • Deploying data sets and models to the environments

(Continued)

TABLE 17.2 (*Continued*)
Sample AIoCF Job Description of Data Engineer (DE)

3	RR03	Optional	• Managing available data resources
			• Managing analytical/integration tools, databases, warehouses
			• Maintenance of the data sets and models to the environments
4	RR04	Optional	• Understand how to optimize data retrieval
			• Understand how to develop dashboards and visualizations
5	RR05	Optional	• Help the other DE understand data sets and data process flow
			• Lead other engineers to implement the data solution
			• Engage with experts & engineers to deliver interpretable results

Section 2: Professional [Data Engineering] Skills

No	Code	Skill Category	Type	Description
1	PS01	Data Engineering	Mandatory	• Data integration process and implementation
				• Data mining process and implementation
				• Data analysis and data visualization
				• Big data process and management
				• Data architecture design and implementation
				• Data warehouse platforms and solutions
				• Data pipeline paradigms and tools
				• Relational and NoSQL database platforms
				• Software engineering and specialized tools
2	PS02	AI Engineering	Optional	• AI and machine learning
				• Computer vision algorithms and models
				• IoT-based data solutions
				• Data visualization tools
3	PS03	Other	Optional	• Cloud computing
				• Blockchain technology
				• Office tools

Section 3: Soft Skills

No	Code	Skill Category	Type	Description
1	SS01	Communication	Mandatory	• Interpersonal communication
				• Teamwork & motivation
2	SS02	Creativity	Optional	• Critical thinking
				• Design thinking
				• Intellectual curiosity
3	SS03	Leadership	Optional	• Proactive problem solving
				• Risk management
				• Leadership

Section 4: Certification and Qualification (Optional)

No	Code	Type	Description
1	SQ01	Optional	International or internal certifications
2	SQ02	Optional	Education background and advanced degree

(*Continued*)

TABLE 17.2 (*Continued*)
Sample AIoCF Job Description of Data Engineer (DE)

		Section 5: Contributions (Optional)	
No	Code	Type	Description
1	SC01	Mandatory	Guided/mentored/coached for junior engineers
2	SC02	Optional	Flight hours as a data engineer in the data ecosystem

17.8 FUTURE SCOPE

Nowadays, the challenges of the AIoCF model when incorporating AI elements into the competency framework to assess employee applicability may not yet be appropriate for some occupations that are less influenced by AI (Khang & Rani et al., 2023). Therefore, when considering the fact that competency assessment systems, especially non-IT enterprises, are extremely ideological and lack a scientific basis, it would be a reasonable decision to propose the requirement for the AI-related assessment criteria to be optional and weighted with the lowest weight to ensure that the competency framework is not dominated by these factors (Snehal et al., 2023).

17.9 CONCLUSION

This model involves analyzing the inevitability of the AIoCF competency framework and the existing assessment model (TCF) in the current human resource management system. Therefore, planning and deploying the AIoCF model on a massive amount of employees and the necessary tools in your organization now necessitate an evaluation of the consent requirements of managers (Babasaheb et al., 2023a). Once everything is ready for implementation, the drawbacks of the existing approaches will be solved by the AIoCF model and its applications. Managers or leaders can reduce relative defects in the competency assessment process and provide predictions on the potential development of career paths for their employees (Babasaheb et al., 2023b).

REFERENCES

AIoCF. (2021). AI Skills-based competency ecosystem in digital economy. *The AI-oriented Competency Model for Digital Economy 5.0*. Retrieved from https://scedex.com/quickstart.htm

Arpita, N., Satpathy, I., Patnaik, B. C. M., Sukanta Kumar, B., & Khang, A. (2023). Impact of artificial intelligence (AI) on talent management (TM): a futuristic overview. *Designing Workforce Management Systems for Industry 4.0: Data-Centric and AI-Enabled Approaches* (1st ed., pp. 32–50). CRC Press, Boca Raton, FL. https://doi.org/10.1201/9781003357070-9

Ashwini, Y. S. & Khang, A. (2023). Challenges faced by marketers in developing and managing contents in workforce development system. *Designing Workforce Management Systems for Industry 4.0: Data-Centric and AI-Enabled Approaches* (1st ed., pp. 332–359). CRC Press, Boca Raton, FL. https://doi.org/10.1201/9781003357070-18

Babasaheb, J., Sphurti, B., & Khang, A. (2023a). Industry revolution 4.0: workforce competency models and designs. *Designing Workforce Management Systems for Industry 4.0: Data-Centric and AI-Enabled Approaches* (1st ed., pp. 14–31). CRC Press, Boca Raton, FL. https://doi.org/10.1201/9781003357070-2

Babasaheb, J., Sphurti, B., & Khang, A. (2023b). Design of competency models in the human capital management system. *Designing Workforce Management Systems for Industry 4.0: Data-Centric and AI-Enabled Approaches* (1st ed., pp. 32–50). CRC Press, Boca Raton, FL. https://doi.org/10.1201/9781003357070-3

Khang, A. (2021). "Material4Studies," Material of Computer Science, Artificial Intelligence, Data Science, IoT, Blockchain, Cloud, Metaverse, Cybersecurity for Studies. Retrieved from https://www.researchgate.net/publication/370156102_Material4Studies

Khang, A., Kali Charan, R., Surabhika, P., Pokkuluri Kiran, S., & Santosh Kumar, P. (2023). Revolutionizing agriculture: exploring advanced technologies for plant protection in the agriculture sector. *Handbook of Research on AI-Equipped IoT Applications in High-Tech Agriculture*, pp. 1–22. IGI Global Press, Hershey, PA. https://doi.org/10.4018/978-1-6684-9231-4.ch001

Khang, A., Muthmainnah, M., Seraj, P. M. I., Yakin, A. A., Obaid, A. J., & Panda, M. R. (2023). AI-aided teaching model for the education 5.0 ecosystem. *AI-Based Technologies and Applications in the Era of the Metaverse* (1st ed., pp. 83–104). IGI Global Press, Hershey, PA. https://doi.org/10.4018/978-1-6684-8851-5.ch004

Khang, A., Rani, S., Gujrati, R., Uygun, H., & Gupta, S. K. (2023). *Designing Workforce Management Systems for Industry 4.0: Data-Centric and AI-Enabled Approaches*. CRC Press, Boca Raton, FL. https://doi.org/10.1201/9781003357070

Khang, A., Shah, V., & Rani, S. (2023). *AI-Based Technologies and Applications in the Era of the Metaverse* (1st ed.). IGI Global Press, Hershey, PA. https://doi.org/10.4018/978-1-6684-8851-5

Khang, A., Shashi Kant, G., Chandra Kumar, D., & Parin, S. (2023). Data-driven application of human capital management databases, big data, and data mining. *Designing Workforce Management Systems for Industry 4.0: Data-Centric and AI-Enabled Approaches* (1st ed., pp. 113–133). CRC Press, Boca Raton, FL. https://doi.org/10.1201/9781003357070-7

Khang, A. (2024). *AI-Oriented Competency Framework for Talent Management in the Digital Economy: Models, Technologies, Applications, and Implementation*. CRC Press, Boca Raton, FL. https://doi.org/10.1201/9781003440901

Pallavi, J., Vandana, T., Ravisankar, M., & Khang, A. (2023). Data-driven AI models in the workforce development planning. *Designing Workforce Management Systems for Industry 4.0: Data-Centric and AI-Enabled Approaches* (1st ed., pp. 179–198). CRC Press, Boca Raton, FL. https://doi.org/10.1201/9781003357070-10

Pooja, A. & Khang, A. (2023). A study on the impact of the industry 4.0 on the employees performance in banking sector. *Designing Workforce Management Systems for Industry 4.0: Data-Centric and AI-Enabled Approaches* (1st ed., pp. 384–400). CRC Press, Boca Raton, FL. https://doi.org/10.1201/9781003357070-20

SCEDEX. (2021). Skills-based competency ecosystem in digital economy. *The AI-oriented Competency Model for Digital Economy 5.0*. Retrieved from https://scedex.com/skills-framework.htm

Shashi, K. G., Khang, A., Parin, S., Chandra Kumar, D., Anchal, P. (2023). Data mining processes and decision-making models in personnel management system. *Designing Workforce Management Systems for Industry 4.0: Data-Centric and AI-Enabled Approaches* (1st ed., pp. 89–112). CRC Press, Boca Raton, FL. https://doi.org/10.1201/9781003357070-6

Snehal, M., Babasaheb, J., & Khang A. (2023). Workforce management system: concepts, definitions, principles, and implementation. *Designing Workforce Management Systems for Industry 4.0: Data-Centric and AI-Enabled Approaches* (1st ed., pp. 1–13). CRC Press, Boca Raton, FL. https://doi.org/10.1201/9781003357070-1

18 Workforce Management
An Endeavour Toward Optimizing Excellence of Higher Education Institutes

Annada Sankar Dash,
Durga Prasad Singh Samanta,
Amarendra Bhuyan, and Udayana Mohanty

18.1 INTRODUCTION

Nowadays, organizations spend huge amounts on the training, development, and career planning of their employees, which is not regarded as a mere expense of the organization. Rather, it is an investment in the development of assets, i.e., the efficiency and effectiveness of the workforce. While talking about work force management, it is treated as a business process solution. Though sound software is essential for the management of the workforce, it should be kept in mind that workforce management is a business process, not a technological solution. Organizational decisions can be satisfiers or dissatisfiers for the workforce. Therefore, the management should try to minimize decisions that can be dissatisfying or optimize satisfaction, irrespective of the amount of compensation. Workforce management today is the focal point in managing an organization. In the contemporary world, organization is an effective man-machine system. Therefore, workforce management should be given utmost importance while taking organizational decisions (Khang & Shah et al., 2023).

Workforce management is the process of strategically allocating people and resources, keeping track of attendance, and fulfilling the compliance of the rules and regulations with the objective of optimizing excellence and reducing the risk of the organization to a minimum. The important aspects of workforce management are setting the target or strategic planning, conducting a proper analysis of the present system, preparing an action plan for developing the workforce management system, implementing the plan, evaluating the system, and, if necessary, reviving the plan (Khang & Muthmainnah et al., 2023).

The higher education sector is an important tool in the developmental process of every nation because the resources, i.e., the experts to manage every sector of nation-building, are supplied or emerged from these institutions. Therefore, the success of higher educational institutes (HEI) is important for nation-building. Generally, the higher educational sector suffers from one important factor or shortcoming, i.e.,

DOI: 10.1201/9781003440901-18

the turnover of employees. Approximately 40% of the employees of HEI change their job for job satisfaction or to meet the change need for any other reason. Therefore, the management of the workforce is important for the performance and excellence of HEI (Khang & Kali et al., 2023).

18.2 LITERATURE REVIEW

There are various definitions of talent management in the literature. Despite these concepts, there is no single, accepted definition of talent management (Morton, 2005). What is frequently linked with talent management methods, however, shares certain commonalities. Talent management, according to Khang (2024), is the capacity to control the flow, supply, and demand of talent through the human capital engine.

The systematic acquisition, identification, development, involvement, retention, and placement of people who are of distinctive value to an organization is what talent management, as defined by Khang (2024), is. In this instance, organizations use talent management as a deliberate way to draw in and keep talent, creating a long-lasting competitive advantage. A plan for developing a framework for managing talent in higher education. 20 (0), a1671 in the SA Journal of Human Resource Management.

Less is known about the uses of performance management systems and the factors that make them effective and useful, despite the fact that numerous texts have concentrated on the advantages and disadvantages of both classic and modern performance management systems. Take a look at the expanding body of literature. It has been discovered that there is little to no attention paid to the actual information that is along with the adoption of these new human resource management (HRM) practices and individual performance management methods in systems for higher education (Ramune, 2016). This is to argue that management in higher education systems has a free option in what they choose to implement. Strategies for personal performance management in order to excel in the field.

Edith Penrose's work, which created the resource-based competitiveness theory. Some HR academics contend that the ideas of competitive advantage and HRM are intertwined. The prices and effectiveness of the company, in Schuler's opinion, were indicators of employee behavior and commercial practices. Management guru Peter Drucker stated in one of his publications that a company's or industry's competitiveness depends on the effective application of both knowledge and information, or knowledge acquisition and application.

18.3 OBJECTIVES

The objectives of this chapter are to discuss:

- Importance of workforce management in HEI.
- The obstacles faced by HEI with regarding proper management of human capital.
- To suggest appropriate measures by which HEI can overcome the challenges of human capital with the help of workforce management.

18.4   RESEARCH METHODOLOGY

The current study is exploratory in nature, and relevant data were acquired from a variety of secondary data sources, including journals, books, websites, papers, etc.

18.4.1   CHALLENGES FACED BY HIGHER EDUCATION

18.4.1.1   Rising of Complexities in Pay and Compliance

The remote worker and the virtual classroom were born out of pandemic restrictions. The option to work and study off-campus has proven to be popular, and many people want to keep it even though many of these restrictions have been removed. In addition to a surge in collective bargaining agreements, the COVID-19 situation also resulted in more job options, increasing the importance of talent retention for institutions. A suitably advanced workforce management solution, like the Workforce Suite, would be completely capable of handling the complexity even if all these aspects have generated additional considerations for HR, payroll, and compliance. These solutions, which are made to be very customizable, can manage even the most stringent collective bargaining agreements and keep scheduling under check.

Higher education can profit from contemporary workforce management even if there are no changes to the employment environment as a result of the epidemic. In order to manage its workforce more effectively, Oregon State University (OSU) opted to adopt the Workforce Suite in 2012. At that time, it was struggling to manage more than 12,000 employees with varied statuses and roles. To accommodate the changing requirements of organizations like OSU, workforce software has kept developing its products. With an average of 300 protected leave cases per month at the institution, keeping compliance with protected leave laws (such as the FMLA) was one challenge they ran into. While precisely and consistently tracking leave, tools like our Absence Compliance Tracker enable organizations like OSU to quickly adjust to changing rules.

18.4.1.2   Cost Control and Budgetary Pressures

Additionally, the state of the market has caused many prospective students to delay or forego their college degree. Due to expensive tuition and worries about student debt, undergraduate enrollment decreased by more than 5% (1 million students) nationally between 2019 and 2018. For some who would normally pursue higher education, rising college expenses have been a major barrier for decades. As a result, higher education funding has decreased, and instructors are under pressure to keep costs in check. The use of labor forecasting and optimization to manage predictable variable costs contributes to the reduction of payroll leakage and unnecessary overtime costs in higher education. To ease the burden on the budget, these savings can be shared among several departments, or they might be delivered directly to the students by way of lower tuition. In their 2022 Value Matrix, Gartner Research has ranked workforce software as No. 1 for our capacity to foster employee engagement, prevent penalties with smart compliance tools, and maximize labor with our labor forecasting solutions.

18.4.1.3   Maintaining, Luring, and Triumphing Superb Talent

Following the global epidemic, remote employment opportunities have skyrocketed. The popularity of remote work has led many employees to demand at least the option of sporadic virtual employment, despite the efforts of many institutions to entice workers and students back to the workplace and the classroom. With more job possibilities than ever before, this has made it difficult to both attract and keep talent. 15% of non-exempt part-time staff in the education industry leave their jobs each year. In another survey, it was discovered that 39% of higher education workers want to leave their positions if they are unable to have a flexible schedule that enables them to work remotely.

Every employee's experience is highly valued in contemporary workforce management, which encompasses all of the interactions an employee has with their employer from the start of their first day to the conclusion of their last. This includes not only just the everyday tasks but also the hiring procedure, training, and interactions with management, HR, and co-workers. According to research, companies that prioritize a pleasant employee experience say that their staff members are happier with their jobs, more engaged, and more likely to feel as though their opinions matter. They are also less inclined to look for alternative employment options (Shah & Khang, 2023).

18.4.2   Meeting the Challenges with Effective Workforce Management

By using efficient workforce management, the obstacles can be overcome, this section examines how strategic WFM has assisted HEIs in creating a future workforce that is sustainable, purpose-driven, and of the highest caliber. Next, it identifies the solutions that effective HRM provides to address the issues that were discussed in the report's earlier chapters, specifically: strategic workforce planning, establishing and implementing efficient performance management strategies, facilitating the growth of leadership, governance, and management, and successfully interacting with employees are all examples of good management (Pooja et al., 2023).

18.4.2.1   Workforce Planning that is Strategically Connected

Although a relatively underdeveloped HRM process in higher education, workforce planning can inspire change as part of an institution's strategic development. As a result, HEIs must ensure that all of their crucial people management procedures, such as workforce planning, reward and performance strategies, learning and development plans, organizational development and culture change plans, employee engagement plans, and development plans for leadership, governance, and management, are completely in line with their chosen strategy. By implementing strategically related workforce initiatives, the HR department may be able to address many of the issues brought up in this research. HEIs (and the workforce) will benefit from these initiatives by becoming more flexible and agile in order to meet new challenges.

18.4.2.2   Facilitation of Leadership, Governance, and Management

Facilitation of leadership, governance, and management development, institutional leadership, governance, and management will face a lot of demands as a result of the global economic slump. The majority of leaders and senior managers have never before witnessed HE changes in the same way and at the same amount as they are now. Senior management teams are faced with considerable challenges as a result of the overall impact of the crises institutions are currently experiencing. Access to high-quality management information will be crucial for providing senior management teams with the tools they need to make wise decisions. Institutions also need to have a swift, flexible decision-making process, be ready to make difficult decisions when necessary, and act promptly with thorough knowledge. Insightful HRM can support the development of management, leadership, etc. (Shyam & Khang, 2023).

18.4.2.3   Optimal Participation of Employees

In order to successfully address some of the biggest issues with people management in high education (HE), such as employee relations or pension's reform, the sector's ability to communicate with its employees is crucial, according to UCEA. The CIPD defines employee engagement as "a blend of dedication to the organization and its objectives and a readiness to assist colleagues" (organizational citizenship). Beyond motivation alone, it goes beyond contentment at work. The employee must provide engagement; it cannot be "imposed" as a requirement of employment. In 2009, the Department of Business, Innovation, and Skills received a study titled Engaging for Success: increasing performance by employee engagement that provided a compelling argument for effective employee engagement as a means of enhancing employee performance.

18.4.2.4   Introduction of Workforce Management System

The HEI will be able to achieve the following:

- Stronger response to business needs.
- Designing the workforce to achieve future capabilities related to talent demand and capacity enhancement.
- Create an appropriate action plan for the future.
- Develop a workforce management team that will cater to the future needs of talent and resource decision.
- Effective methods of recruiting the manpower.
- Improve the employee experience by investing in their development needs.

18.4.2.5   Recommendations for Effective Workforce Planning in Higher Education Institutions

- Changing the mindset to accept the challenge
- Keeping the best talent, which will best suit organizational needs.
- Reduce cost and budget pressure.
- Improving operational efficiency.
- A proper training program should be adopted for the development of the employees.

18.4.3   WFM Strategy's Main Responsibilities

18.4.3.1   Recording of Time

The act of punching a time clock may seem straightforward, but time monitoring involves more than that. Time tracking is not only useful for evaluating productivity but also it may be used to spot trends in overtime or under time. With a time-tracking procedure in place, you can:

- Monitor upcoming work.
- Complement your team with the right quantity of employees.
- Effortful assign duties.

The ability to effectively manage your time and establish a daily routine that is productive in general can both be improved. Using a job management program to track work in real time is the best method to accomplish this. Each team member may independently complete their best work thanks to efficient tools and time management best practices.

18.4.3.2   Organizing Personnel

The correct forecasting and scheduling of staffing requirements are equally as critical as time tracking. This contributes to the development of an effective work-life balance workforce management method. Employee scheduling can be challenging, whether you oversee a team at a lower level of employment or your company has many staffing tiers. To develop a timetable that works for the whole team, you need a system that considers factors like vacation, availability, workload, and absenteeism. The duties of WFM scheduling consist of:

- Controlling employee schedules.
- Dealing with concerns related to too much or too little employees.
- Team growth projections.

Companies can be confident that they won't have too few employees during peak periods of business or too many when things are slow by putting in place a dependable scheduling system.

18.4.3.3   Estimating and Planning a Budget

When predicting how labor demands will alter, historical data analysis can help you balance workloads and improve deployment. For instance, during the holiday season, a retail store might require twice as many employees as it would in the off-season. Among the several methods you can use to forecast using WFM procedures are:

- Addressing high employee turnover rates.
- Taking into account lost time or "shrinkage".
- Making plans for special days or occasions where the workload will be considerably impacted.
- Analyzing historical project information to obtain precise labor and resource estimations.

Workforce planning might be included in your budget as well. Your staff is most likely the most expensive and unpredictable expense for the majority of businesses.

18.4.3.4 Employee Evaluations

Performance evaluations are crucial for assessing team performance, productivity, and overall employee satisfaction, even though they might take a lot of time. Based on team member performance, these measures can also be useful for gauging customer satisfaction.

- Monitoring employee experiences is part of managing team performance.
- Assessing employee productivity
- Measuring client pleasure and experience

Although doing quarterly evaluations is one technique to attain these outcomes, every team has a unique method of managing team performance and job assessment based on skills (Khang & AIoCF.20, 2024), as shown in Figure 18.1.

18.4.3.5 Perks and Pay

Payroll and benefits are a crucial component of a wider personnel management plan, whether or not you have a team specifically dedicated to them. It might also include any additional incentives you give your staff. This includes their regular wages and perks. Payroll is the foundation for a wide range of tasks, such as:

- Considering labor expenses.
- Monitoring paid vacation time (PTO).
- Timesheet tracking.
- Providing bonuses and happy hours as employee incentives.

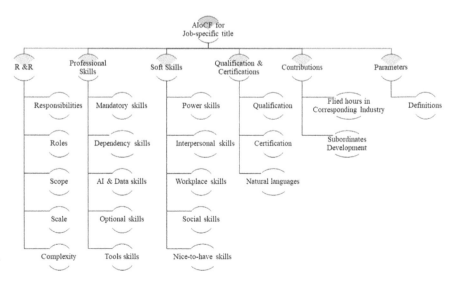

FIGURE 18.1 The list of components of the AIoCF architecture of the assessment tool (Khang, 2021).

Enhancing payroll and benefits can be a wonderful way to boost employee performance and morale beyond just the necessity of processing payments.

18.4.3.6   Control Over Training

A successful workforce must have a training program that is designed for both new and current employees. You must not only produce training materials for new hire orientation but also you must offer chances for ongoing improvement. Among the possible training materials are:

*   Materials for onboarding, no.
*   Assistance with tool management.
*   Market information.
*   Additional sources of information.

It is crucial to arrange materials and documents in a shared location where the entire team can discover information while developing a training program. Using a digital tool or even a shared folder is an option for this.

18.4.3.7   Observing Regulations

Utilizing a workforce management system might make it simpler to abide by wage and labor laws, overtime regulations, and other local employee standards. Make sure your workforce management approach always takes all necessary issues into account by consulting your internal legal team.

18.5   ADVANTAGES OF WORKFORCE MANAGEMENT GREATER PRODUCTIVITY

When tasks are correctly assigned and communicated, productivity is at its peak. Workforce management may boost team productivity by planning workload, keeping track of time, paying fairly, and communicating performance. Your team will experience better work quality as well as more work being produced as a result of higher productivity. Teams and individuals working on it benefit from it in a positive way.

18.5.1   AN EFFICIENT EXCHANGE OF IDEAS

When dealing with various teams, a good workforce management approach helps enhance self-reliance and communication. While the majority of WFM responsibilities might be handled by a single individual, invite important stakeholders from throughout your organization to give their insights and suggestions for improved procedures. You may streamline communication channels and minimize the workload on your internal communications team with a WFM strategy that links tasks, workflows, and systems. In addition to fostering better teamwork, efficient communication can also keep projects on schedule and up to par in terms of quality.

18.5.2   A RISE IN SPIRIT

Several factors—both good and bad—can have an impact on the morale of the team. Morale may create or break a successful team, regardless of the workload, salary and benefits, schedule, or any combination of these elements. Thanks to a staff management strategy, you can thankfully take action to raise morale. Between managers and team members, curiosity and empathy foster an interesting interaction. These straightforward yet effective qualities can enhance the general working atmosphere for your team and boost morale.

18.5.3   RESOURCE OPTIMIZATION

You'll be able to make the best use of your resources with enhanced communication, tracking, and productivity. This will not only enable you to produce more work but also will result in decreased labor expenses. Resources might range from a material good to an intangible one like time; both are precious and shouldn't be wasted.

18.5.4   LOWER COSTS

Automating time tracking, scheduling, and pay calculations not only saves managers and the payroll department's hours of work but also gets rid of expensive human errors and compliance issues. The time spent on these time-consuming chores can subsequently be put to better use. Organizations can use the information provided by workforce management procedures to change staffing numbers as necessary. Companies can cut labor expenditures and unplanned overtime when they can anticipate the peaks and valleys of their workload.

18.5.5   EXCEPTIONAL WORK

The quality you put out ultimately determines what you get back. Therefore, you won't notice a sizable money stream if your quality is subpar. Because of this, enhancing quality is a key component of workforce management. There is always space for development, regardless of whether you think the quality of your work is the greatest it can be. Process improvements could take many different forms, ranging from product modifications to website development and beyond. The options are absolutely endless.

18.6   ESSENTIAL COMPONENTS NEED TO KEEP A HIGH-CALIBER STAFF IN PLACE

- Administration, leadership, and government
- By ensuring and enhancing the caliber of the workforce through professional standards, outstanding research, CPD support, and improving qualification levels.

- Having effective systems and processes in place, such as talent management, performance management, reward programs, career paths for different occupational groups of workers, and a strategic approach to succession planning.

Why the workforce's potential can be completely realized by mainstreaming and actively supporting commitments to equality, diversity, and the health and well-being of all employees.

18.6.1 Governance, Leadership, and Management

Being able to motivate, inspire, and manage personnel is necessary for maintaining a high-level workforce. The unprecedented cost and market constraints, as well as a stringent employment law environment, place extreme expectations on the leaders, governing bodies, and managers of HEIs. The ability to lead and manage cultural change was mentioned by every interviewee as the primary competency required of senior management. While there is still work to be done, several institutions have made major and innovative improvements to their leadership and management capabilities. Every industry needs to figure out how to design development programs for middle- and senior-managers that have a long-term revolutionary impact. The "middle management issue" is what is meant here (Khang & Anuradha et al., 2024).

18.6.2 Establishing a Highly Qualified Workforce Hundred Qualifications

A Ph.D. is the academic staff member degree most usually linked to professors or higher. In England, 55% of academic personnel have doctoral degrees (international staff are more likely to hold a qualification at the Ph.D. level: 70% compared to 50% of UK staff). The percentages vary, nevertheless, depending on the institution type. For instance, 76% of permanent academics in HEIs established prior to 1992 held doctorates, as opposed to 33% among HEIs established following 1992. The topic mix of various sorts of institutions can assist to explain the variations because the proportion of academic personnel with doctorates is much higher in the sciences and engineering and much lower in law, education, and agriculture.

18.6.3 Research Excellence

Despite variations in the importance of research to each institution, virtually all HEIs in England share the goal of excellence in research. It acts as the main motivator for a number of institutional actions, including employing, promoting, and rewarding employees. It is more important than ever to advertise research positions to recent Ph.D. graduates and professionals in the business sector, given the continual growth of the research profession. Even in HEIs founded before 1992, where there is often a bigger proportion of professors involved in scholarly research than in HEIs founded after 1992, many academics still place a high importance on these skills. However, as entrepreneurship activities have increased, there has been a diversification of the skills required (Khang & Rani et al., 2023).

18.6.4 SYSTEMS AND METHODS THAT WORK PERFORMANCE ADMINISTRATION

According to HR directors interviewed for research by Oakleigh Consulting80, managing employee underperformance and managing individual performance in accordance to organizational goals have both increased since 2001. One area where they notably recognized advances was the management of poor performance, which is regarded as having started out with less development than other practices in HRM. Despite this, "addressing the management of poor performance" remains a top concern for HR directors in the present and the future. This has been confirmed by a number of our stakeholders, including the workers' unions and employees, and a number of them have recommended that HEIs construct more comprehensible frameworks and methods for the processes of performance, evaluation, and development.

18.6.5 ISSUES WITH DIVERSITY AND EQUALITY

However, the industry still has significant issues with equity and diversity. Women continue to be underrepresented in senior roles, particularly academic ones, with only 22% of permanent academic staff at the head of department/professor grade being female. Women make up only 10% of professors in STEM fields, a substantially lower proportion than in computer science and engineering. This suggests that these fields are not very diverse. Less males work in administrative and support positions. Senior Black and Minority Ethnic (BME) professionals are still in short supply in HE87 (although numbers are rising). Despite significant challenges and concerns with equal rights, attempts are being made to address both these issues and the culture that surrounds them.

18.6.6 WELLNESS AND HEALTH

The necessity for staff well-being is a new and developing part of institutional practice95, especially in light of the current state of the economy and HEIs' widespread and serious commitment to upholding a high standard of corporate social responsibility. Bettering employee health, engagement, and support through well-being programs can help individuals stay motivated and engaged, respond in a creative and flexible manner, and give their best effort on the job. In the long run, it ought to help HE enterprises create a more adaptive and motivated workforce. The benefits are very much in keeping with the government's aims for the well-being of people of working age 96. Many businesses were investing in workplace programs, according to the most recent government report on the health of the working-age population in Britain (Khang, 2024).

18.7 CONCLUSION

The significant accomplishments of the HE sector, the workforce's most urgent concerns, and the prerequisites for a productive and long-lasting workforce have all been underlined in this chapter. This chapter's objective is to support institutions' strategic planning efforts and future policy decisions with data. In the HE sector in India, it has also brought up a number of difficulties and queries that call for more discussion and research. Given the wide range of HEIs and their diverse perspectives, it is unlikely that there will be straightforward solutions to the questions or even a broad consensus. The following are the main issues we hope the industry will discuss (Khang, 2023).

The payment structure of HEI: there are significant worries about how future salary increases will affect the spending of HEIs in the current economic context. Any decrease in HEIs' income or additional pay hikes would cause major affordability difficulties because staff expenditures typically account for 57% of all institutional spending on average. This has caused several HEIs to wonder if the incremental pay increases (worth about 3% apiece) that roughly two-thirds of HE personnel get annually in addition to the nationally negotiated raises to all points on the pay spine will be sustainable in the future. The best industrial relations strategy for the industry is to forge a true collaboration where the sector's viability and success are fostered (Muthmainnah & Yakin et al., 2023).

REFERENCES

Khang A. (2021). Material4Studies. *Material of Computer Science, Artificial Intelligence, Data Science, IoT, Blockchain, Cloud, Metaverse, Cybersecurity for Studies*. Retrieved from https://www.researchgate.net/publication/370156102_Material4Studies.

Khang, A. (2023). *Advanced Technologies and AI-Equipped IoT Applications in High-Tech Agriculture* (1st ed.). IGI Global Press, Hershey, PA. https://doi.org/10.4018/978-1-6684-9231-4

Khang, A. (Ed.). (2024). *AI and IoT Technology and Applications for Smart Healthcare Systems* (1st ed.). Auerbach Publications. https://doi.org/10.1201/9781032686745

Khang, A., Anuradha Misra, Hajimahmud, V. A., & Litvinova, Eugenia. (2024). Machine Vision and Industrial Robotics in Manufacturing: Approaches, Technologies, and Applications (1st ed.). CRC Press. ISBN: 9781032565972. https://doi.org/10.1201/ 9781003438137

Khang, A., Kali, C. R., Suresh Kumar, S., Amaresh, K., Sudhansu Ranjan, D., & Manas Ranjan, P. (2023). Enabling the future of manufacturing: integration of robotics and IoT to smart factory infrastructure in industry 4.0. *AI-Based Technologies and Applications in the Era of the Metaverse* (1st ed., pp. 25–50). IGI Global Press, Hershey, PA. https://doi.org/10.4018/978-1-6684-8851-5.ch002

Khang, A., Muthmainnah, M., Seraj, P. M. I., Yakin, A. A., Obaid, A. J., & Panda, M. R. (2023). AI-aided teaching model for the education 5.0 ecosystem. *AI-Based Technologies and Applications in the Era of the Metaverse* (1st ed., pp. 83–104). IGI Global Press, Hershey, PA. https://doi.org/10.4018/978-1-6684-8851-5.ch004

Khang, A., Rani, S., Gujrati, R., Uygun, H., & Gupta, S. K. (2023). *Designing Workforce Management Systems for Industry 4.0: Data-Centric and AI-Enabled Approaches*. CRC Press, Boca Raton, FL. https://doi.org/10.1201/9781003357070

Khang, A., Shah, V., & Rani, S. (2023). *AI-Based Technologies and Applications in the Era of the Metaverse* (1st ed.). IGI Global Press, Hershey, PA. https://doi.org/10.4018/978-1-6684-8851-5

Khang, A. (2024). *AI and IoT Technology and Applications for Smart Healthcare Systems*. HBK: 9781032684901, PBK: 9781032679648, EBK: 9781032686745. Taylor and Francis Group, CRC Press, Boca Raton, FL. https://doi.org/10.1201/9781032686745

Khang, A. & AIoCF.17. (2024). Design and modelling of AI-oriented competency framework (AIoCF) for information technology sector. *AI-Oriented Competency Framework for Talent Management in the Digital Economy: Models, Technologies, Applications, and Implementation*. CRC Press, Boca Raton, FL. https://doi.org/10.1201/9781003440901-17

Khang, A. & AIoCF.20. (2024). Implementation of AIoCF model and tools for information technology sector. *AI-Oriented Competency Framework for Talent Management in the Digital Economy: Models, Technologies, Applications, and Implementation.* CRC Press, Boca Raton, FL. https://doi.org/10.1201/9781003440901-20

Morton, L. A. (2005). Take a holistic approach to talent management to manage talent for competitive advantage. *Strategic HR Review,* 4(5), 28–31. https://doi.org/10.1108/14754390580000819

Muthmainnah, M., Khang, A., Seraj, P. M. I., Yakin, A. A., Oteir, I., & Alotaibi, A. N. (2023). An innovative teaching model - the potential of metaverse for English learning. *AI-Based Technologies and Applications in the Era of the Metaverse* (1st ed., pp. 105–126). IGI Global Press, Hershey, PA. https://doi.org/10.4018/978-1-6684-8851-5.ch005

Pooja, K., Babasaheb, J., Ashish, K., Khang, A., & Sagar, K. (2023). The role of blockchain technology in metaverse ecosystem. *AI-Based Technologies and Applications in the Era of the Metaverse* (1st ed., pp. 165–173). IGI Global Press, Hershey, PA. https://doi.org/10.4018/978-1-6684-8851-5.ch011

Ramunė Kasperavičiūtė-Černiauskienė, D. S. (2016). The adoption of ISO 9001 standard within higher education institutions in Lithuania: innovation diffusion approach. *Total Quality Management & Business Excellence,* 29(1), 1–20. https://www.tandfonline.com/doi/abs/10.1080/14783363.2016.1164012

Shah, V. & Khang, A. (2023). Metaverse-enabling IoT technology for a futuristic healthcare system. *AI-Based Technologies and Applications in the Era of the Metaverse* (1st ed., pp. 165–173). IGI Global Press, Hershey, PA. https://doi.org/10.4018/978-1-6684-8851-5.ch008

Shyam, R. S. & Khang, A. (2023). Effects of quantum technology on metaverse. *AI-Based Technologies and Applications in the Era of the Metaverse* (1st ed., pp. 104–203). IGI Global Press, Hershey, PA. https://doi.org/10.4018/978-1-6684-8851-5.ch009

19 Accelerating Student's Creativity in Writing Paragraph through Artificial Intelligence Technology

Eka Apriani, Muthmainnah M, Erfin Wijayanti, Beni Azwar Kons, and Ahmad Al Yakin

19.1 INTRODUCTION

Modern technology is becoming a more significant part of daily life in the twenty-first century, and it has drastically altered how educational institutions operate and how their staff, including teachers and students, live. The usage of technology is widespread throughout the world. They can readily access through social media since technology is a tool that can enhance many other facets of life, including business, economy, education, and health (Khang & Shah et al., 2023).

The use of technological tools, individualized learning materials, and opportunities for advanced learning in education benefit teachers, students, and parents equally. The students have access to the Internet to share new information, communicate, and put their learning into practice. There are many prospects for technology in education today to revolutionize education globally for all age groups and to make available materials for global learning. As a result, for all students to succeed academically, knowing English and using technology are essential, and these skills are a prerequisite for a positive learning environment. The use of technology in the classroom has helped students study more effectively and has also helped them develop multitasking abilities, which is why the educational system has adopted it (Khang & Muthmainnah et al., 2023).

Technology in the classroom can offer relevant content, digital communication, flexible scheduling, and more flexibility of choice (Haris et al., 2017). Technology is an issue that arises regularly, makes it easier for students to understand concepts, and raises the significance of technology in education. Technology can boost students' and teachers' access to materials that can help students learning English as a second language (ESL) or English as a foreign language (EFL) achieve better learning outcomes. Particularly, social networking sites aid in the use of technology in teaching and learning. The core pedagogical ideas that underpin independent language teaching could be forgotten by educational workers as a result of technological advancements (Kranthi, 2017).

In terms of receptive and productive skills, language is still regarded as one of the most crucial components of education and literacy. Writing abilities are one of the most crucial language productive talents, frequently linked to other abilities, and the desirable consequence for students. Writing abilities are highly valued in all educational establishments. When practicing English writing, the learner's motivation is crucial. For successful outcomes, interesting instructional ideas and tactics must be used. The usage of technologies in the classroom may help students establish a positive learning environment. Teachers must use a variety of writing exercises to comprehend the students' strengths and weaknesses (Kawinkoonlasate, 2021).

One of the most crucial communicative language abilities is writing, which begins with the construction of sentences and ends with their integration into a purposeful context. In several systems or majors in education, writing skill becomes a crucial issue to discuss. Additionally, it is a way of clearly and efficiently communicating ideas, thoughts, and emotions through the usage of grammar rules (Martinez et al., 2020). Pre-writing, writing, and post-writing are the three stages of writing, which is a useful talent. Students gather information and draft an outline before beginning to write. They engage in a variety of tasks during the writing period, such as storytelling, report writing, letter writing, and paragraph writing. Teachers review student writing after it has been completed and offer suggestions for improvement (AlSmari, 2019). A variety of writing styles exist. Writing paragraphs serves as the primary organizational unit in works with more than two sentences arranged in it. Topic sentence, information that supports it, and a conclusion make up its three key components. Narrative, explanation, recount information, report, argument, and descriptive paragraphs are among the types that may be employed (Kartawijaya, 2018). Paragraph writing can be challenging for students. Writing may be challenging because students need to become proficient in a variety of sub-skills like writing a topic sentence followed by supporting details and a conclusion. A variety of instructional methods could be among these challenges (Karim et al., 2018).

Students encounter a variety of problems when writing in English. It involves structure and development, a lack of concepts, grammar intuition, punctuation errors, and capitalization. Additionally, researchers and EFL experts have given English writing—particularly paragraph writing—a lot of attention, but many Saudi university students struggle greatly with composing paragraphs (AlTameemy, 2019; AlSmari, 2019; Kartawijaya, 2018). Writing is a challenging endeavor for EFL students, and studies in the field of writing have found that Arab students, in particular, have several difficulties with paragraph writing. With these problems, students must put in more time and effort when writing, especially if they are EFL students. The steps must be followed in order to remedy writing-related concerns. To develop a concept, the students must pay close attention to the paragraph and all of its elements, including the topic sentence, any supporting facts, and the conclusion (Gillet, 2017). To improve students' writing abilities, the aforementioned requirements necessitate a teaching strategy for paragraph writing. Additionally, Ahmadi (2017) noted that in order for teachers to succeed, they must carefully select their teaching techniques. Al-Naibi et al.'s (2018) statement is supported by the fact that teachers frequently employ technology to improve their students' writing. This is due to both the rapid growth of technology and its usefulness in teaching English.

At the current stage of the development of the global educational environment, the use of information and communication technologies (ICT) in education is something that cannot be denied. A combination of ICT, including the use of artificial intelligence (AI) technologies (Popova et al., 2021a; Popova et al., 2021b), online internationalization in universities, and digitalization of education in various configurations, is now possible (Bruhn, 2020; Frolova et al., 2020; Frolova et al., 2021). Thus, ICT becomes more important when it is applied to the education system. Each university can explore more about education or information based on technology, which varies in nations traditionally associated with the "west" and "east." Modern teachers must be able to adjust to employing technology in the classroom for teaching and learning. Students can use technology to better understand the lessons being taught (Muthmainnah et al., 2022).

According to research by Pourhossein (2017), teachers must come up with fun and fascinating ways for students to use technology and activities to practice their language abilities. Therefore, the approach to enhancing learners' English writing abilities may offer appropriate teaching delivery techniques and contemporary teaching media to pique their interest in interacting in the target language and to generate a variety of practice opportunities with the aid of AI. AI is a subfield of computer science that focuses on smart machines that behave and think like people. There are some examples of AI such as speech recognition, natural language processing, image recognition, and other applications (Ilić et al., 2021).

Therefore, according to some theories discussed earlier, the English Study Program at IAIN Curup especially for third-semester students who take paragraph writing major has AI during the teaching and learning process with the lecturer. Students who learn AI are more prepared to deal with today's culture, technology, and environment, according to Muthmainnah et al. (2022). The technology used by the lecturer is ParagraphAI. ParagraphAI is an application in Play Store that helps students build paragraphs in writing. It is a simple application, and when the students type word, the app would show the description of that word. According to Google Play (2023), ParagraphAI is a world-class multilingual AI writing assistant for students and professionals, built by engineers who have worked at Meta, MIT, and OpenAI. ParagraphAI is guaranteed to improve writer writing productivity and quality. This application is based on mobile-assisted language learning (MALL). MALL is "the use of mobile technology in language learning... in MALL there is no need for the learners to sit in a classroom or at a computer to get learning materials". Due to its major influence on the teaching and learning process, Rao (2019) described MALL as an effective and adaptable instrument in education, particularly in ESL/EFL classrooms.

There are some previous research studies regarding the use of AI, the first is Kawinkoonlasate (2021) with the title "*A Study of Using E-Writing Instructional Design Program to Develop English Writing Ability of Thai EFL Learners,*" and the results showed that undergraduate students had a very high degree of satisfaction with this course's instruction, with an average score of 4.34. Additionally, interviews with students suggested that they were happy with the e-writing instructional design program since it helped enhance their writing abilities and foster greater learner autonomy. The second previous study is from Ilić et al. (2021) with the title "Needs and Performance Analysis for Changes in Higher Education and Implementation of Artificial Intelligence, Machine Learning, and Extended Reality," and the findings demonstrated that machine learning (ML) and AI are suitable technologies to be deployed in higher education institutions (HEI) in order to foster student skill development, a collaborative learning environment, and a welcoming setting for research. The third one is from Valizadeh (2022) with the title "Language teachers' perceptions of using Google keyboard in L2 writing," and the findings demonstrated that the teachers considered the inclusion of Gboard in the curriculum to be an efficient intervention that helped improve the students' spelling proficiency.

In order to ascertain students' voices while employing ParagraphAI during the teaching and learning process, the researcher conducted this study so that the title of this research is Accelerating student's creativity in writing through paragraph artificial intelligence technology.

19.2   LITERATURE REVIEW

19.2.1 Paragraph Writing

Writing is a crucial skill for students because it allows them to express their ideas and thoughts in detail. A paragraph is the primary unit of a written piece. A paragraph is made up of several connected sentences that deal with a topic sentence that introduces the paragraph's major idea, followed by supporting details that provide more examples and explanations of the main idea, and a conclusion that restates the paragraph's core idea (Aldera, 2016). Additionally, rationally and coherently moving from one sentence to the next and from one paragraph to the next are both enhanced by correct thought structure in writing. In other words, the ability to arrange a lot of sentences in a specific order and tie them together in specific ways is what it takes to write a paragraph. As a result, from elementary school through college, English language students are given careful instruction in these writing processes (Aldera, 2016).

A paragraph is made up of several sentences that all work to further one core topic (Ireland, Short, & Woollerton, Situngkir, 2019). According to McCloud (2017), the three sentences that make up the growth of a key concept in a paragraph—the topic phrase, the body sentence, and the conclusion—serve as rules for writing effective paragraphs. Situngkir (2019) proposed that each paragraph should contain (a) unified: Each phrase should connect to a single guiding notion; (b) make reference to the paper's main idea, or thesis, in a direct manner; (c) coherent sentences are constructed logically and should follow a predetermined growth plan, and (d) well-established the dominating notion of the paragraph should be sufficiently described and supported by each idea that is included in the paragraph. This is done by using examples and supporting information to do so.

The "germination process," which entails brainstorming, questioning, and taking notes to determine the keyword of the major concept, is where the selection of what to include in paragraph portions first occurs (The Writing Center, 2017). The six laws that govern paragraph development must be followed, and an ideal structural paragraph must be quantifiable and comprehensible (Sarfo, 2015). The first of the six paragraph writing rules is that each sentence must start with a clear, unambiguous reference to what comes before; second, successive phrases that reiterate or demonstrate the same thought should, whenever possible, be formed similarly. This is known as the rule of parallel construction. Third, it is expected that the beginning phrase will prominently reveal the topic of the paragraph, unless it is subtly preliminary. Fourth, the paragraph must be uninterrupted or sequential. Fifth, there should be cohesion within the paragraph, which implies a distinct objective and forbids tangents and irrelevant information. Sixth, an appropriate ratio should exist between the main ideas of the paragraph, just as it did in the sentence.

19.2.2 INFORMATION COMMUNICATION AND TECHNOLOGY (ICT)

All technological tools used to process and transmit information fall under the broad term "information and communication technologies" or ICT. Information technology and communication technology are two related concepts. All aspects of the creation, processing, management, and use of information are covered by information technology. Technology used in communication, however, refers to all methods for handling and transferring data between devices. Information technology and communication technology are thus two interrelated ideas.

According to Asabere and Enguah's definition of ICT in Apriani et al. (2022), ICT refers to the entities, devices, and procedures that provide the physical infrastructure and services for the creation, transmission, processing, archiving, and distribution of all types of information, including speech, text, data, pictures, and video. ICT is suited to all aspects of education, including lesson plans, methodologies, processes, media, content, and assessments (Apriani et al., 2020). There is no denying that every area of language teaching has been impacted by the exponential rise of ICT. In line with Manangsa et al. (2020), lecturers need to be aware of how productive and successful adult learners may be when practicing learning in order to use ICT in higher education. To help each student reach their learning objectives, lecturers should mentor, encourage, and instruct them. ICT is now included in educational activities to improve the learning of foreign languages (Alkamel & Chouthaiwale, 2018). It is possible to foster an environment where students can learn English through ICT. In addition, e-learning is a media that makes use of the Internet to disseminate information and advance user understanding in learning (Apriani et al., 2021). ICT platforms have favorable benefits for students, according to numerous research studies. According to Sanjaya et al. (2020), students have a positive opinion of weblogs based on five indicators provided by the researchers: trust in students' writing, the development of their writing skills, experience and awareness of using ICT, the encouragement of critical thinking, and accessibility.

19.2.3 MOBILE ASSISTED LANGUAGE LEARNING (MALL)

The development of mobile learning (m-learning) and MALL emerged in the 1980s and 1990s as a result of computer-assisted language learning (CALL). Due to the fact that more students utilize mobile devices than desktops or laptops in the ESL/

EFL classroom, it is imperative to consider the MALL approach as a more modern idea. Miangh and Nezarat in Al-Shehab (2020) defined MALL as "the use of mobile technology in language learning… in MALL there is no need for the learners to sit in a classroom or at a computer to get learning materials." The usage of tablets and smartphones for language learning and teaching is referred to as "new technologies" (Kukulska-Hulme et al., 2017). The learning process can be extended by students as needed outside of the classroom. Similar to this, Rao (2019) described MALL as an effective and adaptable instrument in education, particularly in ESL/EFL classrooms due to its major impact on the processes of teaching and learning. Due to the lack of a physical requirement for attendance in the classroom, this approach is more adaptable and practical than previous approaches. In addition, Muthmainnah et al. (2022) said that educators employ collaborative m-learning as a way to raise engagement and academic performance among students in the twenty-first century. Digital learning is compatible with a variety of pedagogical approaches. It means that m-learning or MALL is a technology that supports the education system in the world for now era.

Previous research on the use of MALL has sparked a crucial discussion about ESL/EFL writing classes: While some claim that students are averse to use mobile devices in the classroom, others have contrasted the use of a mobile device with traditional pen and paper while applying new writing approaches (Folk, 2016). Findings indicated that "in comparison with digital writing, traditional paper-and-pencil writing may seem too formal or even boring" (Al-Hamad et al., 2019). Similar to this, several studies on MALL contest its efficacy and adaptability, particularly in the presence of an enormous number of applications for learning–teaching reasons. Although there are many ESL/EFL learning programs available in the digital world, students are unable to determine which applications will best meet their needs. It can be challenging to distinguish between "non-formal and informal learning tools," according to Sendurur (2020), who contends that professors must explicitly demonstrate to students the aims of some applications. This highlights the important role that teachers play in helping students choose applications that are appropriate for their level of language proficiency; students should avoid making decisions about apps based solely on perceived popularity and/or download numbers.

On the other hand, more recent studies on MALL regarding the instruction of writing skills for ESL/EFL learners have shown encouraging outcomes. Al-Hamad et al. (2019) looked at the potential results of MALL in improving adolescent students' writing abilities. Results of the study revealed a significant amount of performance improvement in terms of ideas and content, structure, mechanics, vocabulary, and word choice. This idea was supported by a new study by Jassim and Dzakiria (2019), which used MALL to help Arab EFL students improve their writing abilities. The findings overwhelmingly demonstrate that mobile phones, with their capacity to motivate students, facilitate and enhance the instruction of writing. In terms of content and organization, it was discovered that mobile phones had positive benefits on student writing abilities.

19.2.4 ARTIFICIAL INTELLIGENCE (AI)

Computer science's field of AI is a technology that focuses on creating intelligent machine. This machine is expected to function like humans. Natural language processing, image recognition, audio recognition, and other uses of AI are examples. With the use of various problem-solving algorithms, the term "ML (Machine Learning)" refers to the application of AI to provide systems the capacity to learn from their experience and advance accordingly without explicitly requiring programming. In ML, for instance, computers learn based on the data they analyze rather than on the commands they receive from programs (Rana & Khang et al., 2021).

Vincent and Van der Vlies (2020) stated that AI system is a machine that can think by itself such as making prediction, judgment, or affection like what people do. In addition, different levels of autonomy are built into AI systems. Planning and design, verification and validation, deployment, operation, and monitoring are the stages of an AI system's lifecycle (Vincent & Van Der Vlies, 2020).

The study of AI methods to comprehend or support teaching or learning is known as AI in learning. Over the next 5 years, there will be a rapid increase in the use of AI in education, and by 2025, it is anticipated that global spending will exceed $6 billion. According to Holon IQ's Annual Report for 2019, China will experience the most increase, followed by the United States, which will account for more than half of all spending on AI education globally.

19.2.5 MALL (ParagraphAI) in Writing

Writing is a linguistic skill that is essential to both the growth of language and interpersonal communication. Writing has gotten more attention in language instruction as a result of the rapid development of numerous digital technologies (Li, 2018). A large number of prior research studies have investigated the function of technology in L2 writing classes, including the use of weblogs, Google Docs, and Google Drive (Alharbi, 2020; Ebadi & Rahimi, 2017; Kashani et al., 2013). Collectively, these studies demonstrate how digital technology benefits students' writing. On the other hand, there is a call for experimenting with various technologies that could impact the students' writing (Duman et al., 2015).

Regarding improving the standard of L2 writing, mobile-related technologies have also been extensively investigated. Among the main populations examined in this area

FIGURE 19.1   Layout of ParagraphAI application.

of research are university students and language learners at the school level (Al-Hamad et al., 2019; Chen et al., 2017; Lee, 2020; Yamaç et al., 2020). Yamaç et al. (2020), for instance, investigated the effects of tablet-based second language writing among 96 Turkish primary school children. In comparison with stories written on paper and pencil, the study's findings showed that those written on tablets were of greater quality. Furthermore, Hwang et al. (2014) investigated how 59 students used their mobile devices for writing in pre- and post-tests. The outcomes showed that the experimental group members did noticeably better than the members of the control group.

By using this app, the students just type some words and then let the app build the idea. For example, students search for "Globalization," and the ParagraphAI would show the description of globalization. The lecturer used this application as follows:

- Students must download the application on their phone
- Log in use students' Gmail
- The lecturer gives instruction on how to use the application, such as the example to find the idea in this tool
- The lecturer gives a theme to the students, namely Islamic moderation
- The students are allowed to use ParagraphAI to build the idea about Islamic moderation
- After finding some ideas about Islamic moderation, the students put the idea into their writing project and combined it with their own language
- Students re-arrange the idea to see the good composition of their writing
- The lecturer checks the students' writing

Some examples of the use of ParagraphAI in writing Islamic moderation, tolerance, and nationalist are shown in Figure 19.2, Figure 19.3, and Figure 19.4, respectively.

19.2.6 PERCEPTIONS

Perception has occasionally been used to refer to the awareness of particular concrete objects that are actually in front of our senses. An organism's capacity for perception is how it acquires detailed knowledge of its environment (Febrianara, 2015). The term "perspective" describes an assumption or view regarding something. The viewpoint expressed by MALL in this situation is a judgment regarding the implementation or output of ParagraphAI itself. Perspectives come in two flavors: a positive perspective and a negative one (Zhu & Wang, 2019). Through a process known as perception, people categorize and evaluate the patterns of external inputs (Atkinson & Adolphs, 2005). The first stage in understanding an idea is perception. Information received from the environment is referred to as the perception process. To put it another way, how we perceive the information around us affects how we process it.

19.3 METHODOLOGY

An explanatory quantitative research design was used in this study. The information was gathered from a web form (i.e., Google Form). Sugiyono (2017) claims that the use of questionnaires as a method of data collection involves asking respondents to respond to a list of questions or written inquiries. There are 23 items in

FIGURE 19.2   Example of using ParagraphAI (Islamic moderation).

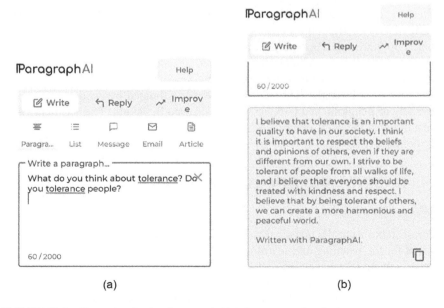

FIGURE 19.3   Example of using ParagraphAI (tolerance moderation).

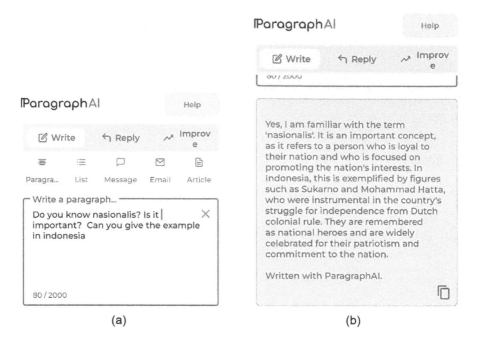

FIGURE 19.4 Example of using ParagraphAI (nationalist moderation).

the questionnaire used. The subjects were 30 students in the English Department at IAIN Curup. The questionnaire was shared for gathering data about students' voice in using ParagraphAI in paragraph writing major (Khang & Kali et al., 2023).

Besides, the researcher also wonders whether the ParagraphAI in paragraph writing major can affect the students' writing achievement. The data gathered were counted by Statistical Package for the Social Sciences (SPSS). Questionnaires are scaled using a Likert scale and then processed or calculated using SPSS or Excel. The Likert scale options are $5 = $ Strongly Agree, $4 = $ Agree, $3 = $ Neutral, $2 = $ Disagree, and $1 = $ Strongly Disagree. The questionnaire was validated by validators from the University of Bengkulu and UIN Fatmawati Sukarno Bengkulu. The researcher also gave a tryout of the questionnaire to 25 students. The result of tryout was that t count was more than t table.

19.4 FINDINGS AND DISCUSSION

19.4.1 FINDING

The researcher used the questionnaire after employing ParagraphAI as a scaffolding tool for the teaching and learning process in writing class to enlighten deeper points in students' voices. Table 19.1 shows the questionnaire results. Thirty students completed the form. The students used their smartphones to gain access to the ParagraphAI. Their responses to the questionnaire statements were presented in percentages.

TABLE 19.1
Students' Voice of ParagraphAI in Writing Class

No	Items	SD	D	N	A	SA	Total
				Percentage			
1	My writing skills have really improved since I started using ParagraphAI	0	0	3.33	33.33	63.33	100
2	My vocabulary has definitely grown since I started using ParagraphAI	0	0	3.33	30	66.67	100
3	My understanding of grammatical structures seems to have improved because to the use of ParagraphAI	6.67	16.67	6.67	33.33	36.67	100
4	I find writing with ParagraphAI to be interesting	0	3.333	0	19.33	73.33	100
5	ParagraphAI increased my motivation to write	0	6.67	0	33.33	60	100
6	I discovered that I was eager to employ ParagraphAI to hone my writing abilities	0	6.67	0	16.67	76.67	100
7	I learned a lot about English grammar through ParagraphAI	0	0	6.67	26.67	66.67	100
8	ParagraphAI contributed greatly to my knowledge of developing my vocabulary in writing	0	0	6.67	30	63.33	100
9	Once the class is over, I will keep using ParagraphAI	0	0	6.67	46.67	46.67	100
10	I would advise other EFL students to use ParagraphAI	0	3.333	6.67	46.67	43.33	100
11	I found it difficult to use ParagraphAI	46.67	43.33	6.67	3.33	0	100
12	I found it difficult to remember using ParagraphAI	50	36.67	6.67	0	6.67	100
13	Using ParagraphAI to learn English was boring for me	66.67	26.67	0	6.67	0	100
14	While using ParagraphAI, I came across familiar words and phrases	0	16.67	16.7	33.33	33.33	100
15	While using ParagraphAI, I learn unfamiliar words and phrases	26.67	26.67	6.67	20	20	100
16	I was pleased to discover that with ParagraphAI, I could improve my writing	0	0	0	33.33	66.67	100
17	I was disappointed when I discovered that I could improve my writing skills by using ParagraphAI	0	6.67	6.67	40	46.67	100
18	Even when I failed at writing, I always wanted to keep using ParagraphAI	0	3.333	6.67	33.33	56.67	100
19	When I failed at writing, I always wanted to stop using ParagraphAI	70	20	3.33	6.67	0	100
20	I found myself using ParagraphAI unconsciously	40	43.33	6.67	6.67	3.33	100
21	I advised my pals to utilize ParagraphAI as I was already using it	0	0	10	33.33	56.67	100
22	I was able to practice my writing more because to ParagraphAI	0	0	6.67	26.67	66.67	
23	I can produce more ideas for writing than I typically can thanks to ParagraphAI	0	0	6.67	20	73.33	100

Examining Table 19.1, the proportion of the questionnaire's items showed that item 1 was students have positive voice about the help of ParagraphAI to improve students' writing skills (96.6%). Item 2 was students have positive voice about the ParagraphAI helped students' vocabulary knowledge (96.6%). Item 3 was students have positive voice about the ParagraphAI helped students to understand grammatical structures easily (70%). Item 4 was students have positive voice about ParagraphAI is interesting to use to write (96.6%). Item 5 was students have positive voice about ParagraphAI increased students' motivation to write (93.3%). Item 6 was students have positive voice about students willing to use ParagraphAI to improve their writing skill (93.3%). Item 7 was students have positive voice about ParagraphAI contributed greatly to students' knowledge of English grammar (93.3%). Item 8 was students have positive voice about ParagraphAI contributed greatly to students' knowledge of developing their vocabulary in writing (93.3%). Item 9 was students have positive voice about the use of ParagraphAI after the lesson (93.3%). Item 10 was students have positive voice about recommend using ParagraphAI to other EFL learners (90%). Item 11 was students have positive voice about students found it difficult to use ParagraphAI (90%). Item 12 was students have positive voice about students found it difficult to remember using ParagraphAI (86.6%). Item 13 was students have positive voice about the use of ParagraphAI to learn writing was boring (93.3%). Item 14 was students have positive voice about using ParagraphAI make students came across familiar words and phrases (66.6%). Item 15 was students have positive voice about using ParagraphAI make students learn unfamiliar words and phrases (53.3%). Item 16 was students have positive voice about students felt well when they could upgrade their writing level while using ParagraphAI (100%). Item 17 was students have positive voice about students felt badly when they could upgrade their writing level while using ParagraphAI (86.6%). Item 18 was students have positive voice about students always wanted to continue using ParagraphAI even though they failed in while writing (90%). Item 19 was students have negative voice about students always wanted to quit using ParagraphAI when they failed in writing (90%). Item 20 was students have positive voice about students found themselves using ParagraphAI unconsciously (83.3%). Item 21 was students have positive voice about students recommended their friends to use ParagraphAI, which they were following (90%). Item 22 was students have positive voice about ParagraphAI could do more practice in writing (93.3%). Lastly, item 23 was students have positive voice that ParagraphAI can generate students' ideas in writing more than students usually do (93.3%), as shown in Figure 19.5.

19.4.2 DISCUSSION

Nineteen statements were deemed to be positive impressions, according to the study's findings, and more than half of the respondents indicated that they agreed or strongly agreed with them. These statements could be examined in relation to students' perspectives on employing ParagraphAI in writing class. To begin with, it was clear that students valued the role of ParagraphAI in assisting them in improving their writing skills and expanding their vocabulary knowledge. Learning with ParagraphAI

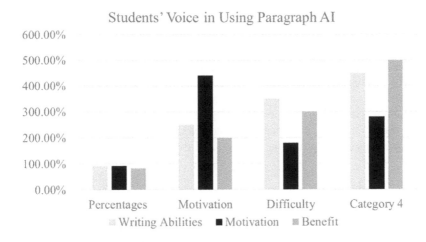

FIGURE 19.5   Students' voice in using ParagraphAI.

provided students with valuable expertise in effectively describing ideas. They may also learn by imitating what they were writing.

Due to the ease with which they could look up the definition of any word by just clicking on it, they also acquired a larger vocabulary. They might also be familiar with and employ a variety of native words that will aid in the expansion of their vocabulary. This result was consistent with research by Alhadiah (2020), who discovered that students thought MALL was a helpful program that was simple to use and that they planned to utilize it in the future. Because of their value both within and outside of the classroom, the use of MALL-based technologies for vocabulary development is strongly advised.

Second, the findings revealed that students improved significantly in terms of motivation, interest, willingness, and the contribution of ParagraphAI to their writing class. They thought it would be fun to employ ParagraphAI as their learning medium. They were also encouraged to learn and practice composing paragraphs more effectively. The use of graph AI helped them learn and improve their vocabulary and grammar skills. In ParagraphAI, the right grammar is always applied for each level or topic. As a result, the students could see how the grammar was used correctly. The students will be given assignments to assess their knowledge and proficiency in applying grammar in sentences after learning the grammar explanation. They can also verify whether their responses are right or wrong. After that, they can fix it for the best grade (Khang & Muthmainnah et al., 2023).

This outcome was consistent with Cukalevska (2020), who claimed that MALL assisted students to cooperate on particular exercises or assisted them to conduct grammar exercises and examinations with a formative intent.

Thirdly, the results demonstrated that students consistently enjoyed using ParagraphAI. They encountered some difficulties with the exercises, but they persisted and continued to use ParagraphAI. They considered learning with ParagraphAI to be enjoyable and satisfying if they were able to respond to the questions and go on to the next level. If they fared poorly in some subjects, they did not want to give up.

Even they meant to tell their pals about ParagraphAI so that they could all work on writing skills together. As a consequence, it was consistent with the study on students in the tenth grade in Buleleng Regency who had a favorable opinion of MALL in regards to learning to write (Indiana et al., 2018).

Finally, the results demonstrated that after learning with ParagraphAI, students increased their writing practice. They may hone their writing abilities more frequently, anywhere, and at any time. Writing abilities in particular, including grammar, vocabulary, fluency, and understanding, could be enhanced by MALL. Through MALL, students can improve their writing abilities by writing more frequently.

19.5   CONCLUSION

According to a majority of academics, MALL can affect students' successful and meaningful language learning. It is accessible to everyone at all times and places. In particular, ParagraphAI was the focus of this study's examination of students' perceptions of MALL. When they learned how to write paragraphs with ParagraphAI, the students demonstrated pleasant thoughts and feelings, according to the results (Khang & Gadirova et al., 2023). They concurred that one of the media that can be employed in writing classes is ParagraphAI. On students' writing abilities and language knowledge, ParagraphAI has a significant influence. When utilizing ParagraphAI independently, students feel considerably more motivated and interested in learning. By using ParagraphAI, students exhibit interest and a positive attitude in their writing. After class, students plan to continue utilizing ParagraphAI. Since they utilize ParagraphAI more frequently, they can improve their writing abilities and feel confident (Khang & Hajimahmud et al., 2023).

Students' motivation and self-confidence in learning English were impacted by their positive perceptions of ParagraphAI. The favorable impression encouraged students to develop their writing abilities. Additionally, they thought that learning to write with ParagraphAI would help them expand their vocabulary. However, further study has to be done on students' attitudes regarding paragraph AI in EFL classes to uncover other factors that might enhance the teaching and learning of English (Khang & Misra et al., 2023).

REFERENCES

Ahmadi, M. R. (2017). The impact of motivation on reading comprehension. *International Journal of Research in English Education*, 2(1), 1–7. https://doi.org/10.18869/acadpub. ijree.2.1.1

Aldera, A. (2016). Cohesion in written discourse: a case study of Arab EFL students. *Arab World English Journal (AWEJ)*, 7(2), 328–341. https://doi.org/10.24093/awej/vol7no2.22

Alhadiah, A. (2020). EFL learners' experience of a MALL-based vocabulary learning tool. *Indonesian Journal of Applied Linguistics*, 10(2), 283–291. https://pdfs.semanticscholar. org/5589/23315c66c7514415c00f394654437fae186e.pdf

Al-Hamad, R., Al-Jamal, D., & Bataineh, R. (2019). The effect of mall instruction on teens' writing performance. *Digital Education Review*, 35, 289–298. https://doi.org/10.1344/ der.2019.35.289-298

Alharbi, M. A. (2020). Exploring the potential of Google Doc in facilitating innovative teaching and learning practices in an EFL writing course. *Innovation in Language Learning and Teaching*, 14(3), 227–242. https://doi.org/10.1080/17501229.2019.1572157.

Alkamel, M. & Chouthaiwale, S. (2018). The use of ICT tools in English language teaching and learning: a literature review. *Veda's Journal of English Language and Literature-JOELL*, 5(2), 29–33. https://www.researchgate.net/profile/Mohammed-Alkamel/publication/330986788_The_Use_of_ICT_Tools_in_English_Language_Teaching_and_Learning_A_Literature_Review/links/5c5efe7d92851c48a9c5f85d/The-Use-of-ICT-Tools-in-English-Language-Teaching-and-Learning-A-Literature-Review.pdf

Al-Naibi, I. et al. (2018). Promoting students' paragraph writing using EDMODO: an action research. *TOJET: The Turkish Online Journal Of Educational Technology*, 17(1), 130–143. https://eric.ed.gov/?id=EJ1165747

Al-shehab, M. (2020). The role of mobile-assisted language learning (MALL) in enhancing the writing skills of intermediate IEP students : expectations vs reality. *Language Teaching Research Quarterly*, 20, 1–17. https://doi.org/10.32038/ltrq.2020.20.01

AlSmari, N. (2019). Fostering EFL students' paragraph writing using EDMODO. *English Language Teaching*, 12(10), 44–54. https://doi.org/10.5539/elt.v12n10p44

AlTameemy, F. (2019). Common paragraph writing errors made by Saudi EFL students' error analysis. *Theory and Practice in Language Studies*, 9(2), 178–187. https://doi.org/10.17507/tpls.0902.07

Apriani, E., Arsyad, S., Supardan, D., & Gusmuliana, P. (2022). ICT platforms for Indonesian EFL students viewed from gender during the COVID-19 pandemic. *Studies in English Language and Education*, 9(1), 187–202. https://jurnal.usk.ac.id/SiELE/article/view/21089

Apriani, E., Supardan, D., & Umami, M. (2020). Independent learning: English teachers' problems in designing a good lesson plan in new normal era at MAN Rejang Lebong [Paper presentation]. *International Conference on the Teaching English and Literature*, Bengkulu, Indonesia. https://ejournal.karinosseff.org/index.php/icotel/article/view/63

Apriani, E., Syafryadin, S., Inderawati, R., Arianti, A., Wati, S., Hakim, I., & Noermanzah, N. (2021). Implementing E-learning training toward English virtual lecturers: the process, perspectives, challenges and solutions. *International Journal of Emerging Technologies in Learning*, 16(4), 240–255. https://doi.org/10.3991/ijet.v16i04.14125

Atkinson, A. P. & Adolphs, R. (2005). Visual emotion perception: mechanisms and processes. *Emotion and Consciousness*, pp. 150–184. Guilford Press, New York. https://www.researchgate.net/profile/Anthony-Atkinson-3/publication/222712280_Visual_emotion_perception_Mechanisms_and_processes/links/00b49524971ae8bb20000000/Visual-emotion-perception-Mechanisms-and-processes.pdf

Bruhn, E. (2020). *Virtual Internationalization in Higher Education*. Springer, Bielefeld. https://doi.org/10.3278/6004797w

Chen, Y., Carger, C. L., & Smith, T. J. (2017). Mobile-assisted narrative writing practice for young English language learners from a funds of knowledge approach. *Language Learning & Technology*, 21(1), 28–41. https://doi.org/10125/44594.

Cukalevska, M. (2020). *Independent Project with Specialization in English the Impact of MALL on English Grammar Learning*. CRC Press, Boca Raton, FL. https://doi.org/10.1201/9781003440901

Duman, G., Orhon, G., & Gedik, N. (2015). Research trends in mobile assisted language learning from 2000 to 2012. *ReCALL*, 27(2), 197–216. https://doi.org/10.1017/S0958344014000287.

Ebadi, S. & Rahimi, M. (2017). Exploring the impact of online peer-editing using Google Docs on EFL learners' academic writing skills: a mixed methods study. *Computer Assisted Language Learning*, 30(8), 787–815. https://doi.org/10.1080/09588221.2017.1363056.

Febrianara, Y. (2015). *Students' Perception of the Implementation of Scaffolding in Public Speaking Clas*s. Sanata Dharma University, Yogyakarta. https://core.ac.uk/download/pdf/127700735.pdf

Folk, M. (2016). Kairotic aurality: audio essays, QR codes, and real audiences. *Mobile Technologies and the Writing Classroom: Reasons for Teachers*, pp. 36–51. The National Council of Teachers of English, US. https://doi.org/10.1201/9781003440901

Frolova, E. V, Rogach, O. V., & Ryabova, T. M. (2020). Digitalization of education in modern scientific discourse: new trends and risks analysis. *European Journal of Contemporary Education*, 9(2), 313–336.

Frolova, E. V., Rogach, O. V., Tyurikov, A. G., & Razov, P. V. (2021). Online student education in a pandemic: new challenges and risks. *European Journal of Contemporary Education*, 10(1), 43–52. https://doi.org/10.13187/ejced.2021.1.43

Gillet, A. (2017). *Academic Writing: Writing Paragraph* [Online]. Retrieved from: www.uefap.com

Haris, M., Yunus, M., & Badusah, J. (2017). The effectiveness of using padlet in ESL classroom. *International Journal of Advanced Research (IJAR)*, 5(2), 783–788. https://doi.org/10.21474/IJAR01/3214

Holon 'IQ's Annual Report on the State of Artificial Intelligence in Global Education. Available online: https://www.holoniq. com/notes/2019-artificial-intelligence-global-education-report/ (accessed on 3 February 2023).

Hwang, W.-Y., Chen, H. S. L., Shadiev, R., Huang, R. Y.-M., & Chen, C.-Y. (2014). Improving English as a foreign language writing in elementary schools using mobile devices in familiar situational contexts. *Computer Assisted Language Learning*, 27(5), 359–378. https://doi.org/10.1080/09588221.2012.733711

Ilić, M. P., Păun, D., Šević, N. P., Hadžić, A., & Jianu, A. (2021). Needs and performance analysis for changes in higher education and implementation of artificial intelligence, machine learning, and extended reality. *Education Sciences*, 11(10), 568. https://doi.org/10.3390/educsci11100568

Indiana, et al. (2018). Tenth grade students' perception toward mobile assisted language learning (MALL). *Learning English in Buleleng Regency in Academic Year 2017/2018*, pp. 1–13.

Jassim, L. L. & Dzakiria, H. (2019). The effects of utilizing mobile on developing English writing skill. *Opción*, 19, 2128–2143.

Karim, S. et al. (2018). Writing strategy instruction to improve writing performance of Bangladeshi EFL learners: a case study. *Journal of Social Sciences and Humanities*, 15(4), 126–136.

Kartawijaya, S. (2018). Improving students' writing skill in writing paragraph through an online technique. *Curricula Journal of Teaching and Learning*, 3(3), 27. https://doi.org/10.22216/jcc.2018.v3i3.3429

Kashani, H., Mahmud, R. B., & Kalajahi, S. A. R. (2013). Comparing the effect of blogging as well as pen-and-paper on the essay writing performance of Iranian graduate students. *English Language Teaching*, 6(10), 202–218. https://doi.org/10.5539/elt.v6n10p202.

Kawinkoonlasate, P. (2021). A study of using E-writing instructional design program to develop English writing ability of Thai EFL learners. *English Language Teaching*, 14(6), 43. https://doi.org/10.5539/elt.v14n6p43

Khang, A., Elmina, G., & Vugar, A. (2023). Role of photochemical reactions in the treatment of water used in the high-tech agriculture. *Advanced Technologies and AI-Equipped IoT Applications in High-Tech Agriculture* (1st ed., pp. 325–326). IGI Global Press, Hershey, PA. https://doi.org/10.4018/978-1-6684-9231-4.ch018

Khang, A., Misra, A., Gupta, S. K., & Shah, V. (2023). *AI-Aided IoT Technologies and Applications in the Smart Business and Production*. CRC Press, Boca Raton, FL. https://doi.org/10.1201/9781003392224

Khang, A., Hajimahmud, V. A., Gupta, S. K., Babasaheb, J., & Morris, G. (2023). *AI-Centric Modelling and Analytics: Concepts, Designs, Technologies, and Applications* (1st ed.). CRC Press, Boca Raton, FL. https://doi.org/10.1201/9781003400110

Khang, A., Muthmainnah, M., Seraj, P. M. I., Yakin, A. A., Obaid, A. J., & Panda, M. R. (2023). AI-aided teaching model for the education 5.0 ecosystem. *AI-Based Technologies and Applications in the Era of the Metaverse* (1st ed., pp. 83–104). IGI Global Press, Hershey, PA. https://doi.org/10.4018/978-1-6684-8851-5.ch004

Khang, A., Kali, C. R., Suresh Kumar, S., Amaresh, K., Sudhansu Ranjan, D., & Manas Ranjan, P. (2023). Enabling the future of manufacturing: integration of robotics and IoT to smart factory infrastructure in industry 4.0. *AI-Based Technologies and Applications in the Era of the Metaverse* (1st ed., pp. 25–50). IGI Global Press, Hershey, PA. https://doi.org/10.4018/978-1-6684-8851-5.ch002

Khang, A., Shah, V., & Rani, S. (2023). *AI-Based Technologies and Applications in the Era of the Metaverse* (1st ed.). IGI Global Press, Hershey, PA. https://doi.org/10.4018/978-1-6684-8851-5

Kranthi, K. (2017). Technology enhanced language learning (TELL). *International Journal of Business and Management Invention*, 6(2), 30–33. Retrieved from https://www.ijbmi.org

Kukulska-Hulme, A., Lee, H., & Norris, L. (2017). Mobile learning revolution: implications for language pedagogy. *The Handbook of Technology and Second Language Teaching and Learning*, pp. 217–233. Wiley-Blackwell, Hoboken, NJ. https://doi.org/10.1002/9781118914069.ch15

Lee, B. J. (2020). Smartphone tapping vs. handwriting: a comparison of writing medium. *EuroCALL Review*, 28(1), 15–21. https://doi.org/10.4995/eurocall.2020.12036.

Li, M. (2018). Computer-mediated collaborative writing in L2 contexts: an analysis of empirical research. *Computer Assisted Language Learning*, 31(8), 882–904. https://doi.org/10.1080/09588221.2018.1465981.

Manangsa, V. A., Gusmuliana, P., & Apriani, E. (2020). Teaching English by using andragogy approach for EFL students. *Teaching English by Using Andragogy Approach for EFL Students*, 4(03), 386–400. https://doi.org/10.33369/jeet.4.3.386-400

Martinez, J. et al. (2020). Using process writing in the teaching of English as a foreign language. *Revista Caribena De Investihacion Educative*, 4(1), 49–61. https://doi.org/10.32541/recie.2020.v41il

McCloud, L. (2017). Writing Effective Paragraph [Online]. Retrieved from: https://writes ite.athabascau.ca/documentation/writing-effective-paragraph

Muthmainnah, M., Cardoso, L., Obaid, A., Al Yakin, J., Jafar, A., & Nurlaila, M. (2022). Expanding on the use of YouMiMe as technology instructional design in learning. *Pegem Egitim ve Ogretim Dergisi*, 13(1), 367–378. https://doi.org/10.47750/pegegog.13.01.40

Muthmainnah, M., Seraj, P. M. I., & Oteir, I. (2022). Playing with AI to investigate human-computer interaction technology and improving critical thinking skills to pursue 21stcentury age. *Education Research International*, 6468995, 1–17. https://doi.org/10.1155/2022/6468995

ParagraphAI: GPT Writer & Chat - Aplikasi di Google Play [Access on February 3rd, 2023]. https://journal.unnes.ac.id/sju/index.php/elt/article/view/64069

Popova, A. V., Balashkina, I. V., Nikitin, P. V., & Prokopovich, G. A. (2021a). Ontology of artificial intelligence: on the dehumanization of social regulators. In E. G. Popkova et al. (Eds.) *Studies in Systems, Decision and Control*. Springer Nature, New York. https://doi.org/10.1007/978-3-030-56433-9_10

Popova, A. V., Gorokhova, S. S., Abramova, M. G., & Balashkina, I. V. (2021b). Legal system and artificial intelligence in modern Russia: goals and tools of digital modernization. In E. G. Popkova et al. (Eds.) *Studies in Systems, Decision and Control*. Springer Nature, New York. https://doi.org/10.1007/978-3-030-56433-9_11

Pourhossein Gilakjani, A. (2017). A review of the literature on the integration of technology into the learning and teaching of English language skills. *International Journal of English Linguistics*, 7(5), 95–106. https://doi.org/10.5539/ijel.v7n5p95

Rana, G., Khang, A., Sharma, R., Goel, A. K., & Dubey, A. K. (2021). *Reinventing Manufacturing and Business Processes through Artificial Intelligence*. CRC Press, Boca Raton, FL. https://doi.org/10.1201/9781003145011

Rao, P. S. (2019). The use of mobile phones in the English classrooms. *Academicia: An International Multidisciplinary Research Journal*, 9(1), 6. https://www.indianjournals.com/ijor.aspx?target=ijor:aca&volume=9&issue=1&article=001

Sanjaya, H. K., Apriani, E., & Edy, S. (2020). Using web blog for EFL students in writing class. *Journal of English Education and Teaching*, 4(04), 516–535. https://doi.org/10.33369/jeet.4.4.516-535

Sarfo, K. (2015). Investigating paragraph writing skills among polytechnic students: the case of Kumasi polytechnic. *International Journal of Language and Linguistics*, 3(3), 145–153. https://article.ijoll.net/pdf/10.11648.j.ijll.20150303.16.pdf

Sendurur, E. (2020). A guide for mobile-assisted language learning in informal setting: pedagogical and design perspectives. In R. Ahmed, A. Al-Kadi, & T. Hagar (Eds.), *Enhancements and Limitations to ICT-Based Informal Language Learning: Emerging Research and Opportunities*, pp. 24–53. IGI Global, Hershey, PA. https://www.igi-global.com/chapter/a-guide-for-mobile-assisted-language-learning-in-informal-settings/242374

Situngkir, D. A. (2019). The Application of process approach in enhancing paragraph writing skills of SMA Negeri 1 Lembang. *Acuity : Journal of English Language Pedagogy, Literature and Culture*, 4(2), 94–127. https://doi.org/10.35974/acuity.v4i2.1043

Sugiyono, F. X. (2017). Neraca Pembayaran: Konsep, Metodologi dan Penerapan (Vol. 4). *Pusat Pendidikan Dan Studi Kebanksentralan (PPSK) Bank Indonesia*. https://www.google.com/books?hl=en&lr=&id=6qLWDgAAQBAJ&oi=fnd&pg=PA1&dq=Neraca+Pembayaran:+Konsep,+Metodologi+dan+Penerapan+(Vol.+4).+Pusat+Pendidikan+Dan+Studi+Kebanksentralan+(PPSK)+Bank+Indonesia.&ots=HRSL0qtSqv&sig=Vh8tRdPJC5N0YUbp6Ink_XFT6so

Valizadeh, M. (2022). Language teachers' perceptions of using Google keyboard in L2 writing. *Journal of Educational Technology and Online Learning*, 5(2), 411–421. https://doi.org/10.31681/jetol.1060950

Vincent-Lancrin, S. & van der Vlies, R. (2020). Trustworthy artificial intelligence (AI) in education: promises and challenges. *OECD Education Working Papers, No. 218*. OECD Publishing, Paris. https://doi.org/10.1787/a6c90fa9-en

Writing Center. (2017). Paragraph Writing [Online]. Retrieved from: www.ad elaide.edu.au/writingcentre. https://eric.ed.gov/?id=EJ1305780

Yamaç, A., Öztürk, E., & Mutlu, N. (2020). Effect of digital writing instruction with tablets on primary school students' writing performance and writing knowledge. *Computers & Education*, 157, 103981. https://doi.org/10.1016/j.compedu.2020.103981

Zhu, C. & Wang, K. (2019). MOOC-based flipped learning in higher education: students' participation, experience and learning performance. *International Journal of Educational Technology in Higher Education*, 16, 1–18. https://doi.org/10.1186/s41239-019-0163-0

20 Implementation of AIoCF Model and Job Assessment Application for Information Technology Sector

Alex Khang

20.1 INTRODUCTION

AIoCF is an extremely fast competency assessment approach because it is a role-based skills assessment. Compared to traditional competency assessment approaches, each skill selects one level through its ranking action, and you can see a corresponding rank of his or her skill on the dashboard. Due to the manner in which AIoCF operates, it is fast in ranking speed with an increase in accuracy (Khang & Shah et al., 2023).

The introduction of AIoCF was a defining moment in the competency assessment area because of its ability to identify employees' skills in real-time and with greater accuracy. The input framework is divided into a grid of S × S cells, each of which is responsible for one item in the selection that defines the level of competency. It also calculates a ranked score for a competency item which also is a measure of how a competency item was qualified. As a consequence, the score has nothing to do with the name of the skill in the dropdown box, but rather with the level of competency.

Entry assessment is the ranking of skills and qualifications between the current role and the new role through various working platforms. AIoCF plays a major role in everyone's workplace. Due to the increase in collaboration, people can communicate with stakeholders from all over the world. New contracts or businesses including projects, services, productions, and new hires make everyone want to have opportunities by enriching their skills and qualifications (Khang & Muthmainnah et al., 2023).

20.2 RELATED WORK

This chapter contains the sequential processes in the design of the competency framework for a Data Engineer (DE) job title, the assessment tool used, and the screen grabs of output, interface, and basic skills prerequisites required to implement

DOI: 10.1201/9781003440901-20

the system properly. The purpose of this work is take into consideration the feasibility of results and feedback in order to provide users with an accurate and reliable system (AIOCF, 2021).

All through the development process of the system, a number of tools were used in implementing different aspects and functionalities of the system. An Excel sheet was used to develop the model. This consisted of various modules such as Self-Assessment by Employee, Assessment by Manager, and Management System, among many others (Khang & Rani et al., 2023).

20.2.1 AIoCF Template

To precisely estimate the rank for a series of connected skills, we have presented a model of the hybrid predictive and fuzzy logic-based control system. The viability and restrictions of using the rate percentage approach to calculate total scores on the job rank state estimate were discussed as shown in Table 20.1.

Details of the competency framework with many components of the job title and their weight percentage are shown in Table 20.2.

TABLE 20.1
Denotes the Evaluation of the Above-Mentioned Existing Methods

Competency-Based Job Description				Weight %
No	Code	Type	Description	Σ = 100%
1	RRxx	Mandatory	Section I: Description for Roles and Responsibilities	30%
2	PSxx	Mandatory	Section II: Description for Professional Skills	35%
3	SSxx	Mandatory	Section III: Description for Soft Skills	15%
4	SQxx	Optional	Section V: Description for Qualification	10%
5	SCxx	Optional	Section IV: Description for Contributions	10%
6	ALxx	Optional	Section VI: Description for Additional Component	0%

TABLE 20.2
Details of the Weight Percentage for Section of Roles and Responsibilities

Section I: Roles and Responsibilities				Weight %
No	Code	Type	Description	Σ = 25%
1	XX01	Mandatory	Description for Responsibilities	10%
2	XX02	Optional	Description for Roles	5%
3	XX03	Optional	Description for Scope [of projects/business activities/technology or platform]	3%
4	XX04	Optional	Description for Scale [of projects/business activities/technology or platform]	2%
5	XX05	Optional	Description for Complexity [of projects/business activities/technology or platform]	1%

20.2.2 AIoCF Weight Percentage

In order to design an assessment for a job title, the focus initially needs to be on what skill categories or skill items are of the highest impact, highest value, and highest contribution to the job role in an employee's career path. Once this has been determined, which covers all knowledge and skills; you can set the right weight percentage of proficiency. An example of a competency framework with many components of a DE job title and their weight percentage is shown in Table 20.3.

20.2.3 AIoCF Job Levels (Job Ranks)

Competency assessment is an important activity in your organization that determines the level of a job title for each employee in the business unit or an organization. Depending on characteristics and the business domain of the job title, the number of the job rank is different as shown in Table 20.4.

TABLE 20.3

Example of the Weight Percentage for Section of DE's Roles and Responsibilities

Section 1: Role and Responsibilities Implementing all the processes from data collection, cleaning, and preprocessing, to training data sets, and deploying them to production, including but not limited to following the responsibilities:				Weight %
No	Code	Type	Description	30%
1	RR01	Mandatory	Collaborating with data scientists and architects on projects	20%
			Data ingests, transforms, and integrates structured data	
			Delivers data to a scalable data warehouse platform	
			Interpreting trends and patterns	
			Data Modeling, database architecture design	
			Analyzing and organizing raw data	
2	RR02	Mandatory	Building data systems and pipelines	5%
			Preparing data for prescriptive and predictive modeling	
			Building algorithms and prototypes	
			Deploying data sets and models to the environments	
3	RR03	Optional	Managing available data resources	3%
			Managing analytical/integration tools, databases, warehouses	
			Maintenance of the data sets and models to the environments	
4	RR04	Optional	Understand how to optimize data retrieval	3%
			Understand how to develop dashboards and visualizations	
5	RR05	Optional	Help the other DE understand data sets and data process flow	1%
			Lead other engineers to implement the data solution	
			Engage with experts & engineers to deliver interpretable results	

(Continued)

TABLE 20.3 (*Continued*)
Example of the Weight Percentage for Section of DE's Roles and Responsibilities

		Section 2: Professional [Data Engineering] Skills			Weight %
No	**Code**	**Skill Category**	**Type**	**Description**	**35%**
1	PS01	Data Engineering	Mandatory	Data Integration Process and Implementation	25%
				Data Mining Process and Implementation	
				Data Analysis and Data Visualization	
				Big Data Process and Management	
				Data Architecture Design and Implementation	
				Data Warehouse Platforms and Solutions	
				Data Pipeline Paradigms and Tools	
				Relational and NoSQL Database Platforms	
				Software Engineering and Specialized Tools	
2	PS02	AI Engineering	Optional	AI and Machine Learning	5%
				Computer Vision Algorithms and Models	
				IoT-Based Data Solutions	
				- Data Visualization Tools	
3	PS03	Other	Optional	Cloud Computing	5%
				Blockchain Technology	
				Office Tools	

		Section 3: Soft Skills			Weight %
No	**Code**	**Skill Category**	**Type**	**Description**	**15%**
1	SS01	Communication	Mandatory	Interpersonal Communication	7%
				Teamwork & Motivation	
2	SS02	Creativity	Optional	Critical Thinking	5%
				Design Thinking	
				Intellectual Curiosity	
3	SS03	Leadership	Optional	Proactive Problem Solving	3%
				Risk Management	
				Leadership	

		Section 4: Certification and Qualification (Optional)		Weight %
No	**Code**	**Type**	**Description**	**10%**
1	SQ01	Optional	International or Internal Certifications	5%
2	SQ02	Optional	Education background and advanced degree	5%

		Section 5: Contributions (Optional)		Weight %
No	**Code**	**Type**	**Description**	**10%**
1	SC01	Mandatory	Guided/Mentored/Coached for Junior Engineers	5%
2	SC02	Optional	Flight hours as a Data Engineer in the Data Ecosystem	5%

TABLE 20.4
Example of Job Levels of Data Engineer Role

No	Code	Job Level	Short Description
1	DE01	Associate Data Engineer (ADE)	Associate Data Engineer performs routine or entry-level engineering tasks. Processes and analyzes data sources, data warehouse, and engineering practices for data products.
2	DE02	[Mid-level] Data Engineer (MDE)	The Data Engineer will build systems for collecting, validating, and preparing high-quality data.
3	DE03	Senior Data Engineer (SDE)	Senior Data Engineer will design, evaluate, develop, and test a data model, maintain a data warehouse and analysis environment, and write scripts for data integration of a wide range of data products, structures, and systems.
4	DE04	Principal Data Engineer (PDE)	Principal Data Engineer will lead the design and development of databases, integration data projects, and change management across a diverse range of clients, and data engineering of industries and technologies.

20.2.4   AIoCF Levels and Template

The AIoCF matrix is an essential part of the assessment process and with the right information, it should help managers to determine whether each job rank is in line with the skill set and whether it is a job the employee actually wants to go up in their career path. Identifying a required proficiency level for a particular rank is important in an organization because these required levels tend to have a major impact on one's job performance as shown in Table 20.5.

Based on input parametric design and taking into consideration the correlation between competency components and job levels in the AIoCF assessment template, a particular proficiency level/responsibility level for each corresponding rank was presented as shown in Table 20.6.

TABLE 20.5
Proficiency Level and Responsibility Level AIoCF Components

No	Description	Job Level 1	Job Level 2	Job Level 3	Job Level 4
1	Section I: Description for Roles and Responsibilities	Responsibility level 1	Responsibility level 2	Responsibility level 3	Responsibility level 4
2	Section II: Description for Professional Skills	Proficiency level 1	Proficiency level 2	Proficiency level 3	Proficiency level 4
3	Section III: Description for Soft Skills	Proficiency level 1	Proficiency level 2	Proficiency level 3	Proficiency level 4
4	Section V: Description for Qualification	Qualification level 1	Qualification level 2	Qualification level 3	Qualification level 4
5	Section IV: Description for Contributions	Contribution level 1	Contribution level 2	Contribution level 3	Contribution level 4

TABLE 20.6

Example of Proficiency Level and Responsibility Level for Data Engineer Title

No	Description	ADE	MDE	SDE	PDE
1	Section I: Description for Roles and Responsibilities	Responsibility level 1	Responsibility level 2	Responsibility level 3	Responsibility level 4
2	Section II: Description for Professional Skills	Proficiency level 1	Proficiency level 2	Proficiency level 3	Proficiency level 4
3	Section III: Description for Soft Skills	Proficiency level 1	Proficiency level 2	Proficiency level 3	Proficiency level 4
4	Section V: Description for Qualification	Qualification level 1	Qualification level 2	Qualification level 3	Qualification level 4
5	Section IV: Description for Contributions	Contribution level 1	Contribution level 2	Contribution level 3	Contribution level 3

Note that the list of responsibility levels as shown in Figure 20.1

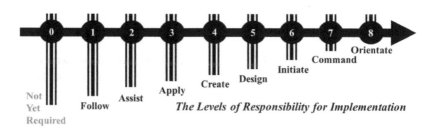

FIGURE 20.1   Nine levels of role and responsibility in the implementation mode.

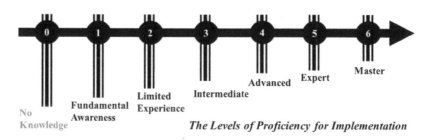

FIGURE 20.2   The eight proficiency levels of professional skills in the implementation mode.

The list of proficiency levels of professional skills are shown in Figure 20.2 and proficiency levels of soft skills are shown in Figure 20.3.

The list of levels of qualifications in below simulation (an example of qualifications):

- 6: Coaching for more than 20 members in 12 months
- 5: Coaching for at least 20 members in 12 months
- 4: Coaching for at least 10 members in 12 months

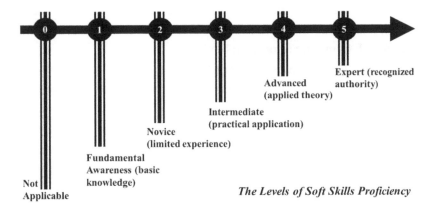

The Levels of Soft Skills Proficiency

FIGURE 20.3 The eight proficiency levels of soft skills in the design and implementation mode.

- 3: Coaching for at least 05 members in 12 months
- 2: Coaching for at least 03 members in 12 months
- 1: Coaching for at least 01 member in 12 months
- 0: No Coaching or Mentoring of any member in 12 months
- 0: Not Yet Required

The list of levels of contributions in below simulation (an example of contributions):

- 4: > TOEIC 700 or Level C1 or Native Proficiency
- 3: >= TOEIC 650 or Level C2 or Near-native or Bilingual proficiency
- 2: >= TOEIC 550 or Level B1 or Full Professional Proficiency
- 1: >= TOEIC 450 or Level B2 or General Professional Proficiency
- 0: No Knowledge or < TOEIC 450
- 0: Not Yet Required

To calculate the total score for each job level, we need to rely on the weight percentage and corresponding level (CL) with the calculation expression as shown in Equation 20.1 and the result as in the following matrix table is shown in Table 20.7 and simulation of score of a data engineer role is shown in Table 20.8:

$$SCORES = \left(CL * Weight\% / MAX(CL)\right) * 100 \tag{20.1}$$

Where,

CL = Responsibility level or Proficiency level or Contribution level or Qualification level

As a simulation of a data engineer job title in Table 20.8, to achieve ADE level, an employee needs to be evaluated by the manager with a total score greater than or equal to 25.8 and less than 51.7, MDE level is greater than or equal to 51.7 and less than 77.5 scores, SDE level is greater than or equal 77.5 and less than 100 scores, and

TABLE 20.7

Score of AIoCF Component for Each Job Level

No	Description	Weight	ADE	MDE	SDE	PDE
1	Section I: Description for Roles and Responsibilities	30%	1*30%/4*100	2*30%/4*100	3*30%/4*100	4*30%/4*100
2	Section II: Description for Professional Skills	35%	1*35%/4*100	2*35%/4*100	3*35%/4*100	4*35%/4*100
3	Section III: Description for Soft Skills	15%	1*15%/4*100	2*15%/4*100	3*15%/4*100	4*15%/4*100
4	Section V: Description for Qualification	10%	1*10%/4*100	2*10%/4*100	3*10%/4*100	4*10%/4*100
5	Section IV: Description for Contributions	10%	1*10%/4*100	2*10%/4*100	3*10%/4*100	3*10%/4*100

TABLE 20.8

Score of AIoCF Component for Each Job Level of Data Engineer Title

No	Description	Weight	ADE	MDE	SDE	PDE
1	Section I: Description for Roles and Responsibilities	30%	7,5	15,0	22,0	30,0
2	Section II: Description for Professional Skills	35%	8,8	17,5	26,3	35,0
3	Section III: Description for Soft Skills	15%	3,8	7,5	11,3	15,0
4	Section V: Description for Qualification	10%	2.5	5.0	7.5	10.0
5	Section IV: Description for Contributions	10%	3.3	6.7	10.0	10.0
Total score(s)		100%	25,8	51,7	77,5	100

PDE level is 100 absolute scores. In case total scores are less than 25.8, the job level could be ranked as Fresher level (Fresher is a candidate with 0 years of professional experience, he or she is still in college or has graduated out of college not more than 2 years ago).

20.2.5 AIoCF RANGE OF SCORES

In fact, to achieve the ADE level, an employee just needs to be evaluated by the manager with a total score less and greater than 25.8 according to a certain ratio or formula ((lower limit number plus upper limit number) divided by 2), total scores of the MDE level is less and greater than 51.7 scores, SDE level is less and greater than 77.5 scores, and PDE is less than or equals 100 scores as shown in Table 20.9.

For example for the data engineer job title in Table 20.9, to achieve the ADE level, an employee needs to be evaluated by the manager with a total score between 12.9 and 38.8; for the MDE level it is between 38.9 and 64.6 scores, for SDE level it is between 64.7 and 88.8 scores, and for PDE it is between 88.9 and 100 scores. In case total scores are less than 12.9 the job level could be ranked as Fresher level.

TABLE 20.9
Range of Scores for Each Job Level of Data Engineer Title

Total	25,8		51,7		77,5		100	
score(s)	12,9	38,8	38,9	64,6	64,7	88,8	88,9	100

20.3 IMPLEMENTATION OF AIoCF JOB ASSESSMENT TOOL

Khang (2024a) designed a competency assessment model and analysis method on the basis of AIoCF architecture. The model was trained with exactly 36,240 job assessments from two big IT companies and then tested with exactly 18,120 employees. He proposed an assessment method for ranking and classifying the competency level using the AIoCF Job Description and the AIoCF Job Assessment.

This model makes many contributions to the challenges of the ranking competency model in the era of AI and the digital economy. In this case study, they used a dataset that included a wide range number of skill sets for the Data Engineer job title acquired with a real context in IT companies as shown in Figure 20.4a.

Data Engineer Job Title

No	Description	Weight	Associate Data Engineer	[Mid-level] Data Engineer	Senior Data Engineer	Principal Data Engineer
1	Section I: Description for Roles and Responsibilities	30%	1: Follow	2: Assist	3: Apply	4: Create
2	Section II: Description for Professional Skills	35%	1: Fundamental Awareness	2: Limited Experience	3: Intermediate	4: Advanced
3	Section III: Description for Soft Skills	15%	1: Novice (limited experience)	2: Intermediate (practical application)	3: Advanced (applied theory)	4: Expert (recognized authority)
4	Section V: Description for Qualification	10%	1: >= TOEIC 450 or Level B2 or General Professional Proficiency	2: >= TOEIC 550 or Level B1 or Full Professional Proficiency	3: >= TOEIC 650 or Level C2 or Near-native or Bilingual proficiency	4: > TOEIC 700 or Level C1 or Native Proficiency
5	Section IV: Description for Contributions	10%	1: Coaching for at least 01 member in 12 months	2: Coaching for at least 03 members in 12 months	3: Coaching for at least 05 members in 12 months	3: Coaching for at least 05 members in 12 months

(a)

(*) Mandatory assessment items. If there is any unsatisfied item, you will not satisfied this rank.

Data Engineer Role - Job Description

Assessment Items	Weight %	Explanation of Assessment Items	Associate Data Engineer	Data Engineer I	Data Engineer II	Senior Data Engineer	Principle Data Engineer
	100%		DSEN01	DSEN02	DSEN03	DSEN04	DSEN05
Roles & Responsibilities	**20%**	**Engineer is required to have the role and responsibility to perform one or more of the following tasks**					
Roles and Responsibilities	15%	Implementing all the processes from data collection, cleaning, and preprocessing, to training data set and deploying them to production, including but not limited to following responsibilities:	1: Follow	2: Assist	3: Apply	4: Create	5. Design
Project Diversity and Complexity	5%	- Lead Data Engineer team to implement the machine learning solution - Engage with domain experts as well as engineers to effectively deliver interpretable results	1. Normal or Common Technologies	2: Medium or Common Technologies	3: Complex or Emerging Technologies	3: Complex or Emerging Technologies	4. Very Complex or Emerging Technologies
Experiences & Contributions	**15%**	**Engineer is required to have one or more of the following professional experiences and industry contributions**					
Data Engineering Experiences (*)	5%	Implementation of Data sets, Data Models, Data Warehouse, Data Lakes for a few AI-Based Industries such as	1: ~1 year experience	2: 2-3 years experience	2: 2-3 years experience	3: 4-6 years experience	4: 7-8 years experience
Subordinate Development	1%	Number of team members or number of people trained/coached	0: Not Required	0: Not Required	1: 1-5 Members or Coaching 1-2 Junior Engineer	2: 5-10 Members or Coaching 3-4 Junior Engineer	3: 10-20 Members or Coaching 1-2 Senior Engineer
Number of Data Science Projects	1%	Number of applications, components, the successful level of project that you have designed, developed, supported, migrated, deployed, managed Data Warehouses, Data Lakes and Data sets maintenanced systems or Data	0: Not Required	1: ~01-03 Products / Components / Ops	2: ~04-05 Products / Components / Ops	3: ~06-10 Products / Components / Ops	4: ~11-15 Products / Components / Ops
SME Community Contributions	2%	Activities to support/training other projects/team, presenter of workshop, contribute IP/solution...	0: Not Required	0: Not Required	1: Training, support other projects	2: Contribute >= 1 IP Solution at project level	3. Contribute >= 1 IP Solution at BU/FSU level
Business Industry Experiences	5%	Banking, Finance, Insurance, Aviation, Automotive, Oil and Gas, Health Care, ERP, Data Warehouse, Logistic, Real estate, Telecom, ...	0: Not Required	0: Fundamental awareness	1: ~1 year experience	2: 2-3 years experience	3: 4-6 years experience
Consulting for Sales/Biding Teams	1%	Technology Stack, Customer Experience, Vision Development, Processes and Activities, Pricing and Positioning Strategies, Drive Product Launch, ...	0: Not Required	0: Fundamental awareness	0: Fundamental awareness	1: ~1 year experience	2: 2-3 years experience
Data Engineering	**42%**	**Engineer is required to have one or more of the following specialized skills to perform assigned tasks**					

(b)

FIGURE 20.4 (a) AIoCF for Data Engineer job title. (b) AIoCF for Data Engineer job title.

Source: AIoCF.Data.Engineer.JD.JA.xlsx (Khang, 2021).

In real-life situations, notwithstanding, job competency assessment stakeholders must fully comprehend the class of benefits and its magnitude in order to affect meaningful measures or assign the required resource in order as shown in Figure 20.4b. Furthermore, there are presently few unified and accessible competency data, resulting in an absence of a widely accepted standard for competency defect recognition of employees in the organization (SCEDEX, 2021).

A skill set like this could be useful in a wide range of implementations. In this case, it was created to provide a collection of skills for use in the data engineering sector that is able to support their employees in planning and monitoring the project in the company. There are three steps in an assessment process, job-based competency evaluated by the employee and job-based competency evaluated by the manager, the comparison report can show the different job levels in the competency framework were correctly evaluated by the employee and the competency framework was correctly evaluated by his or her manager.

20.3.1 SELF-ASSESSMENT BY EMPLOYEE

A set of arrangements and integration of the list groups of AIoCF components are assessed and tested for an employee who was involved in the assessment program as shown in Figure 20.5a and the list of AIoCF components is shown in Figure 20.5b.

The scores of the evaluated level and expected job level as shown in Figure 20.6 are looked up from the established values in the framework (Figure 20.4a) used for comparison and help managers make decisions on choosing a job level for the job title more precisely.

By selecting the expectation of the job level of a Principal Data Engineer; the evaluator can see different levels of competency gaps in the total evaluated score of 87.92 at the required score of 100. This comparison implementation can assist in detecting the competency anomalies that need to be filled in as soon as possible or in planning a more effective learning pathway. The consistent rule is that if the level of any skill in AIoCF changes, the total score will update and the rank of the job title will be reset as shown in Figure 20.7.

20.3.2 ASSESSMENT BY MANAGER

A set of arrangements and integration of AIoCF components to be assessed, tested, and piloted by his or her manager is shown in Figure 20.8.

20.3.3 AIoCF METHOD OF RANGE OF SCORES

In AIoCF, skills play an important role, it is nothing but a required skill designed to complete assigned tasks over a fixed period. The AIoCF model can provide negative consequences if a manager in the workplace misuses them to put a job level for their staff without the transparency assessment. One of the major negative activities that happen in an organization is power bullying. Power bullying a person is a major threat and serious, widely spreading among users without applying AIoCF and especially among middle-level managers. Power bullying action refers to a person's unwanted threats and forces on their subordinates.

(a)

Data Engineer Job Title

Evaluated by Employee: John Smith

No	Description	Weight	Evaluated level	Notes / Evidences	Evaluated Score
1	Section I: Description for Roles and Responsibilities	30%	4: Create	Use Tableau tool to visualize the predicted result of lung diseases	30.0
2	Section II: Description for Professional Skills	35%	3: Intermediate		26.25
3	Section III: Description for Soft Skills	15%	4: Expert (recognized authority)		15.0
5	Section V: Description for Qualification	10%	4: > TOEIC 700 or Level C1 or Native Proficiency		10.0
4	Section IV: Description for Contributions	10%	2: Coaching for at least 03 members in 12 months		6.67
			Evaluated job level	**Senior Data Engineer**	**87.92**

(b)

Data Engineer Role - Job Assessment		CANDIDATE'S SELF-RANKED RESULTS SCORED BY MANAGER / LEADER / SME			CHOOSE EXPECTED RANK Ü TO COMPARE			ACTIONS TO IMPROVE YOUR RANKINGS
[John Smith] - [JSM010]		Not Qualified RANK: Data Engineer I-DSEN02		49.00	DSEN02	36.42		
Assessment Items	Weighted %	Choose competency level for each evaluation criteria	Add remarks or evidences for each evaluation	Actual Score	Required Competency Level for Expected	Required Score	Need to Improve?	Courses or Action(s) to Improve
Roles & Responsibilities	20%			14.50		8.50		
Roles and Responsibilities	15%	4: Create		12.00	2: Assist	6.00	No	
Project Diversity and Complexity	5%	2: Medium or Common Technologies		2.50	2: Medium or Common Technologies	2.50	No	
Experiences & Contributions	15%			9.58		2.75		
Data Engineering Experiences (*)	5%	3: 4-6 years experience		3.75	2: 2-3 years experience	2.50	No	
Subordinate Development	1%	2: 5-10 Members or Coaching 3-4 Junior		0.67	0: Not Required	0.00	No	
Number of Data Science Projects	1%	2: ~04-05 Products / Components / Opt		0.50	1: ~01-03 Products / Components / Opt	0.25	No	
SME Community Contributions	2%	3: Lecturer/Trainer of Technical workshop of		2.00	0: Not Required	0.00	No	
Business Industry Experiences	5%	1: ~1 year experience		1.67	0: Fundamental awareness	0.00	No	
Consulting for Sales/Biding Teams	1%	2: 2-3 years experience		1.00	0: Fundamental awareness	0.00	No	
Data Engineering	42%			12.92		13.17		
Data Integration Process and Implementation (*)	10%	2: Limited Experience		4.00	2: Limited Experience	4.00	No	
Data Mining Process and Implementation (*)	5%	2: Limited Experience		2.00	2: Limited Experience	2.00	No	
Data Analysis and Data Visualization	1%	1: Fundamental Awareness		0.25	1: Fundamental Awareness	0.25	No	

FIGURE 20.5 (a) Data Engineer assessment scenario for self-assessment. (b) Data Engineer assessment scenario for self-assessment (Khang, 2021).

Source: AIoCF.Data.Engineer.JD.JA.xlsx (Khang, 2021).

Data Engineer Job Title

Evaluated by Employee: John Smith

No	Description	Weight	Evaluated Score	Expected level	Required Score
1	Section I: Description for Roles and Responsibilities	30%	22.50	4: Create	30.00
2	Section II: Description for Professional Skills	35%	35.0	4: Advanced	35.0
3	Section III: Description for Soft Skills	15%	7.50	4: Expert (recognized authority)	15.0
5	Section V: Description for Qualification	10%	7.50	4: > TOEIC 700 or Level C1 or Native Proficiency	10.0
4	Section IV: Description for Contributions	10%	3.33	3: Coaching for at least 05 members in 12 months	10.0
			75.83	**Principal Data Engineer**	**100.0**

FIGURE 20.6 Expected level and required scores for a Principal Data Engineer.

Data Engineer Job Title

Evaluated by Employee: John Smith

No	Description	Weight	Evaluated level	Notes / Evidences	Evaluated Score
1	Section I: Description for Roles and Responsibilities	30%	3: Apply	Use Tableau tool to visualize the predicted result of lung diseases	22.50
2	Section II: Description for Professional Skills	35%	4: Advanced		35.0
3	Section III: Description for Soft Skills	15%	2: Intermediate (practical application)		7.50
5	Section V: Description for Qualification	10%	3: >= TOEIC 650 or Level C2 or Near-native or Bilingual proficiency		7.50
4	Section IV: Description for Contributions	10%	1: Coaching for at least 01 member in 12 months		3.33
			Evaluated job level	[Mid-level] Data Engineer	75.83

FIGURE 20.7   The consistent rule of any skill in AIoCF changes.

Data Engineer Job Title

Evaluated for John Smith

Evaluated by Manager: Cruke Smith

No	Description	Weight	Evaluated level	Notes / Evidences	Evaluated Score
1	Section I: Description for Roles and Responsibilities	30%	3: Apply	Use Tableau tool to visualize the predicted result of lung diseases	22.50
2	Section II: Description for Professional Skills	35%	3: Intermediate		26.25
3	Section III: Description for Soft Skills	15%	4: Expert (recognized authority)		15.0
5	Section V: Description for Qualification	10%	3: >= TOEIC 650 or Level C2 or Near-native or Bilingual proficiency		7.50
4	Section IV: Description for Contributions	10%	1: Coaching for at least 01 member in 12 months		3.33
			Evaluated job level	[Mid-level] Data Engineer	74.58

FIGURE 20.8   Mid-level data engineer job assessment scenario for the manager.

By applying AIoCF to the HR management processes, both negative reviews and power bullying can be eliminated to help HR managers reduce the turnover rate. One of the approaches is to apply AIoCF with an approach method of a score boundary (Table 20.9) instead of a specific score as shown in Figure 20.9.

20.3.4   Job Level Making-Decision

Data engineer job-based competency is self-assessed by employees and evaluated by the manager, the results will show the comparison in the competency framework was correctly evaluated by the employee with the best accuracy of 87.92 scores (Figure 20.5) and a job level proposed by the employee is Senior Data Engineer. Meanwhile, the competency framework was correctly evaluated by his or her manager with the best accuracy of 74.58 scores (Figure 20.8). The job level proposed by the manager for the employee would be [Mid-level] Data Engineer.

Based on input parametric design and taking into consideration the correlation between current competency and evaluated selections in assessment results, a

Data Engineer Job Title

Evaluated for John Smith

Evaluated by Manager: Cruke Smith

No	Description	Weight	Evaluated level	Notes / Evidences	Evaluated Score
1	Section I: Description for Roles and Responsibilities	30%	3: Apply	Use Tableau tool to visualize the predicted result of lung diseases	22.50
2	Section II: Description for Professional Skills	35%	3: Intermediate		26.25
3	Section III: Description for Soft Skills	15%	4: Expert (recognized authority)		15.0
5	Section V: Description for Qualification	10%	3: >= TOEIC 650 or Level C2 or Near-native or Bilingual proficiency		7.50
4	Section IV: Description for Contributions	10%	1: Coaching for at least 01 member in 12 months		3.33
			Evaluated job level	Senior Data Engineer	74.58

FIGURE 20.9 Senior data engineer job assessment scenario for the manager.

Data Engineer Job Title　　　　　　　　　　**Assessement Results**

No	Description	Evaluated by Employee	Expected by Employee	Evaluated by Manager	
	John Smith - Assessment Results	Senior Data Engineer	Principal Data Engineer	[Mid-level] Data Engineer	Senior Data Engineer
1	Section I: Description for Roles and Responsibilities	30.0	30.00	22.50	22.50
2	Section II: Description for Professional Skills	26.25	35.00	26.25	26.25
3	Section III: Description for Soft Skills	15.0	15.00	15.0	15.0
5	Section V: Description for Qualification	10.0	10.00	7.50	7.50
4	Section IV: Description for Contributions	6.67	10.00	3.33	3.33
	Total Scores	87.92	100.0	74.58	74.58

FIGURE 20.10 Dashboard of statistic analytics for the new level estimation.

dashboard of statistic analytics for the new level estimation was presented as shown in Figure 20.10.

In order to assess the rank of an employee, two assessment profiles were assigned, each file was evaluated by the employee to show the validity of the self-assessment and the others were evaluated by their manager to show the validity of the managerial assessment. Depending on the point of view of the manager, the scores of evaluated job level and expected job level as well as gaps of skills based on the information help the manager make decisions accurately for the final job level for his or her employee as shown in Figure 20.11 (Khang, 2023).

20.4 LIMITATIONS AND CHALLENGES

Looking toward this massive use of AIoCF, in reality, raises some issues and concerns like feasibility, exactness, and honesty due to the decentralization of evaluation and involvement of reviewers in different fields at the same time (Khang & Kali et al., 2023).

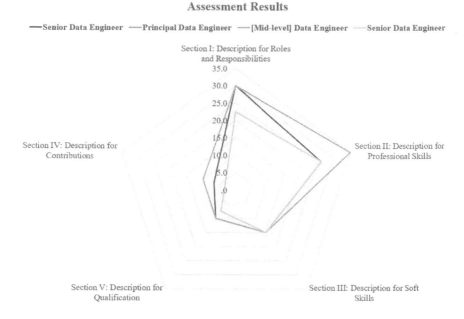

FIGURE 20.11 The final job level for the manager's employee.

AIoCF system requirements are the key modules that a system demands before it can function effectively as it was designed to. The user of the system will have to meet the basic requirements to be able to operate the system successfully. These modules include;

- Self-Assessment by Employee Module
- Assessment by Manager Module
- System Management Module

The model's accuracy is very important as it gives certainty to whether this model can be used to determine job level accurately. To achieve this, simulations and pilots are performed many times on the model for correctness of results (Khang, 2023).

20.5 CONCLUSION

The fundamental goal of the AIoCF competency assessment approach is to identify strong and weak competency skills in order to improve work performance. Looking toward this massive use of AIoCF, in reality, raises some issues and concerns like feasibility, exactness, and honesty due to the decentralization of the evaluation and the involvement of reviewers in different fields at the same time as shown in Figure 20.12.

AIoCF model preprocessing refers to editing skills prior to their use in self-ranking and review stages. This entails specialized knowledge, understanding emerging skills, orienting career pathways, and corrections of rate percentage. In addition, the AIoCF

FIGURE 20.12   Summary of assessment.

model is also used to assess entry competency or interview assessment, the manager can capture many different views of the competencies from the profile of the candidate (Khang & Muthmainnah et al., 2023).

The AIoCF model is a regularization approach for job assessment that reduces the turnover rate. Assessment of fully connected skills is used in competency frameworks to create productivity augmentation, and the AIoCF assessment method is one of the most widely used methods for competency assessment regularization. In this mode, we use a radar chart to illustrate the outcomes of competency between self-assessed job levels by the employee, assessed job levels by the manager, and the expected job level by the employee using the Excel software (Khang, 2024b).

20.6   RESOURCES

Readers can download samples of assessment tools of AIoCF.Data.Engineer. JD.JA.xlsx and AIoCF.xlsx at https://www.researchgate.net/publication/370156102_ Material4Studies.

REFERENCES

AIoCF. (2021). AI Skills-based competency ecosystem in digital economy. *The AI-oriented Competency Model for Digital Economy 5.0.* Retrieved from https://scedex.com/quickstart.htm

Khang, A. (2021). "Material4Studies," Material of Computer Science, Artificial Intelligence, Data Science, IoT, Blockchain, Cloud, Metaverse, Cybersecurity for Studies. Retrieved from https://www.researchgate.net/publication/370156102_Material4Studies

Khang, A. (2023). *Advanced Technologies and AI-Equipped IoT Applications in High-Tech Agriculture.* IGI Global Press, Hershey, PA. https://doi.org/10.4018/978-1-6684-9231-4

Khang, A., Kali, C. R., Suresh Kumar, S., Amaresh, K., Sudhansu Ranjan, D., & Manas Ranjan, P. (2023). Enabling the future of manufacturing: integration of robotics and IoT to smart factory infrastructure in industry 4.0. *AI-Based Technologies and Applications in the Era of the Metaverse* (1st ed., pp. 25–50). IGI Global Press, Hershey, PA. https://doi.org/10.4018/978-1-6684-8851-5.ch002

Khang, A., Muthmainnah, M., Seraj, P. M. I., Yakin, A. A., Obaid, A. J., & Panda, M. R. (2023). AI-aided teaching model for the education 5.0 ecosystem. *AI-Based Technologies and Applications in the Era of the Metaverse* (1st ed., pp. 83–104). IGI Global Press, Hershey, PA. https://doi.org/10.4018/978-1-6684-8851-5.ch004

Khang, A., Rani, S., Gujrati, R., Uygun, H., & Gupta, S. K. (2023). *Designing Workforce Management Systems for Industry 4.0: Data-Centric and AI-Enabled Approaches*. CRC Press, Boca Raton, FL. https://doi.org/10.1201/9781003357070

Khang, A., Shah, V., & Rani, S. (2023). *AI-Based Technologies and Applications in the Era of the Metaverse* (1st ed.). IGI Global Press, Hershey, PA. https://doi.org/10.4018/978-1-6684-8851-5

Khang, A. (2024a). *AI-Oriented Competency Framework for Talent Management in the Digital Economy: Models, Technologies, Applications, and Implementation*. CRC Press, Boca Raton, FL. https://doi.org/10.1201/9781003440901

Khang, A. (2024b). Design and moddeling of AI-oriented competency framework (AIoCF) for information technology industry. *AI-Oriented Competency Framework for Talent Management in the Digital Economy: Models, Technologies, Applications, and Implementation*. CRC Press, Boca Raton, FL. https://doi.org/10.1201/9781003440901-18

SCEDEX. (2021). Skills-based competency ecosystem in digital economy. *The AI-oriented Competency Model for Digital Economy 5.0*. Retrieved from https://scedex.com/skills-framework.htm

21 Artificial Intelligence-Centric Applications in Data Privacy and Cybersecurity for Human Resource Systems

Babasaheb Jadhav, Vikram Barnabas, and Mudassar Sayyed

21.1 INTRODUCTION

We live in a technology-dependent world that generates approximately 329 million terabytes of data daily (Khang & Vugar et al., 2024; Durate, 2023). The IoT (Li et al., 2015) and cloud computing (Velte et al., 2009) are instrumental in generating and gathering such volumes of data. No doubt, human resource (HR) data is important for businesses to survive and flourish, but cyberattacks pose a major challenge in doing so. Cyberattack simply defined as a remote attack on the computing system. The common forms of cyberattacks include ransomware, malware, SQL injection, insider threats, denial of service (DoS), and phishing (Sarker et al., 2020).

Cyberattacks are becoming more common and they affect businesses by causing disruption of services and/or financial losses. According to IBM (IBM, 2022), 83% of the organizations studied have had more than one data breach, 60% of the data breaches increased in price passed on to customers, and USD 9.44 million was the average cost of a data breach in just the US. A McAfee report estimated global cyber-crime losses to exceed USD 1 trillion (McAfee, 2020); furthermore, it is estimated that cybercrime may cost the world USD 10.5 trillion annually by 2025 (Morgan, 2020). This exponential growth of cybercrimes rings a bell for cybersecurity practitioners and researchers (Sarker et al., 2020).

Cybersecurity is the practice to protecting computing systems and data from attack. It involves monitoring the access of the equipment, the program, and the networks and protecting against harm (Kumar, 2019). In Section 21.2, several important terms and concepts related to cybersecurity are briefly discussed. With the rise in more sophisticated cyberattacks, the more conventional cyber defense measures such as encryption, antivirus, access control, user authentication, and firewalls may not be enough (Tapiador et al., 2013). The core issue with such conventional defense mechanisms is that they are operated by very few security practitioners, and the data processing is done to solve a particular problem.

HR is a critically important function of an organization whereby it screens the resumes and candidates, recruits them, and lastly trains them according to the needs of the time. The HR systems are used by organizations for this purpose. The main function of such systems includes, but is not limited to storing employee information and supporting the various HR functions such as payroll, training, etc. HR is all about data and sensitive personal data, if not secured can create havoc. For instance, the financial details of employees should be kept filtered; not all should have access to them. Cybersecurity practitioners need to ensure that such data is kept private and restricted from unauthorized access (Khang & Hahanov et al., 2022).

The cybersecurity applications that are capable of detecting and preventing cyber threats intelligently are at the core of defending individuals, businesses, organizations, and governments against growing cybercrimes. Intelligent tools are the need of the time; as reported by IBM (IBM, 2022), the average cost savings associated with fully deployed security, artificial intelligence (AI), and automation are USD 3.05 million. AI plays an important role in cybersecurity by making the process intelligently automated. The 4th Industrial Revolution is the AI revolution (Fouad, 2019). Along with HR at hand, AI can be used to protect computing systems from cybercrime and keep HR data safe. The notable AI techniques used to solve various cybersecurity issues include NLP, ML, DL, KRR, and rule-based ES (Khang & Misra et al., 2023). Some of these AI techniques are briefly discussed in Section 21.3.

The application areas of AI in data privacy and cybersecurity in HR systems include predicting, detecting, and managing cyberattacks and unauthorized access. Researchers present a comprehensive study on "AI-Centric Applications in Data Privacy and Cybersecurity for Human Resource Systems". The AI-based intelligent models can outperform conventional security systems (Sarker et al., 2021). We also present a workflow that can solve the integrated issues of data privacy and cybersecurity that occur in HR systems. Contributions of this chapter are:

- To provide a brief overview of the related work, the concept of AI-centric data privacy and cybersecurity.
- To present AI-centric data privacy and cybersecurity models. AI-based methods such as DL, NLP, ML, KRR, and ES are also discussed.
- Finally, we present a workflow that can solve the integrated issues of data privacy and cybersecurity that occur in HR systems.

21.2   BACKGROUND AND REVIEW OF THE RELATED WORK

This section introduces terms related to cybersecurity, the concept of AI-centric data privacy and cybersecurity, and a brief overview of the related work.

21.2.1   Data Privacy and Cybersecurity-Related Terms

Information and communication technology (ICT) has dominated our society and industry for more than 50 years. The data is integral to ICT, and as it is driven by global internet connectivity, users knowingly or unknowingly share their data online. The total amount of data is forecast to increase rapidly, reaching 180 zettabytes in 2025 (Taylor, 2022). As computing systems are driven by internet connectivity, data privacy and cybersecurity have gained importance like never before. Thus, detecting and preventing ICT systems from cyber threats has become a major concern for cybersecurity professionals in the present day (Rainie et al., 2014).

Data privacy is understood as the right of individuals to determine who has access to their personal information, what personal information is shared, and the protection of this information from unauthorized parties who should not have access to it (Princess Uche-Awaji, 2022). ICT security means the methods used by an organization to make sure that the computing systems are made available all the time and that integrity and confidentiality are maintained. If, at any point, the system remains unavailable, it may cost the enterprise dearly. The role of ICT in conjunction with the internet is to make things available, while cybersecurity is simply securing the things made available. There are several terms used interchangeably with the term cybersecurity

- Data security
- Information security
- Network security
- Internet security

All the above terms are related to and are frequently used synonymously with cybersecurity. Amongst all the terms, the term cybersecurity is more popular, and its popularity is only increasing, as shown in Figure 21.1.

Cybersecurity refers to understanding the cyber threats and corresponding defense strategies associated with confidentiality, integrity, and availability (Maalem et al., 2020).

21.2.2 CONFIDENTIALITY, INTEGRITY, AND AVAILABILITY

It is also known as the CIA triad and is a common model used to develop cybersecurity systems. The CIA triad with its properties are described below.

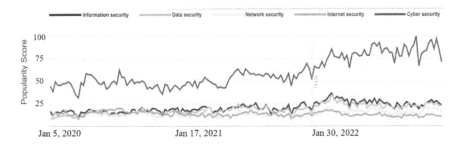

FIGURE 21.1   The comparative popularity score of the cybersecurity terms.

- **Confidentiality**: It refers to the efforts taken by an organization to keep data secret or private. The organization should keep a check on who should and should not have access to the HR data. There are several ways confidentiality can be negotiated. The attacker may gain unlawful access to the HR system, application, or database or use a technique called man-in-the-middle to steal or alter the data they intercept. Some attackers may spy on the network to steal the credentials or may try to gain more system privileges. These are all intentional violations of confidentiality; however, they may happen unintentionally. To ensure confidentiality, restricted data must be classified and labeled, access control policies need to be enabled, multi-factor authentication may be used, and last but not least - all those who are in contact with data should be trained to recognize and avoid the confidentiality breach.
- **Integrity**: It means making sure the data is trustworthy and untampered. Only authentic, accurate, and reliable data passes the integrity check. For example, on an organization's website, the listed information needs to have integrity; otherwise, visitors may feel the organization is not trustworthy. There are certain techniques used to preserve the integrity of the data - like hashing, encryption, digital certificates, or digital signatures.
- **Availability**: The data should be available online when needed. The applications, networks, and system as a whole should be functioning. For example, if an organization has no disaster recovery system in place and if any natural disaster like a flood occurs, its availability may be compromised. Availability can also be compromised because of cyberattacks.

Overall, based on the CIA triad, it can be concluded that confidentiality is all about who is allowed to access the data, integrity is to provide accurate data, and availability is to make this data available when needed.

21.2.3 CYBERSECURITY INCIDENTS AND ATTACKS

Table 21.1 lists several common cyber threats and cyberattacks; this list is not exhaustive but is increasing day by day.

21.2.4 CYBERSECURITY DEFENSE STRATEGIES

A carefully drafted defense strategy is needed to defend the HR system against these threats, it is known as a cybersecurity defense strategy. These strategies are for the overall protection of underlying software, hardware, and data and are responsible for preventing cybersecurity incidents (Khraisat et al., 2019). Below are some of the traditionally used cybersecurity defense strategies.

- Antivirus software (Maimon, 2019) is programmed to scan incoming files on the computing device. It detects and removes any viruses during routine scans.
- Access control (Qi et al., 2018) is a cybersecurity technique that regulates the person who creates, reads, updates, or deletes the resources

TABLE 21.1
Common Cyber-Threats and Cyberattacks

Technical Term	Description	References
Insider threats	Threat occurs from within the organization and is done by a legitimate user	Warkentin and Willison (2009)
Unauthorized access	Accessing information without proper access credentials	Moghimi et al. (2019)
Man-in-the-middle	When someone positions himself in a conversation between a user and an application	Kügler (2003)
Social engineering	The tactic of manipulating, influencing, or deceiving a victim	Jang-Jaccard and Nepal (2014)
Hacking	The attacker gains illegal access to a computer system, network, or device	Alsayed and Bilgrami (2017)
Phishing	The practice of deceiving, pressuring, or manipulating people into sending information or assets to the wrong people	Jang-Jaccard and Nepal (2014)
Data breach	An incident where information is stolen or taken from a system	Shaw (2010)
Computer virus	A type of malicious software or malware	Jang-Jaccard and Nepal (2014)
SQL injection	A cyberattack that injects malicious SQL code into an application	Boyd and Keromytis (2004)
Malware	A class of software including ransomware, viruses, worms, and spyware	Jang-Jaccard and Nepal (2014)
Ransomware	A class of Malware that affects and restricts access to the computer system	McIntosh et al. (2019)
DoS	An attack meant to shut down a machine or network	Jang-Jaccard and Nepal (2014)
Distributed DoS	A cybercrime in which the attacker floods a server with internet traffic	Jang-Jaccard and Nepal (2014)
Zero-day attack	An attack that exploits previously unknown hardware	Alazab et al. (2011)
Trojan horse	A type of malware that downloads onto a computer	Wijayarathne (2022)

- A firewall (Yin, 2016) is a cybersecurity device used to keep a check on incoming and outgoing network traffic based on an organization's previous cybersecurity policies. At its most basic, a firewall is essentially a filter that is placed between a private internal network and the public Internet. Its main purpose is to allow needed traffic in and keep dangerous traffic out.
- Anti-malware (Xue et al., 2017) protects computing systems from malware.
- Multi-factor authentication (MFA) (Williamson & Curran, 2021) forces the extra layer of authentication check.
- SIEM (Irfan et al., 2016) is a security solution that helps organizations recognize potential security threats.

All the above-listed cybersecurity defense strategies have their own benefits and application areas, but they lack or do not fully utilize one very important aspect; the intelligence and context of the application. With the increase in cyberattacks and the novelty of these attacks, the traditional approaches may not be enough to protect the computing systems (Alsayed & Bilgrami, 2017). AI in cybersecurity can automate time-consuming HR tasks done manually by human experts. It scans vast amounts of data and identifies potential threats. This helps human expert's focus on more critical security tasks.

Each cybersecurity defense mentioned above has its own use case, but there is growing interest in AI-centric applications in data privacy and cybersecurity and its intelligence modeling. The concepts of ML, DL, NLP, and KRR are used for AI-centric modeling to solve numerous cybersecurity tasks, such as automatic detection and prevention of phishing, malware, fraud, intrusion, etc. (Khang & Muthmainnah et al., 2023).

21.3 AI-CENTRIC APPLICATIONS IN DATA PRIVACY AND CYBERSECURITY:

As discussed in the previous section, intelligent data privacy and cybersecurity are based on AI.

21.3.1 MACHINE LEARNING-BASED MODELING

ML, along with DL, is an important part of AI and can be applied to build an effective cybersecurity model. In the following section, some of the algorithms are discussed.

- **Supervised Learning**: It is a subcategory of ML and AI. It uses labeled datasets to train algorithms.
- **Unsupervised Learning**: It uses ML algorithms to analyze and cluster unlabeled datasets.
- **Deep Learning**: It is a subset of ML, which is a neural network with three or more layers.

The ML and DL methods discussed above are used to gain insights from the cybersecurity data and can be used to build a data-driven cybersecurity model with applicability to HR systems, as shown in Figure 21.2.

21.3.2 NLP-BASED MODELING

NLP is the branch of AI that enables computers to understand text and words in a natural human way.

Following are the several parts of NLP:

- **Lexical Analysis**: It is the process of identifying and analyzing the structure of words and phrases. For instance, the classification of malicious domains was done by (Kidmose et al., 2018) using lexical analysis. Their study found

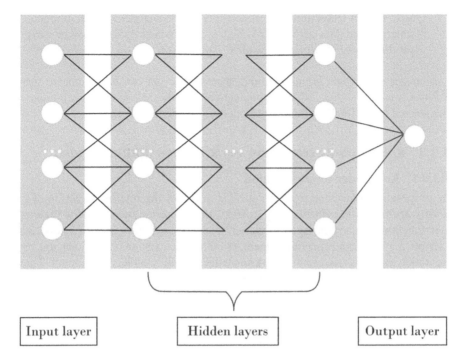

FIGURE 21.2 High-level view of an artificial neural network.

that utilizing lexical features can upsurge the detection efficiency of fraudulent websites with a precision of 0.98.

- **Syntactic Analysis**: In syntactic analysis, the words of a sentence are analyzed for grammar. A parsing (Perera et al., n.d.) algorithm is used to do so; thereby, these words are arranged in a way to show relationships among themselves. This parsing can be used to develop an NLP-based cyberattack prediction model that quickly extracts useful data from a large chunk of data.
- **Identify Software Vulnerabilities**: NLP can be used to make HR software more secure by automating the process of finding bugs in it. The code written by the programmers is usually repetitive because of the strict syntactic rules of a programming language. To identify the bugs in software, the code can be cleaned up to enhance readability and then compared to existing language models prepared by the code conventions of the organization (Saini & Kumar Singh, 2022).
- **Voice-Over Queries**: The security practitioners in an organization look at and report security logs manually (Georgescu, 2020), but NLP can automate this task (Kaur et al., 2021). Just like any other voice command service, security professionals can query and manage security systems with voice commands. Tasks such as troubleshooting or highlighting alerts that require human intervention would also be done by NLP systems shortly (Deng et al., 2019).

- **Quick Threat Detection**: Threat detection needs analysis of large amounts of unstructured data (Hilton et al., 2022), this is where the NLP can be used to produce some meaningful insights.

In summary, if the data is unstructured, threat detection can be automated using NLP-based models. Thus, NLP in conjunction with the ML and DL techniques is instrumental in the intelligent management of cybersecurity threats.

21.3.3 KNOWLEDGE REPRESENTATION AND REASONING (KRR) MODELING

21.3.3.1 Knowledge Representation

The purpose of knowledge representation is to model the intelligent action of the security agent. In the cybersecurity context, this representation benefits the computer function like a security practitioner. There are several types of knowledge: descriptive, structural, procedural, meta, etc. There are also several knowledge representation methods used to build a knowledge-based cybersecurity model. These methods are: logical, semantic network, frame, and production rules (Stephan et al., 2007). In the following, we summarize these.

- Logical representation
- Semantic network representation
- Frame representation
- Production rules

In summary, there are multiple ways in which knowledge can be represented in the HR knowledge-based conceptual model. In the following section, we provide an approach for modeling contextual cybersecurity knowledge to retrieve needed information in the HR system's context.

21.3.3.2 Ontology-Based Cybersecurity Context-Based Knowledge Model

To represent knowledge in cybersecurity is highly complex because of its context-specific nature. Security practitioners, in addition to understanding basic security concepts, also need to know the context of the application. Wen and Katt (2019) argue that context is important for effective cyber defense, as shown in Figure 21.3.

This model is a combination of two fundamental parts: application context and security contextualization, with interaction at scenario and instructions. The application context encompasses the functional area and the platform type, in this case, HR, whereas the security contextualization model encompasses concrete security attacks, concrete security weaknesses, and concrete security practices. This design ensures that the security practitioners understand the cybersecurity-relevant aspects of the underlying application and can also identify potential cyberattacks and cybersecurity threats based on the domain of the application (Khang & Gupta et al., 2023).

21.4 AI-BASED SIEM WORKFLOW

SIEM is a valued component of an organization's cyber defense (Irfan et al., 2016). For cyberattack and cyber-threat detection, SIEM practitioners spend much time and

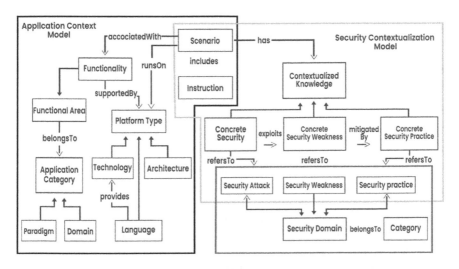

FIGURE 21.3 Ontology-based cybersecurity context-based knowledge mode.

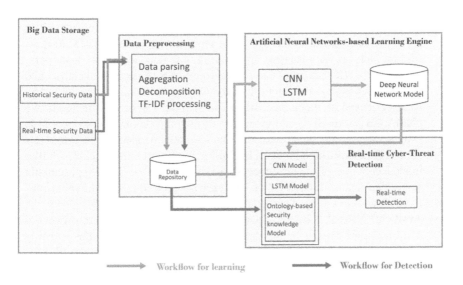

FIGURE 21.4 Workflow incorporating AI-based SIEM and ontology-based security model in the HR systems.

effort differentiating false security alerts from genuine ones. Hence, an AI-based SIEM is proposed in conjunction with the ontology-based security knowledge model to automate the cyberattack and cyber-threat detection in a given HR context, as shown in Figure 21.4.

The logic behind the proposed workflow, as shown in the above figure, is explained as follows. This model employs deep learning methods as well as the data preprocessing mechanism to handle large HR data; both historical and real-time streams. The historical and real-time raw HR data is fed to the deep learning and ontology-based model after it is piped through the data preprocessing mechanism.

Data is parsed, aggregated, and decomposed before being stored in the repository (Khang, 2023).

The resultant cleaned data is later piped to the deep learning model, where the corresponding models like FCNN, CNN, and LSTM can be used. The historical HR data, after being preprocessed and passed through the deep learning models, is piped to the ontology-based model for contextual knowledge. The real-time data is streamed directly to this model, as it can be at the maximum capacity of the stream. Finally, the AI-based model can detect cybersecurity threats in real time.

- **Big data Storage**: Frequently, big data storage is used to collect data for cybersecurity defense as it can collect and process live data and can also be specialized for cyber defense (Lee et al., 2017). Because of these capabilities, we propose a big data platform for our workflow.
- **Data Pre-processing**: ML algorithms need data to train upon; hence, it is necessary to feed the right data to solve the particular problem. The first common step in a deep learning workflow is to pre-process data. It ensures that the data is in a format that is acceptable to artificial neural networks. This pre-processing has two basic advantages; it enhances the desired features and reduces the artifacts that can bias the network. Data pre-processing can also remove noise from the input data. Data pre-processing includes data parsing, aggregation, and decomposition. Data parsing converts HR data from one format to another to better understand and use it. Data parsing takes input from the large dataset and then builds a data structure based on predefined rules. The data parsing process includes two phases: lexical analysis and syntax analysis, both defined in Section 21.2. Data aggregation is the process of reducing many measurements into a few values. Decomposition breaks a big problem into more manageable smaller independent ones. Data pre-processing also includes the TF-IDF processing stage. The cleanses and processed data are now stored in the data repository to be piped into the needed models.
- **Artificial Neural Network-Based Learning Engine**: The pre-processed historical data sitting in the data repository is now fed to the learning engine. Naseer et al. (2018) used CNN, Autoencoder, and RNN with 85% efficiency for intrusion detection. The main advantage of LSTM is the improvement in learning abilities, as they are very efficient in modeling complex sequential data.
- **Real-Time Cyber-Threat Detection**: The knowledge gained using the artificial neural network-based learning engine is then fed to the model running CNN and LSTM in conjunction with the ontology-based cybersecurity context-based knowledge model. Knowledge of cybersecurity is highly complex and context-dependent. The application of ontology-based models will enable cybersecurity practitioners to tackle context-specific security threats. Knowledge can be appreciated fully without knowing the context (Jafari et al., 2008); otherwise, this knowledge is isolated from other relevant knowledge, and as a result it may be limited or even distorted (Goldkuhl & Braf, 2001). Knowledge in itself has little meaning unless it is seen in context, in this case the HR systems.

The knowledge gained from artificial neural network-based learning engines, when fed to the CNN and LSTM models in conjunction with ontology-based security knowledge should prove beneficial for enhanced and automated cyber-threat detection. It can also reduce false positive alerts and automate the response to cyber threats (Khang & Kali et al., 2023).

21.5 CONCLUSION

The field of cybersecurity has gained importance in the recent past and AI is a game changer in many domains, particularly HR. In this chapter, researchers have studied data privacy and cybersecurity in the context of AI in HR systems. The main focus of researchers was to present how AI can be central in intelligent decision-making and present a workflow for AI-centric applications of data privacy and cybersecurity for HR systems (Khang & Hajimahmud et al., 2023).

Researchers presented AI-centric methods such as ML, DL, NLP, etc. Such AI-centric methods can be efficiently used to tackle cybersecurity issues in the context of HR systems with minimal or no human intervention. Later, we briefly discussed various cyber threats and attacks, presented some of the traditional cyber defenses, and discussed how they fall short of securing a computing system (Khang & Rani et al., 2023).

In AI-centric cybersecurity, AI-based modeling discussed in this chapter can automate cyber defenses. Based on the presented material, researchers proposed an AI-based SIEM workflow that learns the threats involving HR data using deep learning methods and, in conjunction with the ontology-based model, can be used to detect and prevent cyber-threats automatically in the HR systems. This workflow's detailed design, development, implementation, and testing are reserved for future research. Researchers believe this chapter can serve as a good beginning for anyone interested in AI-centric data privacy and cybersecurity applications (Khang & Shah et al., 2023).

REFERENCES

Alazab, M., Venkatraman, S., Watters, P., & Alazab, M. (2011). Zero-day malware detection based on supervised learning algorithms of api call signatures. *CRPIT* (Vol. 121). Australian Computer Society, Sydney, NSW. https://www.academia.edu/download/74938048/Zero-day_Malware_Detection_based_on_Supe20211120-8318-944ht9.pdf

Alsayed, A. & Bilgrami, A. (2017). E-banking security: internet hacking, phishing attacks, analysis and prevention of fraudulent activities. *International Journal of Emerging Technology and Advanced Engineering*, 7, 1–12. https://www.researchgate.net/profile/Anwar-Bilgrami/publication/315399380_E-Banking_Security_Internet_Hacking_Phishing_Attacks_Analysis_and_Prevention_of_Fraudulent_Activities/links/59f19d7c0f7e9beabfca5f17/E-Banking-Security-Internet-Hacking-Phishing-Attacks-Analysis-and-Prevention-of-Fraudulent-Activities.pdf

Boyd, S. & Keromytis, A. (2004). SQLrand: preventing SQL injection attacks. *Proceedings of the 2nd Applied Cryptography and Network Security (ACNS) Conference* (Vol. 3089). Springer, Berlin. https://link.springer.com/chapter/10.1007/978-3-540-24852-1_21

Deng, Y., Lu, D., Huang, D., Chung, C. J., & Lin, F. (2019). Knowledge graph-based learning guidance for cybersecurity hands-on labs. *CompEd 2019-Proceedings of the ACM Conference on Global Computing Education*, pp. 194–200. Association for Computing Machinery, Inc., Sichuan, China. https://dl.acm.org/doi/abs/10.1145/3300115.3309531

Durate, F. (2023). Amount of Data Created Daily. https://acp.copernicus.org/articles/23/1511/2023/

Fouad, F. (2019). The fourth industrial revolution is the ai revolution from a business perspective. *International Journal of Information Technology*, 8, 5. https://doi.org/10.1201/9781003440901

Georgescu, T.-M. (2020). Natural language processing model for automatic analysis of cybersecurity-related documents. *Symmetry*, 12(3), 20. https://www.mdpi.com/2073-8994/12/3/354

Goldkuhl, G. & Braf, E. (2001). Contextual Knowledge Analysis - Understanding Knowledge and Its Relations to Action and Communication. https://www.google.com/books?hl=en&lr=&id=MEDSnYpaKr0C&oi=fnd&pg=PA197&dq=).+Contextual+Knowledge+Analysis+-+Understanding+Knowledge+and+Its+Relations+to+Action+and+Communication&ots=B7jXwGM4_b&sig=3M9vBeQUqaUPQbZ60FFbr-Bq3A8

Hilton, K., Siami Namin, A., & Jones, K. S. (2022). Metaphor identification in cybersecurity texts: a lightweight linguistic approach. *SN Applied Sciences*, 4(2), 60. https://link.springer.com/article/10.1007/s42452-022-04939-8

https://ieeexplore.ieee.org/abstract/document/7111524/

IBM. (2022). Cost of a Data Breach Report 2022, pp. 1–59. IBM. https://doi.org/10.1201/9781003440901

Irfan, M., Abbas, H., Sun, Y., Sajid, A., & Pasha, M. (2016). A framework for cloud forensics evidence collection and analysis using security information and event management: SCN-SI-085. *Security and Communication Networks*, 9, 16. https://onlinelibrary.wiley.com/doi/abs/10.1002/sec.1538

Jafari, M., Fathian, M., Jahani, A., & Akhavan, P. (2008). Exploring the contextual dimensions of the organization from a knowledge management perspective. *VINE*, 38, 53–71. https://www.emerald.com/insight/content/doi/10.1108/03055720810870897/full/html

Jang-Jaccard, J. & Nepal, S. (2014). A survey of emerging threats in cybersecurity. *Journal of Computer and System Sciences*, 80(5), 973–993. https://www.sciencedirect.com/science/article/pii/S0022000014000178

Kaur, P., Singh, S. K., Singh, I., & Kumar, S. (2021). Exploring Convolutional Neural Network in Computer Vision-based Image Classification. https://ceur-ws.org/Vol-3080/21.pdf

Khang, A. (2023). *Advanced Technologies and AI-Equipped IoT Applications in High-Tech Agriculture*. IGI Global Press, Hershey, PA. https://doi.org/10.4018/978-1-6684-9231-4

Khang, A., Hahanov, V., Abbas, G. L., & Hajimahmud, V. A. (2022). Cyber-physical-social system and incident management. *AI-Centric smart City Ecosystems: Technologies, Design and Implementation* (1st ed., pp. 2–15). CRC Press, Boca Raton, FL. https://doi.org/10.1201/9781003252542-2

Khang, A., Hajimahmud, V. A., Gupta, S. K., Babasaheb, J., & Morris, G. (2023). *AI-Centric Modelling and Analytics: Concepts, Designs, Technologies, and Applications* (1st ed.). CRC Press, Boca Raton, FL. https://doi.org/10.1201/9781003400110

Khang, A., Gupta, S. K., Rani, S., & Karras, D. A. (2023). *Smart Cities: IoT Technologies, Big Data Solutions, Cloud Platforms, and Cybersecurity Techniques*. CRC Press, Boca Raton, FL. https://doi.org/10.1201/9781003376064

Khang, A., Kali, C. R., Suresh Kumar, S., Amaresh, K., Sudhansu Ranjan, D., & Manas Ranjan, P. (2023). Enabling the future of manufacturing: integration of robotics and IoT to smart factory infrastructure in industry 4.0. *AI-Based Technologies and Applications in the Era of the Metaverse* (1st ed., pp. 25–50). IGI Global Press, Hershey, PA. https://doi.org/10.4018/978-1-6684-8851-5.ch002

Khang, A., Misra, A., Gupta, S. K., & Shah, V. (2023). *AI-Aided IoT Technologies and Applications in the Smart Business and Production*. CRC Press, Boca Raton, FL. https://doi.org/10.1201/9781003392224

Khang, A., Muthmainnah, M., Seraj, P. M. I., Yakin, A. A., Obaid, A. J., & Panda, M. R. (2023). AI-aided teaching model for the education 5.0 ecosystem. *AI-Based Technologies and Applications in the Era of the Metaverse* (1st ed., pp. 83–104). IGI Global Press, Hershey, PA. https://doi.org/10.4018/978-1-6684-8851-5.ch004

Khang, A., Rani, S., Gujrati, R., Uygun, H., & Gupta, S. K. (2023). *Designing Workforce Management Systems for Industry 4.0: Data-Centric and AI-Enabled Approaches*. CRC Press, Boca Raton, FL. https://doi.org/10.1201/9781003357070

Khang, A., Shah, V., & Rani, S. (2023). *AI-Based Technologies and Applications in the Era of the Metaverse* (1st ed.). IGI Global Press, Hershey, PA. https://doi.org/10.4018/978-1-6684-8851-5

Khang, A., Vugar, A., Vladimir, H., & Shah, V. (2024). *Advanced IoT Technologies and Applications in the Industry 4.0 Digital Economy* (1st ed.). CRC Press, Boca Raton, FL. https://doi.org/10.1201/9781003434269

Khraisat, A., Gondal, I., Vamplew, P., & Kamruzzaman, J. (2019). Survey of intrusion detection systems: techniques, datasets, and challenges. *Cybersecurity*, 2, 20. https://link.springer.com/article/10.1186/s42400-019-0038-7

Kidmose, E., Stevanovic, M., & Pedersen, J. (2018). Detection of Malicious Domains Through Lexical Analysis. https://ieeexplore.ieee.org/abstract/document/8560665/

Kügler, D. (2003). *"Man in the Middle" Attacks on Bluetooth* (Vol. 2742). Springer Nature, New York. https://link.springer.com/chapter/10.1007/978-3-540-45126-6_11

Kumar, A. (2019). (ISSN-2349-5162). In JETIREO06119. *Journal of Emerging Technologies and Innovative Research*, 6(3). https://ieeexplore.ieee.org/abstract/document/9965094/

Lee, K.-Y., Kim, K.-H., Kang, J.-J., Choi, S.-J., Im, Y.-S., Lee, Y.-D., & Lim, Y.-S. (2017). Comparison and analysis of linear regression & artificial neural network. *International Journal of Applied Engineering Research*, 12, 9820–9825. https://doi.org/10.1201/9781003440901

Li, S., Xu, L. D., & Zhao, S. (2015). The internet of things: a survey. *Information Systems Frontiers*, 17(2), 243–259. https://link.springer.com/article/10.1007/s10796-014-9492-7

Maalem Lahcen, R. A., Caulkins, B., Mohapatra, R., & Kumar, M. (2020). Review and insight on the behavioral aspects of cybersecurity. *Cybersecurity*, 3(1), 10. https://cybersecurity.springeropen.com/articles/10.1186/s42400-020-00050-w

Maimon, D. (2019). Existing Evidence for the Effectiveness of Antivirus in Preventing Existing Evidence for the Effectiveness of Antivirus in Preventing Cyber Crime Incidents. https://scholarworks.gsu.edu/ebcs_tools/6/

McAfee. (2020). New McAfee Report Estimates Global Cybercrime Losses to Exceed $1 Trillion. https://dl.acm.org/doi/abs/10.1145/3171533.3171535

McIntosh, T., Jang-Jaccard, J., Watters, P., & Susnjak, T. (2019). *The Inadequacy of Entropy-Based Ransomware Detection*, pp. 181–189. Springer, Cham. https://link.springer.com/chapter/10.1007/978-3-030-36802-9_20

Moghimi, D., Eisenbarth, T., & Sunar, B. (2019). MemJam: a false dependency attack against constant-time crypto implementations. *International Journal of Parallel Programming*, 47, 538–570. https://link.springer.com/article/10.1007/s10766-018-0611-9

Morgan, S. (2020). Cybercrime to Cost the World $10.5 Trillion Annually by 2021. https://link.springer.com/article/10.1007/s12103-021-09652-7

Naseer, S., Saleem, Y., Khalid, S., Bashir, M. K., Han, J., Iqbal, M. M., & Han, K. (2018). Enhanced network anomaly detection based on deep neural networks. *IEEE Access*, 6, 48231–48246. https://ieeexplore.ieee.org/abstract/document/8438865/

Perera, I., Hwang, J., Bayas, K., Dorr, B., & Wilks, Y. (n.d.). Cyberattack Prediction Through Public Text Analysis and Mini-Theories. https://ieeexplore.ieee.org/abstract/document/8622106/

Princess Uche-Awaji, A. (2022). Data Privacy and Data Protection: The Rights of Users and the Responsibility of Companies in the Digital World. https://papers.ssrn.com/sol3/papers.cfm?abstract_id=4005750

Qi, H., Di, X., & Li, J. (2018). Formal definition and analysis of access control model based on role and attribute. *Journal of Information Security and Applications*, 43, 53–60. https://www.sciencedirect.com/science/article/pii/S221421261730368X

Rainie, L., Anderson, J., & Connolly, J. (2014). Cyber Attacks Likely to Increase. https://policycommons.net/artifacts/619510/cyber-attacks-likely-to-increase/1600627/

Saini, R. & Kumar Singh, S. (2022). Natural Language Processing Applications in Cyber Security. https://ieeexplore.ieee.org/abstract/document/9817336/

Sarker, I. H., Kayes, A. S. M., Badsha, S., Alqahtani, H., Watters, P., & Ng, A. (2020). Cybersecurity data science: an overview from a machine learning perspective. *Journal of Big Data*, 7(1), 41. https://link.springer.com/article/10.1186/s40537-020-00318-5

Shaw, A. (2010). Data Breach: From Notification to Prevention Using PCI DSS, Vol. 43. https://heinonline.org/hol-cgi-bin/get_pdf.cgi?handle=hein.journals/collsp43&section=24

Stephan, G., Pascal, H., & Andreas, A. (2007). Knowledge representation and ontologies. In R. Studer, S. Grimm, & A. Abecker (Eds.), *Semantic Web Services: Concepts, Technologies, and Applications*, pp. 51–105. Springer Berlin, Heidelberg. https://link.springer.com/chapter/10.1007/978-3-642-02788-8_6

Tapiador, J., Orfila, A., Ribagorda, A., & Ramos, B. (2013). Key-recovery attacks on KIDS, a keyed anomaly detection system. *IEEE Transactions on Dependable and Secure Computing*, 12, 3. https://ieeexplore.ieee.org/abstract/document/6598669/

Taylor, P. (2022). The volume of data/information created, captured, copied, and consumed worldwide from 2010 to 2020, with forecasts from 2021 to 2021. https://doi.org/10.1201/9781003440901

Velte, T., Velte, A., & Elsenpeter, R. (2009). *Cloud Computing, A Practical Approach.* McGraw-Hill, Inc., New York. https://ds.amu.edu.et/xmlui/bitstream/handle/123456789/9207/ Cloud Computing.pdf?sequence=1&isAllowed=y

Warkentin, M., & Willison, R. (2009). Behavioral and policy issues in information systems security: the insider threat. *EJIS*, 18, 101–105. https://www.tandfonline.com/doi/abs/10.1057/ejis.2009.12

Wen, S. F. & Katt, B. (2019). Managing software security knowledge in context: an ontology-based approach. *Information (Switzerland)*, 10(6), 216. https://www.mdpi.com/2078-2489/10/6/216

Wijayarathne, S. (2022). Trojan Horse Malware- Case Study. https://doi.org/10.1201/9781003440901

Williamson, J. & Curran, K. (2021). Best practice in multi-factor authentication. *Semiconductor Science and Information Devices*, 3, 1. https://journals.bilpubgroup.com/index.php/ssid/article/view/3152

Xue, Y., Meng, G., Liu, Y., Tan, T. H., Chen, H., Sun, J., & Zhang, J. (2017). Auditing anti-malware tools by evolving android malware and dynamic loading technique. *IEEE Transactions on Information Forensics and Security*, 1, 1529–1544. https://ieeexplore.ieee.org/abstract/document/7837653/

Yin, J. (2016). Firewall policy management, US Patent 9338134. https://doi.org/10.1201/9781003440901

22 Cybersecurity Techniques in Talent Management and Human Capital Management Systems

Wasswa Shafik and Alex Khang

22.1 INTRODUCTION

In today's digital age, technology has become integral to business operations, including talent management (TM) and human capital management (HCM). However, with the increasing use of these systems, the risk of cyber-attacks and data breaches has also grown significantly (Sheoraj et al., 2023; Jun et al., 2021). This has increased the emphasis on implementing cybersecurity techniques in TM and HCM systems to safeguard sensitive data and ensure business continuity. Cybersecurity techniques in TM and HCM systems include a range of measures, including secure coding practices, access controls, encryption, and regular vulnerability assessments (Ukwandu et al., 2023). By adopting these measures, organizations can protect their critical data, reduce the risk of security breaches, and maintain the trust of their stakeholders. This article will explore some of the most effective cybersecurity techniques used in TM and HCM systems to enhance security and mitigate the risk of cyber-attacks (Alejandro et al., 2023; Budde et al., 2023).

Securing access to sensitive data is one of the most crucial aspects of cybersecurity in TM and HCM systems. Organizations must ensure that only authorized personnel can access and modify critical data. This can be achieved by implementing access controls that limit access to specific data based on user roles and permissions (Allison & Stepney, 2023). Multi-factor authentication (MFA) can also provide an extra layer of security, making it more difficult for cybercriminals to gain unauthorized access. Another essential aspect of TM and HCM systems' cybersecurity is using encryption techniques to protect information and data in transit and stored. Encryption involves converting data into a code that can only be deciphered with a specific key (Fisk et al., 2023).

Regular vulnerability assessments can help organizations identify weaknesses in their TM and HCM systems, allowing them to address these issues before cyber-criminals can exploit them. Organizations can significantly reduce the risk of

DOI: 10.1201/9781003440901-22

cyber-attacks and data breaches by regularly assessing and addressing vulnerabilities (Mahamood et al., 2023). Another vital aspect of cybersecurity in TM and HCM systems is the need to train employees on best practices for cybersecurity. Employees are often the first line of defense against cyber threats and can play a significant role in safeguarding sensitive data. Regular cybersecurity training can help employees understand the importance of cybersecurity, identify potential threats, and take steps to mitigate risks (Chopin & Décary-Hétu, 2023; Zhao et al., 2022).

It is also essential for organizations to have a robust incident response plan in place in case of a security breach. An incident response plan outlines the steps to take during a security incident, including notifying affected parties and mitigating damage (Alshammari, 2023). A well-defined incident response plan can help organizations respond quickly and effectively to security breaches, minimizing the impact on their operations and reputation. It is essential to ensure that third-party vendors with access to TM and HCM systems also have strong cybersecurity measures. Organizations need to assess their vendors' cybersecurity posture, including their policies and procedures, and ensure that they align with the organization's standards (Tran et al., 2023). It is also important to have contracts that clearly outline cybersecurity responsibilities and expectations.

Therefore, cybersecurity is a critical consideration in TM and HCM systems, and it requires a holistic approach that includes a range of techniques, including access controls, encryption, regular vulnerability assessments, employee training, incident response planning, and vendor management (Kucharčíková et al., 2023; Shafik et al., 2022). By implementing robust cybersecurity measures, organizations can protect their critical data and ensure the continuity of their business operations.

22.1.1 A Brief Overview of TM and HCM Systems

Organizations use TM and HCM systems to manage various aspects of their workforce, such as recruitment, onboarding, performance management, training, and development. TM systems typically manage an organization's talent pool, including identifying and attracting top talent, assessing skills, and retaining key employees (Behie et al., 2023). HCM systems, on the other hand, are broader in scope and are designed to manage all aspects of an organization's workforce, including payroll, benefits administration, and compliance with employment laws and regulations. As a result, these systems can help organizations streamline their HR processes, reduce administrative burdens, and provide insights into their workforce (Behie et al., 2023). In addition, with the increasing use of technology in HR, cybersecurity has become a critical consideration in TM and HCM systems to protect sensitive employee data.

TM and HCM systems have become increasingly prevalent in recent years, with organizations of all sizes and industries adopting these platforms to manage their workforce more effectively. As a result, the market for TM and HCM systems has also grown significantly, with a wide range of vendors offering various solutions to meet the needs of different organizations (Afenyo & Caesar, 2023). In addition to the TM and HCM systems' market growth, there has been a growing focus on cybersecurity in these systems. With the increasing frequency and sophistication of cyber-attacks, organizations recognize the importance of implementing strong cybersecurity measures in their TM and HCM systems to protect their critical employee data (Khang & Hahanov et al., 2022).

However, despite the growing emphasis on cybersecurity, many organizations still face challenges in effectively implementing and managing cybersecurity measures in their TM and HCM systems. This is often due to a lack of resources or expertise in cybersecurity or a failure to prioritize cybersecurity as a critical aspect of their HR operations (Khan et al., 2023). Moving forward, it is likely that the importance of cybersecurity in TM and HCM systems will continue to grow as organizations become more reliant on these platforms to manage their workforce. Organizations that prioritize cybersecurity and take proactive steps to enhance their cybersecurity measures will be better positioned to protect their critical data and maintain the trust of their stakeholders (Bukauskas et al., 2023).

22.1.2   IMPORTANCE OF CYBERSECURITY

Cybersecurity is of utmost importance in TM and HCM systems for several reasons, including protecting sensitive employee data, where TM and HCM systems typically contain a vast amount of sensitive employee data, including personal and financial information (Pinto et al., 2023). A cybersecurity breach could compromise this data, leading to identity theft, financial loss, and reputational damage for both the organization and the affected employees. Compliance with regulations where many countries have data protection laws and regulations that organizations must comply with to avoid legal penalties (Allison & Stepney, 2023; HILL et al., n.d.). Non-compliance can result in hefty fines and damage to an organization's reputation. Strong cybersecurity measures are crucial to ensure compliance with these regulations.

Business continuity in such a way that cybersecurity breaches can disrupt an organization's operations, leading to downtime, lost productivity, and financial losses. This can impact an organization's ability to deliver customer services and damage its reputation (Anaam et al., 2023). By implementing strong cybersecurity measures, organizations can reduce the risk of cyber-attacks and ensure the continuity of their business operations. As well as trust and confidence, in today's digital age, organizations must demonstrate that they are trustworthy custodians of sensitive data. By implementing strong cybersecurity measures in TM and HCM systems, organizations can demonstrate their commitment to protecting employee data and building stakeholder trust and confidence. Cybersecurity is critical in TM and HCM systems to protect sensitive employee data, ensure compliance with regulations, maintain business continuity, and build stakeholder trust and confidence (Zhan et al., 2023). Organizations prioritizing cybersecurity in their HR operations will be better positioned to mitigate cyber-attack risks and safeguard their critical data (Lüersmann, 2023). Due to the above, this chapter provides an overview of the TM and HCM systems from a cybersecurity perspective.

The rest of this chapter is structured into six sections. Section 22.2 demonstrates cybersecurity threats in TM and HCM systems, including a brief overview of common threats, for instance, phishing, malware, and insider threats, with examples of past security breaches in TM and HCM systems. Section 22.3 presents the cybersecurity techniques for TM and HCM systems, like encryption and data protection, access control and authentication, network security, and incident response planning. Section 22.4 broadly illustrates some of the best practices for implementing cybersecurity in TM and HCM systems, including employee education and training, regular security audits and assessments, a collaboration between information technology (IT) and HR teams, and compliance with industry regulations (for example, general data protection regulation (GDPR) and health insurance portability and accountability act (HIPAA)). Section 22.5 presents some case studies on the real-world examples of organizations that have successfully implemented cybersecurity in their TM and HCM systems and the analysis of the impact of cybersecurity measures on these organizations. Finally, Section 22.6 avails the conclusion with a summary of the importance of cybersecurity and a few recommendations for organizations looking to improve their cybersecurity posture in these areas.

22.2　CYBERSECURITY THREATS IN TM AND HCM SYSTEMS

Organizations need to implement robust cybersecurity measures to mitigate these cyber threats, including access controls, encryption, regular vulnerability assessments, employee training, incident response planning, and vendor management. By taking a proactive approach to cybersecurity, organizations can reduce the risk of cyber-attacks and safeguard their critical data; cybersecurity threats and some examples in HM and HCM systems are presented in this section (Rani & Chauhan et al., 2021).

22.2.1　CYBERSECURITY THREATS

There are several cybersecurity threats that organizations need to be aware of in TM and HCM systems, including the following.

22.2.1.1　Phishing Attacks

Phishing attacks are a common cyber-attack where an attacker sends fraudulent emails or messages to employees, often disguised as legitimate requests, to trick them into revealing sensitive information or clicking on malicious links (Furstenau et al., 2023; Shokoor et al., 2022). Once attackers have access to this information, they can compromise TM and HCM systems in several ways, like unauthorized access, identity theft, malware installation, and reputational damage. Organizations need to implement robust cybersecurity measures, including employee training, access controls, and incident response planning. Employees should be trained to identify phishing attacks and report suspicious emails or messages. Access controls should be implemented to restrict access to sensitive data and systems. Finally, incident response planning should be in place to enable organizations to respond quickly and effectively to a phishing attack as shown in Figure 22.1. By taking a proactive approach to cybersecurity, organizations can reduce the risk of a successful phishing attack and safeguard their critical TM and HCM data (Khang & Vugar et al., 2024).

22.2.1.2　Malware Attacks

Malware attacks are a type of cyber-attack that involves installing malicious software on an organization's network. These attacks typically involve installing malicious

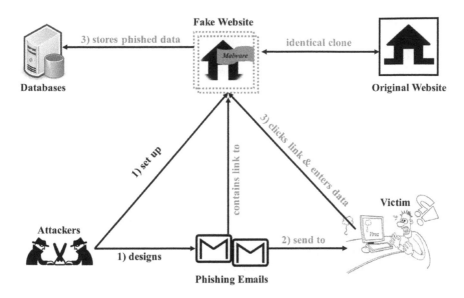

FIGURE 22.1   Phishing emails (Khang, 2021).

software on an organization's network, which can be used to steal sensitive data, disrupt business operations, or enable further attacks on TM and HCM systems (Yao et al., 2021; Shafik, 2023). Some ways in which malware attacks can affect TM and HCM include data theft, system disruption, backdoor access, and ransomware. Organizations must implement strong cybersecurity measures to mitigate the impact of malware attacks on TM and HCM systems, including access controls, regular vulnerability assessments, and incident response planning. Access controls should be implemented to restrict access to sensitive data and systems. Regular vulnerability assessments should be conducted to identify and patch security vulnerabilities. Finally, incident response planning should be in place to enable organizations to respond quickly and effectively to a malware attack.

22.2.1.3   Insider Threats

Insider threats refer to employees or contractors who intentionally or accidentally compromise an organization's security (Yao et al., 2021). This could include employees stealing sensitive data or accidentally exposing it through poor security practices leading to mainly system sabotage, unauthorized access, and data manipulation. To mitigate the impact of insider threats on TM and HCM systems, organizations need to implement robust cybersecurity measures, including access controls, monitoring and auditing, and employee training. First, access controls should be implemented to restrict access to sensitive data and systems. Monitoring and auditing should be conducted to detect and investigate suspicious behavior by insiders. Also, employee training should focus on the risks of insider threats and how to report suspicious behavior.

22.2.1.4 Distributed Denial of Service (DDoS) Attacks

DDoS attacks involve overwhelming an organization's network with traffic to disrupt business operations. This can lead to system downtime, data theft, lost productivity, financial losses, and reputation damage (Tan et al., 2023). To mitigate the impact of DDoS attacks on TM and HCM systems, organizations need to implement robust cybersecurity measures, including network security, traffic monitoring, and incident response planning. Network security measures, such as firewalls and intrusion detection systems, can help detect and prevent DDoS attacks. Traffic monitoring can help detect abnormal traffic patterns and enable organizations to respond quickly to an attack (Wah, 2023). Incident response planning should be in place to enable organizations to respond quickly and effectively to a DDoS attack as shown in Figure 22.2. By taking a proactive approach to cybersecurity, organizations can reduce the risk of a successful DDoS attack and safeguard their critical TM and HCM data.

22.2.1.5 Ransomware Attacks

Ransomware attacks involve encrypting an organization's data and demanding payment for the decryption key. Ransomware attacks can be highly disruptive and can result in the loss of critical data, reputation damage, financial losses, and system downtime (Shafik & Matinkhah, 2021). To mitigate the impact of ransomware attacks on TM and HCM systems, organizations need to implement robust cybersecurity measures, including regular backups, anti-malware software, and employee training. Regular backups can help organizations recover their data in case of a ransomware attack. Anti-malware software can detect and prevent ransomware infections (Yang et al., 2021; Meng et al., 2020). Employee training should focus on recognizing and avoiding phishing attacks, which are common ways for ransomware to infiltrate systems.

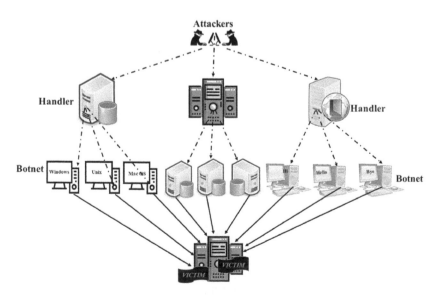

FIGURE 22.2 Distributed DoS attack.

22.2.1.6 Third-Party Risks

TM and HCM systems often involve third-party vendors, such as software providers, data processors, and cloud service providers. These third-party vendors can introduce additional cybersecurity risks, such as data breaches or unauthorized access to sensitive data, compliance violations, and system vulnerabilities (Tyagi et al., 2023; Shafik et al., 2020). Organizations must implement strong cybersecurity measures, including vendor risk management, due diligence, and contractual controls. Vendor risk management should include assessments of third-party partners' security controls and compliance with regulations and industry standards. In addition, due diligence should be conducted before engaging with third-party partners to ensure they meet the organization's security and compliance requirements. Finally, contractual controls should be implemented to ensure third-party partners are held accountable for security and compliance violations.

22.2.1.7 Weak Passwords

Weak passwords can significantly impact the system, or the reuse of passwords across multiple accounts can make it easier for attackers to gain unauthorized access to TM and HCM systems. Additionally, they are easy to guess or crack, making it easier for attackers to gain unauthorized access to systems and data. But most critically, account compromise, data breaches, or even insider threats (Beenish et al., 2023; Ndayishimiye et al., 2023). Organizations are recommended to implement strong password policies and practices, including password complexity requirements, MFA, and employee training. Password complexity requirements, including length, complexity, and regular password changes should be enforced. MFA should be implemented to add a layer of security. Employee training should focus on creating strong passwords and not reusing passwords across multiple accounts.

22.2.1.8 Lack of Encryption

Encryption converts data into a format that can only be read by authorized parties, making it more difficult for attackers to access and steal sensitive data. Sensitive employee data should be encrypted both in transit and at rest to protect it from unauthorized access (Cuong & Thanh, 2023). Without encryption, sensitive data is vulnerable to interception and compromise. There is a need to implement strong encryption practices, including encryption of sensitive data at rest and in transit. Encryption should be used for sensitive employee data, such as personal and financial information, and all data transmitted over public networks. Additionally, organizations should implement access controls to ensure that only authorized parties can access encrypted data.

22.2.1.9 Insider Data Theft

Employees with authorized access to TM and HCM systems can intentionally or unintentionally steal sensitive data for personal gain or malicious purposes. Insider data theft occurs when an employee or other authorized user of the system intentionally steals sensitive data or intellectual property. This leads to the loss of sensitive data, intellectual property theft, and compliance violations (Zhan et al., 2023).

Therefore, organizations must implement strong access controls, employee training, and data loss prevention (DLP) measures. Access controls should limit employees' access to sensitive data and systems to only those who need them to perform their job functions. Employee training should focus on the importance of data protection and the risks of insider data theft. DLP measures should be implemented to monitor and detect the unauthorized transfer of sensitive data (Khanh & Khang et al., 2021).

22.2.1.10   Lack of Regular Software Updates

TM and HCM systems often require regular software updates to patch security vulnerabilities and prevent cyber-attacks. Failing to apply software updates can leave systems vulnerable to attacks and data loss. Organizations need to implement vital patch management processes to mitigate the impact of the lack of regular software updates on TM and HCM systems. This includes regularly assessing software for vulnerabilities, prioritizing updates based on risk, and testing updates before deployment to ensure they do not cause unintended consequences. Additionally, organizations should consider using automated patch management tools that can help streamline the process and ensure updates are applied consistently across all systems. By taking a proactive approach to software updates, organizations can reduce the risk of a successful attack and safeguard their critical TM and HCM data. Thus, TM and HCM systems are vulnerable to various cybersecurity threats, including phishing attacks, malware attacks, insider threats, DDoS attacks, and ransomware attacks (Wah, 2023). Therefore, organizations must implement strong cybersecurity measures and remain vigilant to mitigate these threats and safeguard their critical data.

22.2.2   Examples of Past Security Breaches in TM and HCM Systems

There exist several global breaches.

Target: In 2013, Target suffered a massive data breach that impacted its TM and HCM systems, resulting in the theft of personal information for over 70 million customers, including names, addresses, and credit card information. Yahoo: In 2016, Yahoo revealed that a 2013 data breach had compromised the personal information of all 3 billion user accounts, including their names, email addresses, birth dates, and phone numbers. Equifax: In 2017, Equifax experienced a data breach that exposed the personal information of over 147 million people, including their names, addresses, social security numbers, and birth dates.

Marriott International: In 2018, Marriott International reported a data breach that compromised the personal information of over 383 million customers, including their names, addresses, phone numbers, email addresses, passport numbers, and travel history.

Capital One: In 2019, Capital One suffered a data breach that exposed the personal information of over 100 million customers, including names, addresses, credit scores, and social security numbers.

These examples highlight the significant risk of cyberattacks to TM and HCM systems containing sensitive and personal information. Such breaches can severely affect individuals and organizations, including financial losses, reputational

TABLE 22.1

Selected Security and Data Breach in HCM Systems (https://www.upguard. com/blog/biggest-data-breaches-in-healthcare)

Date	Data Breach	Impact (patients)	Compromised	Solution
September 2011	Tricare	5 million	Prescription information, lab tests, clinical notes, personal health data, phone numbers, addresses, names, social security numbers	Data encryption policies
April–June, 2014	Community Health Systems	4.5 million	Addresses, phone numbers, numbers, social security, birth dates, names	Teach employees to diagnose the threatening traits of malware injection attempts.
July 2015	UCLA Health	4.5 million	Some medical data: identification number, health plan number, Medicaid number, social security numbers, dates of birth, names	Exhaustive investigation
August 2013	Advocate Health Care	4.03 million	Health insurance information, clinical information, demographic information, dates of birth, addresses, names	Physical security controls
July 2015	Medical Informatics Engineering	3.9 million	Birth statistics, names of children, medical conditions, doctor names, disability codes, diagnosis, lab results, social security numbers, dates of information, and security questions	Monitor for sensitive data
May 2020	Trinity Health	3.3 million	Certain financial information, claim information, medications, lab results, medical record numbers, vaccination types,	The consciousness of the ransomware attack sequence

damage, and legal and regulatory penalties. Therefore, organizations need to implement robust cybersecurity measures to protect against such threats and stay vigilant in the face of constantly evolving cyber threats; Table 22.1 presents some data breaches that happened in different years and the affected sensitive data of patients.

22.3 CYBERSECURITY TECHNIQUES FOR TM AND HCM SYSTEMS

Within this section, we present some of the cybersecurity techniques that can be used to protect TM and HCM systems from cyber threats, and these include:

22.3.1 ENCRYPTION AND DATA PROTECTION

These techniques help safeguard sensitive employee data by converting it into a secure format that can only be accessed with the proper decryption key. This can prevent unauthorized access and theft of employee data, reduce the risk of data breaches, and protect employee privacy (Chingoriwo, 2022). Data protection measures also help organizations comply with privacy regulations and standards. As a result, organizations can better protect their TM and HCM systems and safeguard critical employee data by implementing encryption and data protection techniques.

22.3.2 ACCESS CONTROL AND AUTHENTICATION

These techniques ensure that only authorized individuals can access sensitive employee data and perform specific actions within the system. Access control measures limit access to sensitive data based on user roles and permissions, while authentication techniques verify the identity of users before granting access (Zhan et al., 2023). Substantial access control and authentication measures can help prevent unauthorized access to TM and HCM systems, reduce the risk of data breaches, and protect employee privacy. It also helps organizations comply with privacy regulations and standards.

22.3.3 NETWORK SECURITY

Network security is a critical technique for supporting the security of TM and HCM systems. This technique involves implementing measures to protect the TM and HCM system's network from cyber threats, including firewalls, intrusion detection and prevention systems (IDPS), and virtual private networks (VPNs) (Anaam et al., 2023). Network security helps prevent unauthorized access to the TM and HCM system's network, reduces the risk of data breaches, and protects employee privacy. As a result, organizations can better protect their TM and HCM systems against cyber threats or cybercrimes and safeguard critical employee data by implementing strong network security measures.

22.3.4 INCIDENT RESPONSE PLANNING

This technique involves developing and implementing a plan to respond to cyber-security incidents, such as data breaches or cyber-attacks that may compromise the security of the TM and HCM systems. An effective incident response plan can help organizations quickly identify and contain security incidents, minimize the impact of security breaches, and restore the system to normal operations as soon as possible (Mahamood et al., 2023). In addition, by implementing a robust incident response plan, organizations can better protect their TM and HCM systems against cyber threats and safeguard critical employee data.

22.3.5 MFA

The MFA technique requires users to provide multiple forms of authentication before accessing the TM and HCM systems, such as a password and a security token or a biometric factor like a fingerprint (Alejandro et al., 2023). As a result, MFA helps prevent unauthorized access to the system and reduces the risk of data breaches, making it much harder for attackers to gain access even if they have compromised a user's password. As a result, by implementing MFA, organizations can better protect their TM and HCM systems against cyber threats and safeguard critical employee data.

22.3.6 Regular Software Updates

A critical technique for supporting the security of TM and HCM systems against cyber threats or cybercrimes. Software updates often include security patches that address vulnerabilities that cybercriminals can exploit (Mahamood et al., 2023). By ensuring that software is up-to-date with the latest security patches, organizations can reduce the risk of cyberattacks and data breaches, protect sensitive employee data, and safeguard the integrity of the TM and HCM systems. Regular software updates also help organizations comply with security and privacy regulations and standards.

22.3.7 Access Controls

Access controls involve implementing policies, procedures, and technologies to limit access to the TM and HCM systems to authorized users. By restricting access to sensitive employee data, organizations can reduce the risk of data breaches and insider threats. Access controls also help ensure that only authorized individuals can perform specific actions on the TM and HCM systems, such as modifying employee records or changing access privileges (Fisk et al., 2023). By implementing strong access controls, organizations can better protect their TM and HCM systems against cyber threats and safeguard critical employee data.

22.3.8 Employee Training

Employees are often the weakest link in the security chain, and they can unwittingly become targets of cybercriminals through phishing attacks or social engineering techniques. Organizations can help employees recognize and avoid common cyber threats, such as phishing emails or malware attacks, by providing regular security awareness training (Fisk et al., 2023). This training can also educate employees about how to securely handle sensitive employee data and follow the best password management and access control practices. By investing in employee training, organizations can improve their security posture and reduce the risk of cyberattacks and data breaches.

22.3.9 Data Loss Prevention

DLP involves implementing policies and technologies to prevent the unauthorized disclosure of sensitive employee data. This can include monitoring data movement within and outside the organization, blocking the transmission of sensitive data through unauthorized channels, and enforcing access controls to ensure that only authorized individuals can access and modify employee data (Allison & Stepney, 2023).

By implementing strong DLP measures, organizations can reduce the risk of insider threats, accidental data leaks, and data breaches, safeguarding the confidentiality, integrity, and availability of the TM and HCM systems.

22.3.10 REGULAR BACKUPS

By backing up critical employee data regularly, organizations can protect against the risk of data loss or corruption due to cyberattacks, system failures, or natural disasters. Regular backups can help organizations recover quickly from a cyber incident and minimize the impact on employee data and business operations (Ukwandu et al., 2023; Deepa, 2023). In addition, backups can be used to test incident response and disaster recovery plans (DRP), ensuring that the organization is prepared to respond to cyber incidents effectively. By implementing regular backups, organizations can ensure the continuity and availability of the TM and HCM systems and protect critical employee data.

Organizations should implement various cybersecurity techniques to ensure the security of TM and HCM systems against cyber threats. These include encryption and data protection, access control and authentication, network security, incident response planning, MFA, regular software updates, access controls, employee training, DLP, and regular backups. These techniques help to safeguard employee data, prevent unauthorized access or disclosure, detect and respond to cyber incidents, and ensure the continuity and availability of the TM and HCM systems. By implementing a layered approach to cybersecurity, organizations can better protect against the evolving threat landscape and reduce the risk of cybercrime. As the TM and HCM systems get more involved in security protection approaches, some best practices exist during the implementation, as discussed in the next section.

22.4 BEST PRACTICES FOR IMPLEMENTING CYBERSECURITY IN TM AND HCM SYSTEMS

Cybersecurity is a critical aspect of TM and HCM systems. Implementing best practices for cybersecurity can help organizations protect their sensitive data, prevent unauthorized access, and safeguard their systems against cyber threats, even though sometimes it has some weaknesses. In this context, organizations need to adopt a proactive approach to cybersecurity by implementing robust security measures and regularly updating them. In this article, we will discuss some of the best practices for implementing cybersecurity in TM and HCM systems, which can help organizations mitigate the risks associated with cyber threats and ensure the security of their systems and data.

22.4.1 CONDUCT A THOROUGH RISK ASSESSMENT

Before implementing any cybersecurity measures, conduct a comprehensive risk assessment to identify potential vulnerabilities and threats in the TM and HCM systems. A thorough risk assessment is a crucial best practice for implementing cybersecurity in TM and HCM systems. The weakness of not conducting a risk assessment is that it leaves the system vulnerable to attacks that were not considered (Anaam et al., 2023). By conducting a thorough risk assessment, organizations can identify potential threats and vulnerabilities, determine the likelihood of an attack, and evaluate the potential impact. This information can help organizations prioritize and allocate resources to address the most significant risks. The possible benefits of conducting a thorough risk assessment include enhanced security, better decision-making, increased efficiency, and reduced costs associated with mitigating attacks.

22.4.2 ESTABLISH A CYBERSECURITY POLICY

Develop and implement a cybersecurity policy outlining procedures and guidelines for managing and protecting sensitive data. Establishing a cybersecurity policy is another crucial best practice for implementing cybersecurity in TM and HCM systems. The weakness of not having a policy is that it can lead to inconsistent security practices and confusion among employees (Mahamood et al., 2023). A cybersecurity policy can help establish clear expectations for employees, ensure compliance with regulations, and provide a framework for incident response. The possible benefits of establishing a cybersecurity policy include increased security awareness, improved compliance, reduced legal liability, and increased trust with customers and stakeholders. By establishing a cybersecurity policy, organizations can ensure that security is a top priority and everyone works together toward the same goals.

22.4.3 LIMIT ACCESS TO SENSITIVE DATA

One of the best practices for implementing cybersecurity in TM and HCM systems is limiting access to sensitive data. The weakness of not implementing access controls is that it makes the system vulnerable to insider threats and unauthorized access. When employees have access to sensitive data, they can compromise the system's security unintentionally or maliciously (Bukauskas et al., 2023). On the other hand, implementing access controls limits access to sensitive data to only those who require it to perform their duties. This helps to reduce the risk of insider threats, accidental data breaches, and unauthorized access. It also provides a detailed record of who accessed the data, when, and what they did with it, which can aid in identifying the source of a data breach.

22.4.4 USE MFA

MFA is a cybersecurity best practice involving multiple methods to verify a user's identity before granting access to sensitive data or systems. The weakness of not implementing MFA is that passwords can be compromised, or employees can unknowingly give away their credentials to cybercriminals (Alejandro et al., 2023). This leaves sensitive data and systems vulnerable to unauthorized access. On the other hand, the benefits of using MFA are substantial, including increased security, decreased risk of data breaches, and protection against unauthorized access. Additionally, MFA can help to comply with regulatory requirements and improve overall security posture, making it an essential practice for securing TM and HCM systems.

22.4.5 Implement Data Encryption

Implementing data encryption is a best practice to ensure that sensitive data is protected from unauthorized access. The weakness of data encryption is that it can impact system performance and require additional resources. However, the benefits of data encryption outweigh the risks (Fisk et al., 2023). Encryption helps prevent data breaches, as even if an attacker gains access to the data, they cannot read it without the decryption key. Additionally, data encryption helps organizations comply with data protection regulations and standards. By implementing data encryption in TM and HCM systems, organizations can ensure that sensitive data is secure and protected from cyber threats.

22.4.6 Regularly Update Software and Security Patches

Regularly updating software and security patches is an essential best practice for implementing cybersecurity in TM and HCM systems. Failure to do so leaves the systems open to exploitation by cybercriminals who can exploit vulnerabilities in the software. The main weakness of this best practice is that it can be time-consuming and costly to ensure that all software and security patches are updated regularly, especially in large organizations with many systems (Khan et al., 2023; Pinto et al., 2023). However, the benefits of regularly updating software and security patches far outweigh the cost and effort involved. Regular updates ensure that the systems are protected against the latest cyber threats and that any vulnerabilities are fixed promptly, minimizing the risk of data breaches and other security incidents.

22.4.7 Firewall Implementation

One weakness of implementing a firewall in TM and HCM systems is that it may not provide complete protection against all cyber threats. A determined attacker could potentially find ways to bypass or penetrate the firewall. Additionally, if the firewall is not configured correctly or regularly updated, it could create vulnerabilities for attackers to exploit. However, the benefits of implementing a firewall in TM and HCM systems as a best practice for implementing cybersecurity cannot be overstated (Budde et al., 2023; Dao et al., 2023). A firewall provides an essential layer of protection by monitoring and controlling incoming and outgoing network traffic. It can block unauthorized access attempts and prevent the spread of malware within the network. By implementing a firewall, organizations can significantly reduce the risk of cyber-attacks and better protect their sensitive data and systems.

22.4.8 Regularly Data Backup

Regularly backing up data is a crucial best practice for implementing cybersecurity in TM and HCM systems. The weakness of not doing so is that in the event of a cyberattack or system failure, essential data may be lost or unrecoverable, causing significant damage to the organization. On the other hand, regularly backing up data ensures that critical information is recoverable during a cyberattack, system failure, or accidental loss (Furstenau et al., 2023; Mohapatra et al., 2023). Backing up data also provides an opportunity to test and validate the recovery process, helping organizations to identify any weaknesses and make necessary improvements. Regular backups also enable organizations to comply with regulatory requirements related to data retention and protection.

22.4.9 Conduct Regular Security Awareness Training

One feebleness of conducting regular security awareness training for TM and HCM systems employees is that it can be time-consuming and costly. However, the benefits outweigh the drawbacks. Regular training can educate employees on identifying and responding to potential cyber threats, reducing the likelihood of successful attacks (Cuong & Thanh, 2023; Wah, 2023). It can also promote a culture of security consciousness and encourage employees to protect the organization's data actively. Additionally, employees can learn the best data handling and password management practices, preventing accidental data breaches. Overall, regular security awareness training can effectively mitigate human error, one of the most common causes of cybersecurity incidents in TM and HCM systems.

22.4.10 Conduct Periodic Security Audits

Periodic security audits is an essential best practice for implementing cybersecurity in TM and HCM systems. These audits help to identify vulnerabilities in the system, including weaknesses in hardware, software, and personnel practices. One weakness is that conducting these audits can be time-consuming and costly, especially for larger organizations (Beenish et al., 2023). However, the benefits of these audits far outweigh the costs. Audits help organizations to stay on top of evolving threats and to ensure that their security measures remain effective. By identifying vulnerabilities, audits allow organizations to address them before malicious actors can exploit them proactively. Additionally, audits can help organizations demonstrate compliance with industry regulations and standards, which can help to build trust with customers and partners.

22.4.11 Use IDPS

This provides an additional layer of defense to detect and prevent unauthorized access to TM and HCM systems. These systems monitor network traffic and system activity for signs of suspicious behavior, such as known attack signatures or anomalies that indicate potential attacks. One weakness of IDPS is that they may generate false positives, causing unnecessary alerts that may reduce the efficiency of security teams (Mohapatra et al., 2023). Moreover, IDPS should not be relied on as a sole defense mechanism. However, the benefits of using IDPS in TM and HCM systems include enhanced visibility and real-time monitoring of network activity and the ability to detect and prevent attacks before they cause significant damage. This can help organizations respond more quickly and effectively to security incidents, minimizing the impact of cyber threats on their systems and data.

22.4.12   IMPLEMENT A DRP

This is another best practice for ensuring business continuity during a cybersecurity incident. A weakness of not having a DRP is that organizations are unprepared to recover from an unexpected disruption, leading to data loss, downtime, and reputational damage. However, the benefits of having a DRP in place for TM and HCM systems include minimizing the impact of a cyberattack, ensuring the continuity of critical business functions, reducing downtime, and preventing data loss (Furstenau et al., 2023). The DRP should include a clear action plan, including regular backups, data recovery procedures, communication plans, and testing protocols. By implementing a DRP, organizations can be better prepared to respond to any cybersecurity incident and quickly recover from disruptions.

22.4.13   MONITOR NETWORK ACTIVITY

Monitoring network activity is a crucial best practice for implementing cybersecurity in TM and HCM systems. It involves continuously observing and analyzing network traffic to identify any malicious activity or potential security breaches. One weakness of this practice is that it requires a significant investment in network monitoring tools and skilled personnel to manage and analyze the data collected (Fisk et al., 2023). However, the benefits of monitoring network activity outweigh the costs, as it allows for early detection and response to potential cyber threats, reducing the risk of data breaches and protecting sensitive information. Additionally, network monitoring can provide valuable insights into network performance, enabling organizations to optimize their systems and improve efficiency.

22.4.14   PERFORM REGULAR VULNERABILITY ASSESSMENTS

Regular vulnerability assessments are a crucial best practice for implementing cybersecurity in TM and HCM systems. One potential weakness is that assessments may not always catch every vulnerability, exposing the system to potential attacks. However, the benefits of regular vulnerability assessments are numerous. They allow organizations to identify and prioritize vulnerabilities, which can be addressed before attackers exploit those (Tran et al., 2023). This can prevent the loss of sensitive data and damage the system's reputation. Additionally, assessments provide valuable insights into the organization's security measures' effectiveness, allowing for adjustments and improvements. By performing regular vulnerability assessments, organizations can stay ahead of potential threats and ensure the ongoing security of their TM and HCM systems.

22.4.15   REGULARLY REVIEW AND UPDATE ORGANIZATION CYBERSECURITY MEASURES

Regularly reviewing and updating cybersecurity measures is a crucial best practice for implementing cybersecurity in TM and HCM systems. The weakness of not doing this is that cyber threats and attacks constantly evolve, and if the organization's security measures remain static, the organization may become vulnerable to new threats. The possible benefits of regularly reviewing and updating organization cybersecurity measures include staying ahead of the latest threats and responding quickly to and mitigating any potential breaches (Ndayishimiye et al., 2023; Tran et al., 2023). In addition, reviewing and updating organization measures can ensure they are still relevant and effective in protecting the organization's TM and HCM systems. This best practice requires ongoing effort and resources, but keeping TM and HCM systems secure is a necessary investment.

22.4.16 COMPLIANCE WITH INDUSTRY REGULATIONS

Compliance with industry regulations is an essential best practice for implementing cybersecurity in TM and HCM systems. While industry regulations may vary across different sectors, complying with these regulations can help organizations meet legal requirements for protecting sensitive data and preventing cyber-attacks. However, one weakness of compliance with industry regulations is that it may lead to a false sense of security (Pinto et al., 2023). Complying with regulations does not necessarily mean an organization is fully protected from cyber threats. Additionally, regulations can be slow to adapt to new and emerging threats, which may leave organizations vulnerable to new attack methods.

However, the benefits of compliance with industry regulations are significant. Compliance can help organizations establish a framework for addressing cybersecurity risks and vulnerabilities. It can also help them identify and mitigate potential system and process weaknesses. Compliance can also help organizations build trust with their customers and stakeholders (Fisk et al., 2023). By demonstrating a commitment to protecting sensitive data and preventing cyber-attacks, organizations can establish a reputation for reliability and trustworthiness.

Conversely, non-compliance can lead to fines, legal fees, and damage to an organization's reputation. Therefore, compliance with industry regulations is a critical best practice for implementing cybersecurity in TM and HCM systems (Allison & Stepney, 2023; Fisk et al., 2023). While there are weaknesses to relying solely on compliance, the benefits of establishing a framework for addressing cybersecurity risks and building trust with stakeholders make compliance an essential component of any cybersecurity strategy.

22.5 CASE STUDIES

Some real-world examples exist of organizations that have successfully implemented cybersecurity in their TM and HCM systems and with an advanced analysis of the impact of cybersecurity measures on these organizations are discussed below.

22.5.1 MICROSOFT

Microsoft is a global technology company implementing a robust cybersecurity program in its TM and HCM systems. To safeguard its systems from cyber threats, the company leverages a combination of advanced technology, rigorous policies, and effective procedures. For instance, Microsoft has implemented robust access controls, end-to-end data encryption, and MFA to protect its sensitive data from unauthorized access or theft. Moreover, Microsoft employs a comprehensive suite of cybersecurity measures, such as regular vulnerability assessments, IDPS, and security information and event management tools. These measures help detect, analyze, and respond to potential security incidents promptly, minimizing the risk of data breaches or other security threats.

22.5.2 CISCO

A multinational technology company incorporates cybersecurity features in its TM/HCM systems to ensure data security and prevent cyber threats. These features include MFA, data encryption, regular software updates, and intrusion detection systems. They also limit access to sensitive data, conduct regular security awareness training, and perform periodic security audits. Cisco also has a DRP for security breaches or data loss. Additionally, they comply with industry regulations and standards, such as the GDPR, HIPAA, and the International Organization for Standardization (ISO) 27001. These cybersecurity measures help protect Cisco's TM/HCM systems against cyber threats and maintain their sensitive data's confidentiality, integrity, and availability.

22.5.3 INTERNATIONAL BUSINESS MACHINES CORP (IBM)

IBM has incorporated cybersecurity features in its TM and HCM systems to ensure data security and privacy. One such feature is the IBM Cloud Pak for security, an open and integrated platform enabling security teams to detect and respond to threats across cloud and on-premises environments. The platform provides security intelligence and analytics, threat management, identity and access management, and data protection capabilities. IBM has also integrated machine learning and artificial intelligence algorithms into its security solutions, enabling them to analyze and predict potential security threats and vulnerabilities.

22.5.4 DELOITTE

A global consulting firm, Deloitte incorporates various cybersecurity features in its TM and HCM systems. For example, they use MFA to ensure only authorized personnel access sensitive data. Additionally, they have implemented a role-based access control system that limits access to data to only individuals who need it to perform their duties. Deloitte also conducts regular security awareness training to ensure its employees know the latest cybersecurity threats and how to prevent them. In addition, they regularly review and update their cybersecurity measures to keep up with the constantly evolving cybersecurity landscape. Moreover, they conduct periodic security audits to assess their system's vulnerability to cyber threats and identify improvement areas. Lastly, Deloitte has implemented a DRP to recover from any cyber-attack or data breach quickly.

22.5.5   GENERAL ELECTRIC (GE)

GE has implemented cybersecurity features in its TM and HCM systems to protect critical data. One of the key features is continuous monitoring of the systems to identify any potential security breaches. The company also has a comprehensive incident response plan that outlines the steps to be taken in the event of a security incident. GE has implemented MFA and role-based access controls to ensure only authorized personnel can access sensitive data. In addition, the company regularly conducts vulnerability assessments and penetration testing to identify and address any potential security weaknesses. GE has also established partnerships with leading cybersecurity companies to protect its systems against the latest threats (Khang, 2023).

22.5.6   WALMART

Walmart has implemented several cybersecurity measures in its TM and HCM systems to ensure employee data security. They have established a comprehensive cybersecurity policy and use MFA for access control. Walmart also implements data encryption and regularly updates software and security patches to prevent potential vulnerabilities. They also conduct regular security awareness training and periodic security audits to identify and address security weaknesses. Walmart has implemented a DRP to ensure business continuity during security breaches or other disruptions. Additionally, they monitor network activity and perform regular vulnerability assessments to identify potential threats and address them promptly (Khang & Muthmainnah et al., 2023).

22.5.7   JPMORGAN CHASE

JPMorgan Chase, one of the largest financial institutions in the world, places a high priority on cybersecurity measures in their TM and HCM systems. They have implemented a multi-layered approach to security, including regular security audits and vulnerability assessments, as well as encryption and access controls to protect sensitive data. They also utilize a security information and event management (SIEM) system to monitor network activity and detect potential threats in real-time. JPMorgan Chase also has a dedicated team of cybersecurity experts who continuously review and update their cybersecurity measures to ensure they stay ahead of evolving threats. Additionally, they provide regular security awareness training to all employees to ensure they understand their role in maintaining a secure environment. JPMorgan Chase's approach to cybersecurity in their TM and HCM systems reflects a commitment to proactively identifying and mitigating risks to protect their customers' information and financial assets (Khang & Kali et al., 2023).

22.5.8   AUTOMATIC DATA PROCESSING (ADP)

This is a leading provider of HCM solutions. To ensure the security of their TM and HCM systems, ADP employs a multi-layered approach to cybersecurity that includes several features. One of the critical features is MFA, which requires employers to provide multiple forms of identification to access sensitive information. The company also utilizes a network of firewalls and IDPS to monitor and protect against cyber threats. ADP's systems are regularly updated with the latest security patches and undergo continuous vulnerability assessments. Additionally, ADP has implemented data encryption and secure file transfer protocols to protect sensitive information in transit. These features, along with regular security awareness training for employees and periodic security audits, help ADP maintain the highest level of cybersecurity for its TM and HCM systems (Khang & Rani et al., 2023).

22.6 CONCLUSION

Cybersecurity is critical in managing talent and human resources, and organizations must implement proactive measures to secure their TM and HCM systems. Best practices such as conducting risk assessments, limiting access to sensitive data, implementing firewalls, and regularly backing up data can improve an organization's cybersecurity posture. Compliance with industry regulations and standards can also protect against legal and financial consequences.

Successful implementation of cybersecurity in TM and HCM systems by organizations such as Cisco, IBM, Deloitte, General Electric, Walmart, and JPMorgan Chase demonstrates that effective cybersecurity is achievable through advanced technology, robust policies, and proactive employee training. Organizations must take a holistic approach, involving all relevant stakeholders to ensure comprehensive and integrated cybersecurity. Regular vulnerability assessments and security audits can identify and address potential weaknesses, and investing in employee training can build a strong culture of cybersecurity (Khang & Shah et al., 2023).

REFERENCES

Afenyo, M. & Caesar, L. D. (2023). Maritime cybersecurity threats: Gaps and directions for future research. *Ocean & Coastal Management*, 236, 106493, https://doi.org/10.1016/j.ocecoaman.2023.106493

Alejandro, M. C., Andrés, G. H., & Ricardo, V. F. (2023). Constructing an architecture-based cybersecurity solution for a system. *MethodsX*, 10, 102010, https://doi.org/10.1016/j.mex.2023.102010

Allison, J. & Stepney, O. (2023). *Cyber Security in English Secondary Education Curricula: A Preliminary Study*, pp. 193–199. ACM, New York. https://doi.org/10.1145/3545945.3569758

Alshammari, F. H. (2023). Design of capability maturity model integration with cybersecurity risk severity complex prediction using bayesian-based machine learning models. *Service Oriented Computing and Applications*, 17(1), 59–72. https://doi.org/10.1007/s11761-022-00354-4

Anaam, E., Ghazal, T. M., Haw, S. C., Alzoubi, H. M., Alshurideh, M. T., & Al Mamun, A. (2023). Utilization of blockchain technology in human resource management. *2023 IEEE 2nd International Conference on AI in Cybersecurity (ICAIC)*, Houston, TX, USA, pp. 1–5. https://ieeexplore.ieee.org/abstract/document/10044181/

Beenish, H., Javid, T., Fahad, M., Siddiqui, A. A., Ahmed, G., & Syed, H. J. (2023). A novel markov model-based traffic density estimation technique for intelligent transportation system. *Sensors*, 23(2), 768. https://www.mdpi.com/1424-8220/23/2/768

Behie, S. W., Pasman, H. J., Khan, F. I., Shell, K., Alarfaj, A., El-Kady, A. H., & Hernandez, M. (2023). Leadership 4.0: the changing landscape of industry management in the smart digital era. *Process Safety and Environmental Protection*, 172, 317–328. https://doi. org/10.1016/j.psep.2023.02.014

Budde, C. E., Karinsalo, A., Vidor, S., Salonen, J., & Massacci, F. (2023). Consolidating cybersecurity in Europe: a case study on job profiles assessment. *Computers & Security*, 127, 103082. https://doi.org/10.1016/j.cose.2022.103082

Bukauskas, L., Brilingaitė, A., Juozapavičius, A., Lepaitė, D., Ikamas, K., & Andrijauskaitė, R. (2023). Remapping cybersecurity competences in a small nation state. *Heliyon*, 9, e12808. https://doi.org/10.1016/j.heliyon.2023.e12808

Chingoriwo, T. (2022). Cybersecurity challenges and needs in the context of digital development in Zimbabwe. *British Journal of Multidisciplinary and Advanced Studies*, 3(2), 77–104. https://doi.org/10.37745/bjmas.2022.0046

Chopin, J. & Décary-Hétu, D. (2023). Dark web pedophile site users' cybersecurity concerns: a lifespan and survival analysis. *Journal of Criminal Justice*, 86, 102060. https://doi. org/10.1016/j.jcrimjus.2023.102060

Cuong, N. T. & Thanh, H. C. (2023). Mediatized infrapolitics and government accountability in vietnam. *The Dragon's Underbelly: Dynamics and Dilemmas in Vietnam's Economy and Politics*, pp. 234–261. ISEAS Publishing, Singapore. https://doi.org/ 10.1355/9789815011401-010

Dao, Q. T., Dang, T. K., Nguyen, T. P. H., & Le, T. M. C. (2023). VNLES: a reasoning-enable legal expert system using Ontology modeling-based method: a case study of vietnam criminal code. *2023 17th International Conference on Ubiquitous Information Management and Communication*, pp. 1–7. Seoul, Korea. https://doi.org/10.1109/ IMCOM56909.2023.10035590

Deepa, R. (2023). The application of blockchain in talent supply chain management. *Blockchain in a Volatile-Uncertain-Complex-Ambiguous World*, pp. 121–139. Elsevier, Amsterdam. https://doi.org/10.1016/B978-0-323-89963-5.00015-0

Fisk, N., Kelly, N. M., & Liebrock, L. (2023). Cybersecurity communities of practice: strategies for creating gateways to participation. *Computers & Security*, 132, 103188. https:// doi.org/10.1016/j.cose.2023.103188

Furstenau, L. B., Leivas, P., Sott, M. K., Dohan, M. S., López-Robles, J. R., Cobo, M. J., Bragazzi, N. L., & Choo, K.-K. R. (2023). Cybersecurity communities of practice: strategies for creating gateways to participation. *Digital Communications and Networks*, 132, 103188. https://doi.org/10.1016/j.dcan.2023.03.005

Hill, D., Keller, K. M., Rico, M., Shostak, M., & Matthews, M. (n.d.). *Talent Management and Diversity, Equity, and Inclusion in Private-Sector Organizations*. RAND Corporation, Santa Monica, CA. https://www.rand.org/pubs/research_reports/RRA988-4.html

Jun, Y., Craig, A., Shafik, W., & Sharif, L. (2021). Artificial intelligence application in cybersecurity and cyberdefense. *Wireless Communications and Mobile Computing*, 2021, 1–10. https://doi.org/10.1155/2021/3329581

Khang, A. (2023). "Material4Studies," *Material of Computer Science, Artificial Intelligence, Data Science, IoT, Blockchain, Cloud, Metaverse, Cybersecurity for Studies*. Retrieved from https://www.researchgate.net/publication/370156102_Material4Studies

Khang, A. (2023). *Advanced Technologies and AI-Equipped IoT Applications in High-Tech Agriculture*. IGI Global Press, Hershey, PA. https://doi.org/10.4018/978-1-6684-9231-4

Khang, A., Hahanov, V., Abbas, G. L., & Hajimahmud, V. A. (2022). Cyber-physical-social system and incident management. *AI-Centric Smart City Ecosystems: Technologies, Design and Implementation* (1st ed., pp. 2–15). CRC Press, Boca Raton, FL. https://doi. org/10.1201/9781003252542-2

Khang, A., Kali, C. R., Suresh Kumar, S., Amaresh, K., Sudhansu Ranjan, D., & Manas Ranjan, P. (2023). Enabling the future of manufacturing: integration of robotics and IoT to smart factory infrastructure in industry 4.0. *AI-Based Technologies and Applications in the Era of the Metaverse* (1st ed., pp. 25–50). IGI Global Press, Hershey, PA. https://doi.org/10.4018/978-1-6684-8851-5.ch002

Khang, A., Muthmainnah, M., Seraj, P. M. I., Yakin, A. A., Obaid, A. J., & Panda, M. R. (2023). AI-aided teaching model for the education 5.0 ecosystem. *AI-Based Technologies and Applications in the Era of the Metaverse* (1st ed., pp. 83–104). IGI Global Press, Hershey, PA. https://doi.org/10.4018/978-1-6684-8851-5.ch004

Khang, A., Rani, S., Gujrati, R., Uygun, H., & Gupta, S. K. (2023). *Designing Workforce Management Systems for Industry 4.0: Data-Centric and AI-Enabled Approaches.* CRC Press, Boca Raton, FL. https://doi.org/10.1201/9781003357070

Khang, A., Shah, V., & Rani, S. (2023). *AI-Based Technologies and Applications in the Era of the Metaverse* (1st ed.). IGI Global Press, Hershey, PA. https://doi.org/10.4018/978-1-6684-8851-5

Khan, S. K., Shiwakoti, N., Stasinopoulos, P., & Warren, M. (2023). Modelling cybersecurity regulations for automated vehicles. *Accident Analysis & Prevention*, 186, 107054. https://doi.org/10.1016/j.aap.2023.107054

Khang, A., Vugar, A., Vladimir, H., & Shah, V. (2024). *Advanced IoT Technologies and Applications in the Industry 4.0 Digital Economy* (1st ed.). CRC Press, Boca Raton, FL. https://doi.org/10.1201/9781003434269

Khanh, H. H. & Khang, A. (2021). The role of artificial intelligence in blockchain applications. *Reinventing Manufacturing and Business Processes through Artificial Intelligence* (2nd ed., pp. 20–40). CRC Press, Boca Raton, FL. https://doi.org/10.1201/9781003145011-2

Kucharčíková, A., Mičiak, M., Tokarčíková, E., & Štaffenová, N. (2023). The investments in human capital within the human capital management and the impact on the enterprise's performance. Sustainability, 15(6), 5015. https://doi.org/10.3390/su15065015

Lüersmann, A. (2023). Artificial intelligence in human resource management: personnel marketing and recruiting. *Developments in Information and Knowledge Management Systems for Business Applications* (Vol. 6, pp. 607–626). Springer, New York. https://doi.org/10.1007/978-3-031-27506-7_23

Mahamood, A. K., Malik, M., Ruhani, A. B., & Zolkipli, M. F. (2023). Cybersecurity strengthening through penetration testing: emerging trends and challenges. *Borneo International Journal*, 6(1), 44–52. https://majmuah.com/journal/index.php/bij/article/view/341

Meng, H., Shafik, W., Matinkhah, S. M., & Ahmad, Z. (2020). A 5 g beam selection machine learning algorithm for unmanned aerial vehicle applications. *Wireless Communications and Mobile Computing*, 2020, 1–16. https://doi.org/10.1155/2020/1428968

Mohapatra, L. M., Kamesh, A. V. S., & Roul, J. (2023). Challenges and path ahead for artificial intelligence-aided human resource management. *The Adoption and Effect of Artificial Intelligence on Human Resources Management, Part A*, pp. 107–121. Emerald Publishing Limited, Bingley. https://doi.org/10.1108/978-1-80382-027-920231006

Ndayishimiye, C., Dubas-Jakóbczyk, K., Holubenko, A., & Domagala, A. (2023). Competencies of hospital managers-a systematic scoping review. *Frontiers in Public Health*, 11, 1095. https://doi.org/10.3389/fpubh.2023.1130136

Pinto, S. J., Siano, P., & Parente, M. (2023). Review of cybersecurity analysis in smart distribution systems and future directions for using unsupervised learning methods for cyber detection. *Energies*, 16(4), 1651. https://doi.org/10.3390/en16041651

Rani, S., Chauhan, M., Kataria, A., & Khang, A. (2021). IoT equipped intelligent distributed framework for smart healthcare systems. *Networking and Internet Architecture*, 2, 30. https://doi.org/10.48550/arXiv.2110.04997

Shafik, W. (2023). Cyber security perspectives in public spaces: drone case study. *Handbook of Research on Cybersecurity Risk in Contemporary Business Systems 2023*, pp. 79–97. IGI Global Press, Hershey, PA. https://doi.org/10.4018/978-1-6684-7207-1.ch004

Shafik, W. & Matinkhah, S. M. (2021). Unmanned aerial vehicles analysis to social networks performance. *The CSI Journal on Computer Science and Engineering*, 18(2), 24–31. https://doi.org/10.11591/cjcse

Shafik, W., Matinkhah, M., & Sanda, M. N. (2020). Network resource management drives machine learning: a survey and future research direction. *Journal of Communications Technology, Electronics and Computer Science*, 2020, 1–15. https://doi.org/10.22385/jctecs.v30i0.312

Shafik, W., Matinkhah, S. M., & Shokoor, F. (2022). Recommendation system comparative analysis: internet of things aided networks. *EAI Endorsed Transactions on Internet of Things*, 8(29), e5. https://doi.org/10.4108/eetiot.v8i29.1108

Sheoraj, Y. & Sungkur, R. K. (2022). Using AI to develop a framework to prevent employees from missing project deadlines in software projects-case study of a global human capital management (HCM) software company. *Advances in Engineering Software*, 170, 103143. https://doi.org/10.1016/j.advengsoft.2022.103143

Shokoor, F., Shafik, W., & Matinkhah, S. M. (2022). Overview of 5G & beyond security. *EAI Endorsed Transactions on Internet of Things*, 8(30), e2. https://doi.org/10.4108/eetiot.v8i30.1622

Tan, L. N., Gupta, N., & Derawi, M. (2023). H∞ control for oscillator systems with event-triggering signal transmission of internet of things. *IEEE Access*, 11, 8938–8949. https://doi.org/10.1109/ACCESS.2023.3239665

Tran, T. M., Beuran, R., & Hasegawa, S. (2023). Gamification-based cybersecurity awareness course for self-regulated learning. *International Journal of Information and Education Technology*, 13(4), 724–730. https://www.ijiet.org/vol13/IJIET-V13N4-1859.pdf

Tyagi, P., Chilamkurti, N., Grima, S., Sood, K., & Balusamy, B. (2023). *The Adoption and Effect of Artificial Intelligence on Human Resources Management, Part A*. Emerald Publishing Limited, Bingley. https://doi.org/10.1108/978-1-80382-027-920231015

Ukwandu, E., Okafor, E. N., Ikerionwu, C., Olebara, C., & Ugwu, C. (2023). Assessing cyber-security readiness of Nigeria to industry 4.0. *Proceedings of the International Conference on Cybersecurity, Situational Awareness and Social Media, Wales*, pp. 355–374. https://doi.org/10.1007/978-981-19-6414-5_20

Wah, N. K. S. (2023). Integration of AI/Ml in 5G technology toward intelligent connectivity, security, and challenges. *Machine Learning Algorithms and Applications in Engineering*, pp. 239. CRC Press, Boca Raton, FL. https://doi.org/10.1201/9781003104858-14

Yang, Z., Jianjun, L., Faqiri, H., Shafik, W., Talal Abdulrahman, A., Yusuf, M., & Sharawy, A. M. (2021). Green internet of things and big data application in smart cities development. *Complexity*, 2021, 1–15. https://doi.org/10.1155/2021/4922697

Yao, J., Craig, A., Shafik, W., & Sharif, L. (2021). Artificial intelligence application in cyber-security and cyberdefense. *Wireless Communications & Mobile Computing*, 2021, 1–10. https://doi.org/10.1155/2021/3329581

Zhan, J., Deng, J., Xu, Z., & Martínez, L. (2023). A three-way decision methodology with regret theory via triangular fuzzy numbers in incomplete multi-scale decision information systems. *IEEE Transactions on Fuzzy Systems*, 31(8), 2773–2787. https://doi.org/10.1109/TFUZZ.2023.3237646

Zhao, L., Zhu, D., Shafik, W., Matinkhah, S. M., Ahmad, Z., Sharif, L., & Craig, A. (2022). Artificial intelligence analysis in cyber domain: a review. *International Journal of Distributed Sensor Networks*, 18(4). https://doi.org/10.1177/15501329221084882

23 Role of Augmented Reality and Virtual Reality in the Post-Covid-19 Scenario of Employee Training

B. C. M. Patnaik, Ipseeta Satpathy, Alex Khang,
Rashmi Gujrati, and Hayri Uygun

23.1 INTRODUCTION

In augmented reality (AR), digital objects can be overlaid onto real-world objects (augmented components) in the presence of technology (Azuma, 1997; Akçayır & Akçayır, 2017). It was Azuma (1997), who pointed out three key characteristics of AR: blending the real and virtual worlds, allowing real-time interactions, and accurately registering virtual objects in 3D (Azuma, 1997). As a result of these AR features, users are able to observe, interact with, and create digitally enhanced reality together or individually (Khang & Kali et al., 2023).

As smart phones and wearables become more affordable and accessible, AR technology has become more widely available. Its use in educational settings has increased in popularity as a result. Despite the importance of lifelong learning and a knowledge-based society, AR has been scarce and exploratory in professional training contexts, despite its gradual adoption and increasing investigation in "K–16 education (K–12 and higher education)" (Andersson, 2001; Marin, 2013).

As a result of training, employees develop the skills necessary to perform a specific job with greater proficiency (Liu et al., 2017). Professional training has three distinctive features. Rather than improving performance, it aims to develop employees (e.g., skill acquisition and knowledge growth). Furthermore, its target audience is not degree-seeking students but employees and professional staff (Khanh & Khang, 2021).

Furthermore, it is often conducted in workplaces rather than classrooms since it is job-specific and highly contextual. Many professional training programs are offered in the healthcare and medical fields, engineering fields, service sectors, manufacturing industries, and teacher training programs.

The purpose of continuing medical education is to develop the practical skills of medical residents and interns in clinical diagnosis (Davis et al., 1999), and many countries require educators to undergo in-service teacher education (Korthagen, 2004). Conferences, lectures, workshops, and traineeships are the most common types of professional training (Mazmanian & Davis, 2002).

DOI: 10.1201/9781003440901-23

Employee engagement can be enhanced through training. Businesses can save money and time by using virtual reality (VR) to simulate different scenarios and tasks. Currently, we can observe the growth of VR and AR in the training arena, especially in the corporate world (Goulding et al., 2012). An immersive learning environment can be created using VR, which provides a safe, cost-effective, and measurable learning environment.

The two technologies are also slowly spreading throughout the business world, including human resources management and development (HRMD). Employee learning opportunities can be revolutionized so that they facilitate the acquisition of knowledge, skills, and performance goals for employees. In this process, new employees are trained and specialized skills are developed. Training and development of new employees are two HRMD areas where such technologies can have a tremendous impact (Petrock, 2017).

Successful social interaction requires soft skills such as customer service, negotiation, sales pitching, and business networking. A prospective employee should have these enterprise skills in order to function effectively in a business. These skills are becoming increasingly relevant as automation increases.

The purpose of the study is to understand the role of AR and VR in the post-Covid-19 on managerial skills, to study the role of AR and VR in the post-Covid-19 on technical skills, and to decode the role of AR and VR in the post-Covid-19 on life skills. The present study was confined to 5 government organizations and 5 private sector organizations (Khang & Sivaraman, 2022). The respondents include both male and female groups, and they consist of lower-level, middle-level and higher-level employees working in the study units.

23.2 RATIONALITY OF THE STUDY

With the invention of new technology, it is important to be more relevant to equip with the changing scenario, so also the employees of both the government and private sectors of different organizations in Odisha. The findings of the present study will definitely give an opportunity to explore other organizations and implement the same for the betterment of the organization (Babasaheb et al., 2023).

23.3 RESEARCH GAPS

After going through various reviews, it was found that very limited research has been undertaken in the post-Covid-19 scenario related to managerial skills, technical skills, and life skills in the study areas for the lower-level, middle-level, and higher-level employees of both the government and private sectors.

23.4 RESEARCH DESIGN

A combination of primary and secondary data is used in the present study. Snowball sampling was used to collect the primary data. A total of 162 samples were collected, and for these, 293 questionnaires were distributed. Initially, 18 variables were identified, but after the pilot study, the attributes were restricted to only 14

(Rani & Chauhan et al., 2021). The sample respondents include 3 levels of employees, such as lower-level, middle-level, and higher-level employees. In this direction, we considered 1–2 years of experience as entry level, 3–5 years as lower-level, 5–10 years as middle-level, and above 10 years as higher-level (Khang & Shah et al., 2023).

23.5   SAMPLE SIZE DETERMINATION

This study used a ratio of 1:4–1:10 to calculate sample size. A minimum sample size of 4 items and a maximum sample size of 10 items should be calculated according to the above method. Since 14 items were used for the study, a minimum and maximum sample size of 56 and 140 are required, respectively. To ensure a better representation, we included all 162 responses received in the present case. Our study will be adequate if we take the rule of (Bhambri et al., 2022) into consideration.

In Table 23.1, the government's lower-level, middle-level, and higher-level employees include 32, 28, and 18; overall, the total respondents include 78. Similarly, the private sector lower-level, middle-level, and higher-level include 19, 34, and 31, and overall includes 84 respondents (Figure 23.1).

With reference to Table 23.2, the maximum possible weight for the government lower level (GLL) with respect to managerial skills, technical skills, and life skills is 960, 640, and 640, and the same for the government middle level (GML) is 840, 560, and 560. Similarly, for the government higher level (GHL), the same is 540, 360, and 360, respectively.

TABLE 23.1
Sampling Frame

Name of the Organizations	Lower Level	Middle Level	Higher Level	Total
Government Organizations				
Odisha Mining Corporation (OMC)	8	6	3	17
National Aluminum Company (NALCO)	7	5	4	16
Odisha Power Generation Corporation (OPGC)	7	6	4	17
Mahanadi Coalfield Ltd (MCL)	5	4	3	12
Odisha Hydro Power Corporation (OHPC)	5	7	4	16
Total	32	28	18	78
Private Organizations				
Indian Metal and Ferro Alloys Ltd. (IMFA)	3	8	7	18
FALCON Marine	4	7	8	19
Pro-Mineral Private Ltd. (PMPL)	4	7	6	17
Mayfair	3	6	5	14
Odisha Stevedores Ltd. (OSL)	5	6	5	16
Total	19	34	31	84

Table 23.1 presents lower-level, middle-level and higher-level employees.
Source: Authors own compilation.

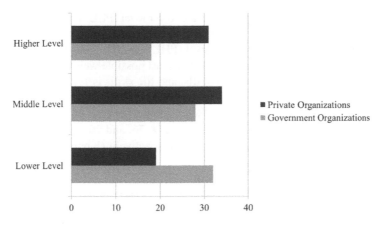

FIGURE 23.1 Sampling frame of lower-level, middle-level and higher-level employees.

Source: Authors own compilation.

TABLE 23.2
Computation of Maximum Possible Weight and Least Possible Weight

Category	Managerial Skills	Technical Skills	Life Skills
Government Lower Level (GLL)			
Maximum possible weight	$32 \times 6 \times 5 = 960$	$32 \times 4 \times 5 = 640$	$32 \times 4 \times 5 = 640$
Least possible weight	$32 \times 6 \times 1 = 192$	$32 \times 4 \times 1 = 128$	$32 \times 4 \times 1 = 128$
Government Middle Level (GML)			
Maximum possible weight	$28 \times 6 \times 5 = 840$	$28 \times 4 \times 5 = 560$	$28 \times 4 \times 5 = 560$
Least possible weight	$28 \times 6 \times 1 = 168$	$28 \times 4 \times 1 = 112$	$28 \times 4 \times 1 = 112$
Government Higher Level (GHL)			
Maximum possible weight	$18 \times 6 \times 5 = 540$	$18 \times 4 \times 5 = 360$	$18 \times 4 \times 5 = 360$
Least possible weight	$18 \times 6 \times 1 = 108$	$18 \times 4 \times 1 = 72$	$18 \times 4 \times 1 = 72$
Private Lower Level (PLL)			
Maximum possible weight	$19 \times 6 \times 5 = 570$	$19 \times 4 \times 5 = 380$	$19 \times 4 \times 5 = 380$
Least possible weight	$19 \times 6 \times 1 = 114$	$19 \times 4 \times 1 = 76$	$19 \times 4 \times 1 = 76$
Private Middle Level (PML			
Maximum possible weight	$34 \times 6 \times 5 = 1,020$	$34 \times 4 \times 5 = 680$	$34 \times 4 \times 5 = 680$
Least possible weight	$34 \times 6 \times 1 = 204$	$34 \times 4 \times 1 = 136$	$34 \times 4 \times 1 = 136$
Private Higher Level (PHL)			
Maximum possible weight	$31 \times 6 \times 5 = 930$	$31 \times 4 \times 5 = 620$	$31 \times 4 \times 5 = 620$
Least possible weight	$31 \times 6 \times 1 = 186$	$31 \times 4 \times 1 = 124$	$31 \times 4 \times 1 = 124$

Table 23.2 presents the maximum possible weight for GLL with respect to managerial skills, technical skills, and life skills.
Source: Authors' own compilation.

The maximum possible weight for private lower level (PLL) with respect to managerial skills, technical skills, and life skills is 570, 380, and 380, and for private middle level (PML) and private higher level (PHL) are 1020, 680, 680, 930, 620, and 620, respectively, as shown in Table 23.3. The least possible weight for the GLL with respect to managerial skills, technical skills, and life skills is 192, 128, and 128, and the same for the GML is 168, 112, and 112. Similarly, for the GHL, the same is 108, 72, and 72, respectively. The least possible weight for the PLL with respect to managerial skills, technical skills, and life skills is 114, 76, and 76, and for the PML and PHL, it is 204, 136, 136, 186, 124, and 124, respectively (Khang & Muthmainnah et al., 2023).

TABLE 23.3
Data Analysis

Attributes	GLL	GML	GHL	PLL	PML	PHL
Managerial Skills						
AR and VR helps in improving Stress Management.	142	120	78	84	154	144
AR and VR facilitates in Fast Learning.	140	127	81	81	148	131
AR and VR trains the employees about the Adaptability.	149	131	81	85	148	135
AR and VR enhance the Quick Thinking.	142	121	77	80	149	141
AR and VR help to train about Conflict Management.	139	125	85	84	146	134
AR and VR facilitate ability to Work Independently.	136	134	74	79	137	136
Total score	848	758	476	413	882	721
Maximum possible score	960	840	540	570	1,020	930
Least possible score	192	168	108	114	204	186
% of total score to maximum possible score	88.33	90.24	88.15	72.46	86.47	77.53
Average score	83.86					
Technical Skills						
AR and VR improve the Data-Driven Decision of the employees.	140	131	75	87	141	137
AR and VR help to improve the knowledge about ICT Tools.	144	130	74	84	147	142
AR and VR facilitate Solution Delivery.	142	130	79	81	147	133
AR and VR trains about Budget Planning.	139	134	79	82	138	136
Total score	562	525	307	334	573	548
Maximum possible score	640	560	360	380	680	620
Least possible score	128	112	72	76	136	124
% of total score to maximum possible score	87.81	93.75	85.28	87.89	84.26	88.39
Average score	87.90					

(Continued)

TABLE 23.3 (*Continued*)
Data Analysis

Attributes	GLL	GML	GHL	PLL	PML	PHL
	Life Skills					
AR and VR facilitates in improving Self-Awareness.	143	127	81	83	141	136
AR and VR trains Empathy.	139	129	77	81	137	133
AR and VR develop Critical Thinking.	141	127	77	81	146	141
AR and VR provide Creative Thinking.	139	133	71	76	131	146
Total score	562	516	306	321	555	556
Maximum possible score	640	560	360	380	680	620
Least possible score	128	112	72	76	136	124
% of total score to maximum possible score	87.81	92.14	85	84.47	81.38	89.68
Average score	86.78					

Table 23.3 presents the question related to managerial skills: the percentage of total skills to maximum possible skills.

Source: Annexure- A, B, C, D, E, and F.

23.6 RESULTS AND DISCUSSION

With reference to Table 23.3, answering the question related to managerial skills, the percentage of total skills to maximum possible skills is 88.33, 90.24, 88.15, 72.46, 86.47, and 77.53, respectively, for the GLL, GML, GHL, PLL, PML, and PHL, respectively, and over all, the average is 83.86%. Similarly, for the technical skills, the percentage of total skills to maximum possible skills is 87.81, 93.75, 85.28, 87.89, 84.26, and 88.39, and the average weight is 87.90 (Shah & Khang, 2023).

Finally, for life skills, the weights are 87.81, 92.14, 85, 84.47, 81.38, and 89.68, respectively; GLL, GML, GHL, PLL, PML, and PHL for the average weight are 86.78. This shows that the actual scores are better than the total maximum possible score, so this leads to the conclusion that the various variables identified are relevant for employee training (Khang et al., 2022).

23.7 CONCLUSION

The various skills for the training, such as managerial, technological, and life skills, are very much relevant for the post-Covid-19 scenario. AR and VR play a crucial role for facilitating the same. With the advancement of technology, every organization tries adopt the same for transparency and speed of doing the skills with accuracy.

The various managerial skills improve stress management, fast learning, adaptability, quick thinking, etc. Similarly, technical skills facilitate data-driven decisions, improve ICT tools, solution delivery, and budget planning. The life skills also help in improving self-awareness, training empathy, developing critical thinking, and providing creative thinking. These are the needs of the hour for the achievement of organizational goals, and this will enable the employees to discharge their duties without any difficulties (Rana & Khang et al., 2021).

23.8 SOURCE PRIMARY DATA

23.8.1 ANNEXURE-A: GOVERNMENT LOWER LEVEL (GLL)-32 RESPONSES

Annexure-A- Government Lower Level (GLL)-32 Responses						
	CA	A	N	DA	CDA	
Variables	5	4	3	2	1	**Score**
Managerial Skills						
AR and VR helps in improving Stress Management.	23	4	2	2	1	142
AR and VR facilitates in Fast Learning.	24	3	1	1	2	140
AR and VR trains the employees about the Adaptability.	26	2	3	1	0	149
AR and VR enhance the Quick Thinking.	23	3	3	3	0	142
AR and VR help to train about Conflict Management.	21	4	4	3	0	139
AR and VR facilitate ability to Work Independently.	22	3	2	3	2	136
Technical Skills						
AR and VR improve the Data-Driven Decision of the employees.	21	5	4	1	1	140
AR and VR help to improve the knowledge about ICT Tools.	22	6	3	0	1	144
AR and VR facilitate Solution Delivery.	23	4	2	2	1	142
AR and VR trains about Budget Planning.	22	2	3	4	0	139
Life Skills						
AR and VR facilitates in improving Self-Awareness.	23	4	2	3	0	143
AR and VR trains Empathy.	22	3	3	4	0	139
AR and VR develop Critical Thinking.	22	4	3	3	0	141
AR and VR provide Creative Thinking.	21	5	2	4	0	139

23.8.2 Annexure-B: Government Middle Level (GML)-28 Responses

Annexure-B-Government Middle Level (GML)-28 Responses						
	CA	A	N	DA	CDA	
Variables	5	4	3	2	1	Score
Managerial Skills						
AR and VR helps in improving Stress Management.	19	3	2	2	1	120
AR and VR facilitates in Fast Learning.	21	4	1	1	1	127
AR and VR trains the employees about the Adaptability.	22	3	3	0	0	131
AR and VR enhance the Quick Thinking.	18	5	2	2	1	121
AR and VR help to train about Conflict Management.	21	3	2	0	2	125
AR and VR facilitate ability to Work Independently.	23	2	3	0	2	134
Technical Skills						
AR and VR improve the Data-Driven Decision of the employees.	22	4	1	1	0	131
AR and VR help to improve the knowledge about ICT Tools.	22	3	2	1	0	130
AR and VR facilitate Solution Delivery.	23	2	1	2	0	130
AR and VR trains about Budget Planning.	24	2	2	0	0	134
Life Skills						
AR and VR facilitates in improving Self-Awareness.	21	3	2	2	0	127
AR and VR trains Empathy.	22	2	3	1	0	129
AR and VR develop Critical Thinking.	22	2	1	3	0	127
AR and VR provide Creative Thinking.	24	2	1	1	0	133

23.8.3 Annexure-C: Government Higher Level (GHL)-18 Responses

Annexure-C- Government Higher Level (GHL)-18 responses						
	CA	A	N	DA	CDA	
Variables	5	4	3	2	1	Score
Managerial Skills						
AR and VR helps in improving Stress Management.	12	3	1	1	1	78
AR and VR facilitates in Fast Learning.	13	2	2	1	0	81
AR and VR trains the employees about the Adaptability.	12	4	1	1	0	81
AR and VR enhance the Quick Thinking.	13	2	1	0	1	77

(Continued)

AR and VR help to train about Conflict Management.	14	3	1	0	0	85
AR and VR facilitate ability to Work Independently.	12	2	2	2	0	74
Technical Skills						
AR and VR improve the Data-Driven Decision of the employees.	11	2	2	3	0	75
AR and VR help to improve the knowledge about ICT Tools.	10	3	2	2	1	74
AR and VR facilitate Solution Delivery.	11	3	4	0	0	79
AR and VR trains about Budget Planning.	12	2	3	1	0	79
Life Skills						
AR and VR facilitates in improving Self-Awareness.	13	2	2	1	0	81
AR and VR trains Empathy.	12	2	1	3	0	77
AR and VR develop Critical Thinking.	11	3	2	2	0	77
AR and VR provide Creative Thinking.	10	3	2	3	0	71

23.8.4   ANNEXURE-D: PRIVATE LOWER LEVEL (PLL)-19 RESPONSES

Annexure-D- Private Lower Level (PLL)-19 Responses

	CA	A	N	DA	CDA	
Variables	**5**	**4**	**3**	**2**	**1**	**Score**
Managerial Skills						
AR and VR helps in improving Stress Management.	12	3	2	2	0	84
AR and VR facilitates in Fast Learning.	12	2	3	2	0	81
AR and VR trains the employees about the Adaptability.	13	2	4	0	0	85
AR and VR enhance the Quick Thinking.	11	3	3	2	0	80
AR and VR help to train about Conflict Management.	14	2	0	3	0	84
AR and VR facilitate ability to Work Independently.	12	2	3	0	2	79
Technical Skills						
AR and VR improve the Data-Driven Decision of the employees.	14	2	3	0	0	87
AR and VR help to improve the knowledge about ICT Tools.	12	3	4	0	0	84
AR and VR facilitate Solution Delivery.	11	4	2	2	0	81
AR and VR trains about Budget Planning.	12	2	4	1	0	82
Life Skills						

(Continued)

Annexure-D- Private Lower Level (PLL)-19 Responses (*Continued*)					
	CA	A	N	DA	CDA	
Variables	**5**	**4**	**3**	**2**	**1**	**Score**
AR and VR facilitates in improving Self-Awareness.	12	3	3	1	0	83
AR and VR trains Empathy.	11	4	2	2	0	81
AR and VR develop Critical Thinking.	12	3	1	3	0	81
AR and VR provide Creative Thinking.	12	2	2	3	0	76

23.8.5 ANNEXURE-E: PRIVATE MIDDLE LEVEL (PML)-34 RESPONSES

Annexure-E-Private Middle Level (PML)-34 Responses					
	CA	A	N	DA	CDA	
Variables	**5**	**4**	**3**	**2**	**1**	**Score**
Managerial Skills						
AR and VR helps in improving Stress Management.	26	3	2	3	0	154
AR and VR facilitates in Fast Learning.	24	2	4	4	0	148
AR and VR trains the employees about the Adaptability.	25	2	3	4	0	148
AR and VR enhance the Quick Thinking.	23	4	4	3	0	149
AR and VR help to train about Conflict Management.	22	6	2	2	2	146
AR and VR facilitate ability to Work Independently.	24	2	3	5	0	137
Technical Skills						
AR and VR improve the Data-Driven Decision of the employees.	23	2	2	2	2	141
AR and VR help to improve the knowledge about ICT Tools.	23	3	5	2	1	147
AR and VR facilitate Solution Delivery.	24	2	3	5	0	147
AR and VR trains about Budget Planning.	22	2	4	2	4	138
Life Skills						
AR and VR facilitates in improving Self-Awareness.	21	4	3	5	1	141
AR and VR trains Empathy.	20	3	6	3	3	137
AR and VR develop Critical Thinking.	22	4	4	4	0	146
AR and VR provide Creative Thinking.	21	2	2	2	4	131

23.8.6 ANNEXURE-F: PRIVATE HIGHER LEVEL (PHL)-31 RESPONSES

Annexure-F- Private Higher Level (PHL)- 31 Responses

Variables	CA 5	A 4	N 3	DA 2	CDA 1	Score
Managerial Skills						
AR and VR helps in improving Stress Management.	24	3	4	0	0	144
AR and VR facilitates in Fast Learning.	22	2	2	2	3	131
AR and VR trains the employees about the Adaptability.	21	3	4	3	0	135
AR and VR enhance the Quick Thinking.	22	4	5	0	0	141
AR and VR help to train about Conflict Management.	21	3	3	4	0	134
AR and VR facilitate ability to Work Independently.	20	4	6	1	0	136
Technical Skills						
AR and VR improve the Data-Driven Decision of the employees.	22	2	5	2	0	137
AR and VR help to improve the knowledge about ICT Tools.	23	4	3	1	0	142
AR and VR facilitate Solution Delivery.	21	2	4	4	0	133
AR and VR trains about Budget Planning.	20	5	4	2	0	136
Life Skills						
AR and VR facilitates in improving Self-Awareness.	21	3	5	2	0	136
AR and VR trains Empathy.	22	2	2	4	1	133
AR and VR develop Critical Thinking.	24	2	3	2	0	141
AR and VR provide Creative Thinking.	25	3	2	0	0	146

REFERENCES

Akçayır, M. & Akçayır, G. (2017). Advantages and challenges associated with augmented reality for education: a systematic review of the literature. *Educational Research Review*, 20, 1–11. https://doi.org/10.1016/j.edurev.2016.11.002

Andersson, E. P. (2001). Continuing education in Sweden-to what purpose? *The Journal of Continuing Education in Nursing*, 32(2), 86–93. https://doi.org/10.3928/0022-0124-20010301-08

Azuma, R. T. (1997). A survey of augmented reality. *Presence: Teleoperators and Virtual Environments*, 6(4), 355–385. https://doi.org/10.1162/pres.1997.6.4.355

Bhambri, P., Rani, S., Gupta, G., & Khang, A. (2022). *Cloud and Fog Computing Platforms for Internet of Things*. CRC Press, Boca Raton, FL. https://doi.org/ 10.1201/9781032101507

Davis, D., O'Brien, M. A., Freemantle, N., Wolf, F. M., Mazmanian, P., & Taylor-Vaisey, A. (1999). Impact of formal continuing medical education. *JAMA*, 282(9), 867. https://doi.org/10.1001/jama.282.9.867

Goulding, J., Nadim, W., Petridis, P., & Alshawi, M. (2012). Construction industry offsite production: a virtual reality interactive training environment prototype. *Advanced Engineering Informatics*, 26(1), 103–116. https://doi.org/10.1016/j.aei.2011.09.004

Khang A., Chowdhury, S., & Sharma, S. (Eds.). (2022). *The Data-Driven Blockchain Ecosystem: Fundamentals, Applications, and Emerging Technologies* (1st ed.). CRC Press, Boca Raton, FL. https://doi.org/10.1201/9781003269281

Khang, A., Kali, C. R., Suresh Kumar, S., Amaresh, K., Das, S. R., & Panda, M. R. (2023). Enabling the future of manufacturing: integration of robotics and IoT to smart factory infrastructure in industry 4.0. *AI-Based Technologies and Applications in the Era of the Metaverse* (1st ed., pp. 25–50). IGI Global Press, Hershey, PA. https://doi.org/10.4018/978-1-6684-8851-5.ch002

Khang, A., Muthmainnah, M., Seraj, P. M. I., Yakin, A. A., Obaid, A. J., & Panda, M. R. (2023). AI-aided teaching model for the education 5.0 ecosystem. *AI-Based Technologies and Applications in the Era of the Metaverse* (1st ed., pp. 83–104). IGI Global Press, Hershey, PA. https://doi.org/10.4018/978-1-6684-8851-5.ch004

Khang, A., Rani, S., & Sivaraman, A. K. (2022). *AI-Centric Smart City Ecosystems: Technologies, Design and Implementation* (1st ed.). CRC Press, Boca Raton, FL. https://doi.org/10.1201/9781003252542

Khang, A., Shah, V., & Rani, S. (2023). *AI-Based Technologies and Applications in the Era of the Metaverse* (1st ed.). IGI Global Press, Hershey, PA. https://doi.org/10.4018/978-1-6684-8851-5

Khanh, H. H. & Khang, A. (2021). The role of artificial intelligence in blockchain applications. *Reinventing Manufacturing and Business Processes through Artificial Intelligence*, pp. 20–40. CRC Press, Boca Raton, FL. https://doi.org/10.1201/9781003145011-2

Korthagen, F. A. J. (2004). In search of the essence of a good teacher: Towards a more holistic approach in teacher education. *Teaching and Teacher Education*, 20(1), 77–97. https://doi.org/10.1016/j.tate.2003.10.002

Liu, Q., Sang, Z., Chen, H., Wang, J., & Zhang, H. (2017). An efficient algorithm for complex-valued neural networks through training input weights. *Neural Information Processing*, 2017, 150–159. https://doi.org/10.1007/978-3-319-70093-9_16

Most-believe-automation-will-shape-future-workforce. (n.d.). SAGE Business Researcher. https://doi.org/10.1177/237455680329.n3

Marin, S. (2013). The educational systems and teacher training in the knowledge-based society. *Procedia - Social and Behavioral Sciences*, 93, 1039–1044. https://doi.org/10.1016/j.sbspro.2023.09.326

Mazmanian, P. E. & Davis, D. A. (2002). Continuing medical education and the physician as a learner. *JAMA*, 288(9), 1057. https://doi.org/10.1001/jama.288.9.1057

Petrock, V. (2017). When is Training in Virtual Reality a Good Investment? eMarketeer. https://parc.ipp.pt/index.php/iirh/article/view/2865

Rana, G., Khang, A., Sharma, R., Goel, A. K., & Dubey, A. K. (2021). *Reinventing Manufacturing and Business Processes through Artificial Intelligence*. CRC Press, Boca Raton, FL. https://doi.org/10.1201/9781003145011

Rani, S., Chauhan, M., Kataria, A., & Khang, A. (2021). IoT equipped intelligent distributed framework for smart healthcare systems. *Networking and Internet Architecture*, 2, 30. https://doi.org/10.48550/arXiv.2110.04997

Shah, V. & Khang, A. (2023). Metaverse-enabling IOT technology for a futuristic healthcare system. *AI-Based Technologies and Applications in the Era of the Metaverse* (1st ed., pp. 165–173). IGI Global Press, Hershey, PA. https://doi.org/10.4018/978-1-6684-8851-5.ch008

24 Designing Artificial Intelligence-Enabled Training Approaches and Models for Physical Disabilities Individuals

Chabi Gupta and Alex Khang

24.1 INTRODUCTION

24.1.1 CONCEPTUAL FRAMEWORK AND LITERATURE REVIEW

Conceptual framework of this research is as follows (also depicted in Figure 24.1):

- Introduction to AI-enabled approaches for workforce training,
- Definition of AI-enabled training systems,

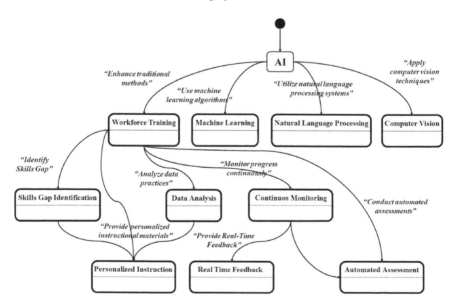

FIGURE 24.1 Opportunities galore integrating AI into existing workforce training methodologies.

DOI: 10.1201/9781003440901-24

- Importance of inclusivity in workforce training,
- Overview of the research objective and scope,
- Challenges in training systems for physically challenged people,
- Identifying the unique challenges faced by physically challenged individuals in training,
- Accessibility and assistive technologies,
- Ensuring equal opportunities and addressing biases,
- Benefits of AI-enabled training systems,
- Improved accessibility and adaptability for physically challenged individuals,
- Enhanced personalization and individualized learning experiences,
- Increased engagement and motivation,
- Role of adaptive learning in workforce training,
- Exploring adaptive learning algorithms and techniques, and
- Customizing learning paths and content based on individual needs.

24.2 INTRODUCTION TO AI-ENABLED APPROACHES FOR WORKFORCE TRAINING

This is a crucial aspect of modern educational practices. In recent years, the integration of artificial intelligence (AI) into various fields has been gaining significant attention and traction, and the realm of workforce training is no exception. This emerging trend encompasses the use of AI technologies such as machine learning algorithms, natural language processing systems, and computer vision techniques to enhance traditional methods employed in employee education (Iverson, 2000; Jiang et al., 2016; Karwal & Tandon, 2021). By harnessing the potential offered by AI-powered solutions in training programs, organizations can leverage sophisticated algorithms to analyze data patterns, identify skill gaps among employees more effectively, and subsequently provide personalized instructional materials that cater to individual needs (Kim, 2006).

Moreover, these AI-enabled approaches enable trainers to monitor progress continuously through automated assessments while providing ongoing feedback based on real-time performance metrics (Kumar & Suresh, 2018; Li et al., 2021). Thus, integrating technology-driven advancements like AI into existing workforce training methodologies offers tremendous opportunities for improving overall productivity levels within an organization's human resources pool, as shown in Figure 24.1.

24.3 DEFINITION OF AI-ENABLED TRAINING SYSTEMS

AI-enabled training systems refer to educational platforms or technologies that utilize AI algorithms and techniques to enhance the learning experience. These systems are designed to provide more comprehensive and immersive training by utilizing data-driven insights, machine learning capabilities, natural language processing, and other AI technologies ("Impact of workforce diversity on employees' performance," 2020; Liu et al., 2023). With AI-enabled training systems, learners can benefit from personalized instruction tailored to their specific needs.

The advanced algorithms in these systems analyze user behavior patterns and adapt the content delivery accordingly. This allows for a more interactive and adaptive approach that caters to individual learning styles. Moreover, AI technology enables these training systems to collect vast amounts of data on learner progress and performance. This wealth of information is then used by educators or trainers for continuous improvement purposes, enabling them to make data-informed decisions about instructional strategies. In summary, AI-enabled training systems harness the power of AI techniques to offer a more personalized, adaptive, and effective approach toward education (Kim, 2006; Kumar & Suresh, 2018).

24.4   IMPORTANCE OF INCLUSIVITY IN WORKFORCE TRAINING

In a rapidly changing global economy, organizations must recognize the value of creating diverse and inclusive work environments. By promoting inclusivity, employers can foster innovation and creativity within their teams. Inclusivity in workforce training involves providing equal opportunities for individuals from all backgrounds to access professional development programs and resources ("Impact of workforce diversity on employees' performance," 2020; Li et al., 2021). It requires acknowledging different perspectives, experiences, and skillsets that each employee brings to the table. This approach allows organizations to tap into a wider pool of talent and encourages collaboration among team members.

Furthermore, Dev et al. (2021) and Liu et al. (2023) in their research highlight that including employees with diverse backgrounds enhances problem-solving abilities by introducing varied viewpoints and solutions. When people from different cultures or demographics come together in an inclusive environment during training sessions, they are more likely to challenge conventional thinking patterns leading to creative breakthroughs. It is essential for companies not only to prioritize diversity but also actively cultivate an atmosphere where everyone feels welcomed—regardless of factors such as race, gender identity/expression, disability status, etc.,—through supportive policies ensuring inclusion at every level (Dev et al., 2021; Li et al., 2021).

24.5   OVERVIEW OF RESEARCH OBJECTIVES AND SCOPE

24.5.1   RESEARCH OBJECTIVES

- To investigate AI-based strategies and models that can be used to develop workforce training systems specifically for individuals with physical disabilities.
- To identify the challenges and ethical concerns associated with incorporating AI in workforce training for people with physical disabilities.
- To develop progressive approaches and frameworks driven by AI that can empower individuals with physical disabilities by improving their employability prospects and fostering inclusivity within workforce training.

24.5.2 SCOPE OF THE RESEARCH

The research will focus on the following areas:

- Effective strategies and real-life examples of AI-based training platforms for people with physical disabilities.
- Obstacles and ethical concerns associated with incorporating AI in workforce training, such as mitigating biases, ensuring accessibility, and safeguarding privacy.
- The integration of AI-powered solutions in training programs for physically challenged individuals, such as personalized learning algorithms, predictive analytics, and adaptive content delivery systems.

The study employs a multidimensional approach to explore the subject matter thoroughly. It incorporates diverse methodologies such as an extensive literature review, in-depth case studies, and insightful interviews with leading authorities in the field of AI and workforce training. By utilizing this multifaceted approach, valuable insights can be obtained regarding the intersection of AI technology and workforce development for individuals with physical disabilities. The research endeavors to give rise to practical implications that can guide stakeholders involved in designing and implementing AI-based systems for enhancing workforce training opportunities specifically tailored for people facing physical disabilities (Shah & Khang et al., 2023).

24.6 CHALLENGES IN TRAINING SYSTEMS FOR PHYSICALLY CHALLENGED PEOPLE

Today, one of the key concerns is how to effectively train physically challenged individuals. This has become a pressing issue as it requires specialized systems and strategies that cater to the unique needs of these individuals. To address this challenge, innovative training methods must be developed with depth and complexity. One potential solution involves incorporating virtual reality (VR) technology in training programs for physically challenged people. Eun et al. (2022) and Sooraj et al. (2020) discuss that by creating immersive environments, VR can provide realistic simulations of various scenarios where physical movements are required. For instance, an individual with limited mobility could virtually practice navigating through obstacles or performing tasks requiring dexterity as shown in Figure 24.2.

Moreover, personalized training plans should be implemented to ensure maximum effectiveness for everyone based on their specific challenges and abilities (Dev et al., 2021; Gorski et al., 2022). These tailored plans would consider factors such as the type and severity of disability while accounting for personal goals and preferences. Additionally, collaboration among interdisciplinary teams comprising healthcare professionals, engineers, and educators is crucial in developing comprehensive training systems that address different aspects—from understanding physical limitations to designing inclusive environments (Kim, 2006; Kumar & Suresh, 2018). The situation necessitates going beyond traditional approaches by embracing innovative technologies like VR and tailoring interventions according

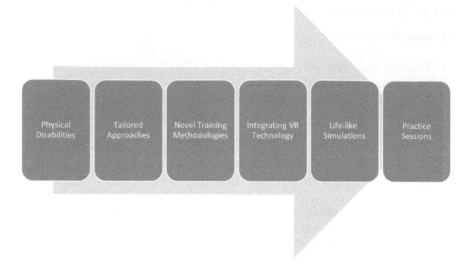

FIGURE 24.2 Challenges in training systems for physically challenged people.

to individual requirements alongside fostering collaborative efforts among experts as shown in Figure 24.2.

Identifying the distinct difficulties encountered by physically challenged individuals in training requires a comprehensive understanding of their specific needs and limitations. These individuals often face additional obstacles that can hinder their ability to fully engage in various aspects of training programs (Dev et al., 2021; Li et al., 2021). One significant challenge is related to accessibility.

Caner & Bhatti (2020), Sadiku et al. (2020), and Xu & Babaian (2021) discuss in their study about physically challenged individuals who may encounter difficulties in accessing training facilities, which leads to limited participation opportunities. Moreover, traditional equipment and learning materials might not be tailored or accommodated for their diverse physical abilities and requirements (Eun et al., 2022; Sooraj et al., 2020).

Furthermore, there exists a psychological component as well—disabled trainees might battle negative self-perceptions or societal stereotypes that limit their confidence levels when attempting new exercises or techniques during each session. Southworth et al. (2023) suggest overcoming such mental barriers becomes crucial for creating an inclusive training environment where these individuals can flourish. To address these issues effectively, it's important for trainers and organizations to implement inclusive practices that focus on adapting infrastructure, providing specialized equipment if needed, offering modified teaching methods based on individual capabilities ensuring equal access opportunities.

24.6.1 ACCESSIBILITY AND ASSISTIVE TECHNOLOGIES

In the ever-evolving technological landscape, there has been a notable surge in the development of AI solutions aimed at assisting individuals with mental, physical, visual, or hearing impairments (Dev et al., 2021; Gorski et al., 2022).

These AI-powered tools have proven to be invaluable in enabling these individuals to navigate through a spectrum of tasks spanning from everyday routines to more intricate situations. Notably, they are designed not just for individual use but also facilitate seamless collaboration between differently abled individuals and their non-disabled counterparts (Eun et al., 2022; Sooraj et al., 2020).

Henceforth, it is imperative that technology be leveraged as an aid in training and fostering the professional growth of disabled employees on par with their able-bodied colleagues. Effective workforce training and development programs for individuals with physical disabilities necessitate the incorporation of accessibility features and assistive technologies (Kirongo et al., 2022; Svetlana et al., 2022). These essential components play a crucial role in ensuring equal access to training opportunities, promoting inclusivity, and empowering disabled individuals to excel in their chosen fields.

Accessibility encompasses various elements that make information, resources, and facilities readily accessible to people with physical challenges. This includes providing wheelchair ramps or elevators for those with mobility impairments and implementing captioning or sign language interpretation services for individuals who are deaf or hard of hearing. Additionally, the use of assistive technologies further enhances the learning experience by compensating for specific limitations faced by physically challenged learners (Eun et al., 2022; Kirongo et al., 2022; Sooraj et al., 2020; Svetlana et al., 2022).

For instance, screen readers enable visually impaired trainees to access online courses through audio output, while adaptive keyboards allow those with limited dexterity to type more easily. By recognizing these factors as integral parts of workforce training programs addressing physical disabilities, organizations can foster an inclusive environment that promotes personal growth and professional success.

In addition to the AI tools, there are several other technologies that businesses can utilize to enhance their workflow. One such technology is robotic process automation, which can automate repetitive tasks. By delegating mundane and time-consuming activities to Robotic process automation (RPA) systems, employees can minimize manual interventions and allocate more of their time toward high-value work (American Hospital Association, 2017).

Furthermore, AI-integrated recommendation systems offer organizations the opportunity to retrieve personalized suggestions from vast sets of data. This not only enhances decision-making processes but also enables businesses to provide tailored recommendations based on individual preferences (Farrow, 2022; Ishii et al., 2020; Wilson & Daugherty, 2019).

Another valuable application of AI lies in computer vision technologies (American Hospital Association, 2017). With advanced object detection algorithms and image recognition techniques, AI-powered computer vision applications have emerged as a powerful tool for identifying and interpreting visual data accurately. This includes facial recognition capabilities that prove useful in industries like security or retail marketing where it's important to detect individuals or analyze customer emotions through facial expressions. Moreover, by integrating AI into cybersecurity practices, businesses gain a proactive approach toward threat detection and elimination in real-time instances (American Hospital Association, 2017).

24.6.2   ENSURING EQUAL OPPORTUNITIES AND ADDRESSING BIASES

To ensure equal opportunities and address biases in workforce training programs, it is crucial to implement technology with built-in tools that specifically target the needs of physically challenged employees. Eun et al. (2022) and Gorski et al. (2022) discuss in their study that by incorporating these specialized features into training programs, employers can create a more inclusive environment where all individuals have an equal chance to succeed.

One example of such technology is assistive devices or software that accommodate different physical disabilities. These tools could include screen readers for visually impaired individuals, ergonomic keyboards for those with limited hand dexterity, or voice recognition software for individuals with mobility impairments.

Additionally, VR simulations can help bridge the gap between theoretical learning and practical application by providing accessible experiences tailored to everyone's needs (Caner & Bhatti, 2020; Kalyanaraman et al., 2022). By leveraging these technologies within workforce training initiatives, organizations demonstrate their commitment to fostering diversity and inclusion while also ensuring that every employee has an opportunity to thrive professionally (Kirongo et al., 2022; Sooraj et al., 2020).

AI voice technologies have revolutionized communication for individuals with impairments, proving to be invaluable in enhancing accessibility. Prominent examples of these innovations include Siri, Alexa, and Echo, which effectively overcome barriers faced by people with visual disabilities. These advanced programs excel at describing text and images to those who are visually impaired as shown in Figure 24.3.

Additionally, the integration of text-to-speech or speech-to-text technologies aids individuals with brain injuries in comprehending information more easily or expressing themselves effectively (Caner & Bhatti, 2020; Kalyanaraman et al., 2022; Southworth et al., 2023).

Notably, Google's Parrotron app stands out as an AI tool specifically designed to address the needs of speech-impaired individuals. By intelligently translating distorted speech patterns into coherent conversations, it empowers such persons and facilitates their engagement within social interactions.

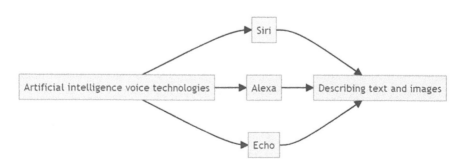

FIGURE 24.3   Prominent examples of these innovations include Siri, Alexa, and Echo.

24.7 BENEFITS OF AI-ENABLED TRAINING SYSTEMS

To foster a comprehensive and cooperative work environment that embraces employees irrespective of their disabilities, businesses can leverage AI tools. Gorski et al. (2022) and Li et al. (2021) in their study discuss that by equipping individuals with efficient speech transcription tools, business leaders facilitate equal participation in collaborative settings and the exchange of ideas among different departments and teams. This inclusive approach ensures that every employee's voice is heard without any barriers or limitations imposed by their disabilities. The utilization of such advanced technologies supports an atmosphere where diverse perspectives are valued, enriching discussions and problem-solving processes (Brynjolfsson & McAfee, 2017; Sadiku et al., 2020).

In the contemporary business landscape, there is growing recognition of the potential that AI holds for revolutionizing employee learning experiences. Kirongo et al. (2022) and Sooraj et al. (2020) talk about the business leaders who are increasingly seeking ways to tap into this transformative power by leveraging AI-powered solutions. These innovative tools have demonstrated their ability to significantly enhance and enrich employee training and development programs. Brynjolfsson and McAfee (2017) and Sadiku et al. (2020) conclude that by harnessing AI technology, organizations can elevate traditional learning methods to new heights.

Through sophisticated algorithms and machine learning capabilities, AI-powered systems have the capacity to analyze vast amounts of data and provide personalized recommendations for individuals' professional growth. This tailored approach ensures that employees receive targeted content that aligns with their specific needs and aspirations, leading to more effective skill-building outcomes. Furthermore, these advanced solutions enable interactive and engaging learning experiences through immersive technologies like VR or augmented reality (AR) (American Hospital Association, 2017; Kirongo et al., 2022; Svetlana et al., 2022). By simulating real-life scenarios in a safe environment, employees can actively participate in hands-on exercises or problem-solving activities that foster critical thinking skills while maximizing knowledge retention. Overall, incorporating AI-powered solutions into an organization's training strategy enhances not only the efficiency but also the effectiveness of its workforce's ongoing education journey (Sooraj et al., 2020).

24.8 IMPROVED ACCESSIBILITY AND ADAPTABILITY FOR PHYSICALLY CHALLENGED INDIVIDUALS

AI is playing a crucial role in empowering disabled individuals to lead independent and fulfilling personal as well as professional lives. Kirongo et al. (2022) and Svetlana et al. (2022) advocate the advent of AI technology which has introduced various programs that assist businesses in streamlining their operations, ultimately resulting in more efficient workflows. Take for instance virtual assistants like Google Assistant or Amazon Alexa, which are powered by AI algorithms: these intelligent software solutions can effortlessly schedule meetings, set reminders, and automate several voice-related tasks.

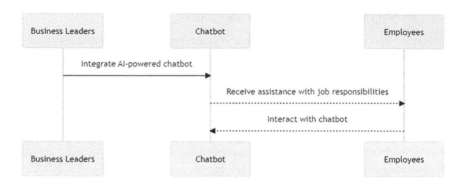

FIGURE 24.4   Case diagram for an AI-enabled interactive interface.

Moreover, business leaders now can integrate AI-powered chatbots into various work-related training applications. This allows employees to receive assistance with diverse job responsibilities through an interactive interface as shown in Figure 24.4. For example, Microsoft Bing's collaboration with Skype illustrates this integration effectively; you can simply pose a question using Bing on Skype and instantly obtain a relevant answer (Brynjolfsson Erik & McAfee Andrew, 2017; Kaur & Mohammad Aslam, 2020).

By harnessing the potential of AI technologies such as these invaluable assistant tools or chatbots embedded within workplace apps for different organizational requirements., both employers and employees benefit from enhanced productivity levels (Akerkar, 2019; Zohuri, 2020).

24.9   ENHANCED PERSONALIZATION AND INDIVIDUALIZED LEARNING EXPERIENCES

In addition to personalized learning paths, leaders can leverage AI tools within their organizations as a means of identifying skill gaps among employees and gaining insight into everyone's competencies (American Hospital Association, 2017; Ayandibu & Kaseeram, 2019). This information then enables the development of targeted training programs aimed at promoting upskilling opportunities throughout the company as shown in Figure 24.5.

VR and AR also play a crucial role in revolutionizing employee training by offering interactive and immersive learning experiences. American Hospital Association (2017) and Ayandibu and Kaseeram (2019) point out that these technologies create engaging scenarios where employees can actively participate in real-life simulations or visualizations directly related to their roles. Importantly, providing personalized learning experiences is especially pertinent for individuals with disabilities within an organization. Through tailored approaches supported by AI technologies, businesses empower these employees with inclusive opportunities designed specifically around acquiring the necessary skills and knowledge required for success in their daily responsibilities (Dell'Era & Rahova, 2019; Levukova, 2022).

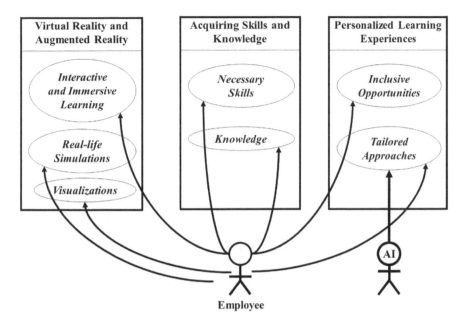

FIGURE 24.5 Personalized learning paths.

24.10 INCREASED ENGAGEMENT AND MOTIVATION

In today's rapidly changing and interconnected world, establishing meaningful connections and networking is more imperative than ever. Kirongo et al. (2022) and Svetlana et al. (2022) in their study advocate the AI technology that has emerged as a powerful tool to facilitate the inclusion of people with disabilities in society. For instance, Microsoft's virtual assistant Cortana uses AI capabilities such as voice control to aid visually impaired individuals in navigating their devices effortlessly. Also, innovative programs like RogerVoice and Ava leverage AI algorithms to transcribe group conversations for those with hearing impairments. These cutting-edge solutions not only capture spoken words accurately but also include essential features such as punctuation marks, the sender's name, and even important terms from the user's personal dictionary.

Moreover, businesses can integrate these AI-powered virtual assistants into their operations to help disabled employees participate fully in meetings and manage schedules efficiently like any other employee would do. Additionally, employers benefit from translation tools empowered by AI that facilitate communication among multilingual team members who speak different languages. This can be of help in designing appropriate workforce training and development programs.

Particularly when it comes to individuals with disabilities in the workforce, fostering an enthusiastic and motivated environment enables them to effectively expand their network and contribute positively within their spheres (American Hospital

Association, 2017; Ayandibu & Kaseeram, 2019). This can result in numerous benefits such as improved collaboration among colleagues and increased prospects for sharing knowledge and acquiring new skills, as well as cultivating a work culture that fosters inclusivity.

24.11 ROLE OF ADAPTIVE LEARNING IN WORKFORCE TRAINING

Adaptive learning plays a crucial role in workforce training programs that are designed using AI tools, particularly for individuals with disabilities. The integration of adaptive learning methods allows these programs to cater to the unique needs and requirements of differently abled employees, enhancing their overall learning experience (Healey & Hodgkinson, 2009; Kahane, 2012; Skulimowski, 2013). By leveraging AI technology, such training programs can adapt and customize the content delivery based on an individual's strengths and weaknesses. This personalized approach ensures that each employee receives tailored instruction and support, maximizing their potential for skill development and knowledge acquisition as shown in Figure 24.6.

Moreover, adaptive learning enables trainers to track learners' progress more effectively by collecting data on their performance throughout the program (Healey & Hodgkinson (2009); Kahane, (2012); Skulimowski, (2013)). Using this data-driven feedback loop, trainers can provide timely interventions or modifications

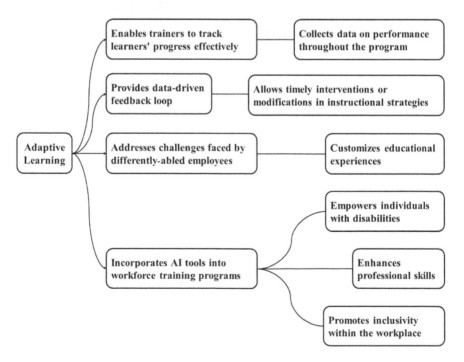

FIGURE 24.6   Role of adaptive learning integrated with AI for workforce training.

in instructional strategies to address specific challenges faced by differently-abled employees. Thus, incorporating adaptive learning into workforce training programs through AI tools empowers individuals with disabilities by providing them with customized educational experiences. Such initiatives not only enhance their professional skills but also promote inclusivity within the workplace.

24.12   EXPLORING ADAPTIVE LEARNING ALGORITHMS AND TECHNIQUES

The significance and relevance of using adaptive learning algorithms and techniques cannot be overstated when it comes to designing AI-enabled training and development programs for physically challenged individuals in the workforce. These innovative approaches are crucial in helping organizations effectively integrate this segment of the workforce into their operations. Dell'Era and Rahova (2019) and Khandelwal & Upadhyay (2021) suggest that by utilizing adaptive learning algorithms, AI technology can adapt its instruction methods to the requirements of physically challenged individuals. This personalized approach ensures that each employee receives tailored support, enhancing their overall learning experience.

Also, incorporating these techniques into training programs enables organizations to overcome traditional barriers faced by physically challenged employees. It promotes inclusivity within the workplace by creating an environment where all members have equal opportunities for professional growth (Levukova, 2022; Zohuri, 2020). In conclusion, leveraging adaptive learning algorithms as part of AI-enabled training programs plays a pivotal role in integrating physically challenged workers into businesses successfully.

24.13   CUSTOMIZING LEARNING PATHS AND CONTENT BASED ON INDIVIDUAL NEEDS

There are several ways that can customize learning paths and content based on individual needs when using AI in workforce systems for physically challenged people. Here are a few examples: use AI to assess individual learning needs. AI can be used to assess the individual learning needs of physically challenged people by gathering data on their prior knowledge, skills, and abilities. This data can then be used to create customized solutions.

One potential application of AI is the provision of instantaneous feedback to learners as they progress through their educational journeys. By leveraging AI technology, educators can deliver real-time feedback that helps students identify and rectify mistakes or misconceptions in a timely manner during the learning process. The utilization of AI algorithms enables continuous monitoring and assessment, allowing for personalized guidance tailored to each learner's individual needs. This approach shifts away from traditional methods where feedback is typically given after completion or periodic intervals, providing an immediate response mechanism that promotes active engagement with course materials. This feedback can help learners to identify areas where they need more support and to adjust their learning strategies (Dell'Era & Rahova, 2019; Khandelwal & Upadhyay, 2021).

The application of AI in tailoring content according to individual requirements has become increasingly prevalent. AI offers the capability to customize learning material and cater specifically to the unique needs of each learner. An illustrative instance is how AI can generate transcriptions for videos or audio recordings, making them accessible for individuals with hearing impairments such as deafness or auditory difficulty. With AI's ability to adapt content, learners who face challenges due to hearing disabilities can benefit from equivalent access opportunities when engaging with multimedia resources. AI can also be used to provide alternative text for images or graphics for learners who are blind or visually impaired.

Use AI to track progress and provide recommendations. Dell'Era & Rahova (2019) and Khandelwal & Upadhyay (2021) suggest that AI can be used to track the progress of learners and to provide recommendations for additional resources or activities. This data can help learners to stay on track and to identify areas where they need more support. Akerkar (2019) and Zohuri (2020) also talk about that; by using AI to customize learning paths and content, the organizations can create more inclusive and effective workforce training systems for physically challenged people. This will help to ensure that all people, regardless of their disabilities, can learn and succeed in the workforce.

Be aware of the different types of disabilities that people may have. Some disabilities, such as visual impairments or hearing impairments, may require specific accommodations in the learning environment.

To enhance the learning experience, it is crucial to employ diverse and multifaceted methods. It is important to acknowledge that individuals differ in their preferred ways of processing information. For instance, certain individuals may find reading as the most effective method for comprehension, while others may benefit more from auditory stimuli or hands-on activities.

- Real-time feedback and assessment for personalized progress tracking,
- AI-enabled workforce management systems,
- Leveraging AI for workforce planning and management,
- Opportunity marketplaces for matching skills with tasks,
- Enhancing agility and flexibility in the workforce,
- Design principles for AI-enabled training systems,
- User-centered design for inclusive training experiences,
- Ethical considerations and bias mitigation in AI algorithms,
- Designing intuitive and user friendly interfaces,
- Data-centric approaches in workforce management
- Utilizing data analytics for workforce optimization,
- Predictive analytics for future workforce needs,
- Leveraging AI for talent acquisition and retention,
- Visualization and simulation in training systems,
- Importance of visualization in training content delivery,
- Simulated environments for practical training experiences,
- VR and AR applications,

- AI and data analytics in training systems,
- Collecting and analyzing training data for insights,
- Predictive modeling for identifying training gaps,
- Continuous learning and improvement through AI feedback loops,
- Personalization and customization in training programs,
- Adapting training programs to individual preferences and capabilities,
- AI-powered content recommendations and learning pathways, and
- Addressing accessibility needs and providing tailored accommodations.

Provide opportunities for collaboration. Learning with others can be a great way to support each other and to learn from each other's experiences. Be patient and understanding. It may take people with disabilities longer to learn new things. Be patient and understanding and offer them the support they need to succeed as shown in Figure 24.7.

There have been several notable case studies concerning the implementation of AI in the development of workforce training systems for individuals with physical disabilities. An exemplary initiative is Microsoft's "AI for Accessibility" program, which has successfully utilized AI to create innovative tools and technologies aimed at enhancing independence and productivity among disabled populations.

An important application stemming from this initiative is the groundbreaking "Seeing AI" app. Advanced AI algorithms in smartphone application empowers visually impaired individuals by providing them with a digital representation of their

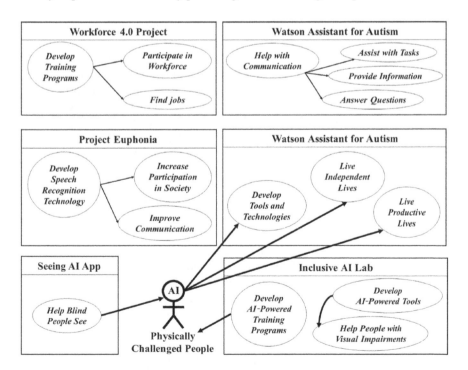

FIGURE 24.7 Training systems for individuals with physical disabilities.

surroundings. By leveraging machine vision technology, "Seeing AI" app enables blind people to effectively navigate daily activities that may otherwise prove challenging or inaccessible without sight. Also, these advancements in building intelligent solutions demonstrate how technology can significantly contribute toward creating an inclusive society where opportunities are accessible to everyone, irrespective of their physical limitations (Brynjolfsson Erik & McAfee Andrew, 2017; Kaur & Mohammad Aslam, 2020).

An intriguing example is Google's "Project Euphonia", which utilizes AI to enhance speech recognition technology specifically tailored for people with speech impairments. The primary objective behind this endeavor is to facilitate improved communication and greater societal involvement among individuals facing such challenges. By utilizing advanced AI algorithms, Project Euphonia aims to strengthen the accuracy and understanding of speech patterns exhibited by those with impaired speech capabilities.

IBM's "Watson Assistant for Autism" is a prime example of how cutting-edge research and technological advancements are being utilized to address the unique challenges faced by individuals with autism spectrum disorder. This undertaking can be regarded as an innovative case study focused on leveraging AI-enabled approaches and models in developing workforce training systems tailored explicitly for people with physical disabilities, particularly those living with autism spectrum disorder (ASD). By harnessing the immense capabilities of AI, this groundbreaking assistant seeks to revolutionize communication channels and cultivate better interaction skills among autistic individuals.

Brynjolfsson Erik and McAfee Andrew (2017) and Kaur and Mohammad Aslam (2020) discuss in their study that through its advanced algorithms and machine learning techniques, Watson Assistant has the potential to elevate not only the quality of life but also improve socio-cognitive development outcomes for those affected by ASD. By using Watson Assistant, individuals can benefit from various features such as obtaining answers to queries, receiving informative content, scheduling appointments efficiently, and managing financial matters effectively. One significant advantage of utilizing AI technology like Watson Assistant in supporting individuals with autism is its ability to provide personalized assistance tailored to everyone's unique needs.

The advanced algorithms powering this system allow it to learn patterns, and IBM's "Watson Assistant for Autism" offers a notable illustration of how AI can be leveraged to design workforce training systems specifically catered toward individuals with physical challenges. The aim of this assistant is to enhance communication and facilitate interaction for people diagnosed with ASD. By employing AI, the system enables users to have their queries answered, access relevant information, and even receive assistance in managing daily tasks like scheduling appointments or handling financial matters.

Akerkar (2019), Kaur & Mohammad Aslam (2020), and Zohuri (2020) discuss about the integration of AI technology into such initiatives which demonstrate the potential benefits it holds for marginalized communities. These advancements not only improve quality of life but also promote inclusivity by providing equal opportunities for engagement within the organization. Consequently, more comprehensive research studies focusing on integrating AI-enabled approaches should be conducted as they hold significant promise in transforming workforce development programs targeted at underserved populations.

24.14 REAL-TIME FEEDBACK AND ASSESSMENT FOR PERSONALIZED PROGRESS TRACKING

Several other notable case studies have emerged in the realm of designing workforce training systems for physically challenged individuals using AI-enabled approaches and models. One such example is the "Workforce 4.0" project in India, which leverages AI to develop tailored programs that cater specifically to individuals with disabilities. The primary objective behind this initiative is to facilitate greater employment opportunities for disabled persons, enabling them to actively engage in various work environments.

Also, within university research settings, initiatives like the "Inclusive AI Lab" at the University of California, Berkeley are striving toward creating inclusive and accessible AI technologies for people with disabilities. This prominent lab undertakes a range of projects aimed at enhancing accessibility and inclusion through advancements in AI-driven tools designed specifically for visually impaired individuals as well as developing comprehensive AI-powered training programs catering toward diverse disability requirements. Incorporating real-time feedback and assessment mechanisms for personalized progress tracking is paramount in the design of training and development programs that aim to integrate individuals with diverse abilities into any organization's workforce (Vineeta et al., 2023).

Leveraging AI-enabled technologies can be especially beneficial in this endeavor (Zohuri, 2020). By utilizing these advanced tools, organizations can obtain a comprehensive understanding of everyone's unique needs, strengths, and areas for improvement. Real-time feedback allows trainers and instructors to provide timely guidance, support, and corrective measures to ensure maximum effectiveness in skill acquisition (Brynjolfsson Erik & McAfee Andrew, 2017). Additionally, a well-designed assessment system enables accurate monitoring of progress over time while also facilitating the identification of specific areas where further intervention or adaptation may be required. By utilizing such strategies as part of their broader inclusion efforts, organizations not only promote diversity but also foster an inclusive culture that values every employee's contribution.

24.14.1 RESEARCH METHODOLOGY

- **Data Sample**: The data sample includes a diverse group of physically challenged people from different backgrounds and with different levels of disabilities. This helps to ensure that the research findings are generalizable to a wider population of physically challenged people.
- Demographic profile: The demographic profile of the data sample includes information in the following Table 24.1.
- **Work Experience**: The research also collects information on the work experience of the participants, including the types of jobs they have held,

TABLE 24.1
Demographic Profile of Respondents

Factor	Demographic Profile		
Age	25–45 years – 21%	45–65 years - 69%	65 and above years – 10%
Gender	Male – 75%	Female – 25%	
Race/ethnicity	South Asian – 92%	Others – 8%	
Disability type	Low – 10%	Medium – 75%	High – 15%
Level of Education	Graduate – 60%	Postgraduate – 20%	Professionals – 20%
Employment status	Public Sector – 25%	Private Sector – 65%	Others – 10%
Geographic location	Asia and Nearby Regions – 80%	Europe – 10%	Others – 10%

their level of job satisfaction, and the challenges they have faced in the workplace.

- **Training Needs**: The research also collects information on the training needs of the participants, including the types of training they would like to receive, their preferred learning methods, and their expectations for the benefits of training. Here are some specific questions that were asked to collect this information as shown in Figure 24.8.

A diverse set of approaches such as surveys, interviews, and focus groups were employed to gather the data sample. To detect trends and patterns in the data, statistical methods were utilized during the analysis phase. The conclusions derived from this research played a crucial role in formulating actionable suggestions for constructing workforce training systems that embrace inclusivity. The questionnaire survey included specific questions to analyze further the issues, challenges, and feasibility

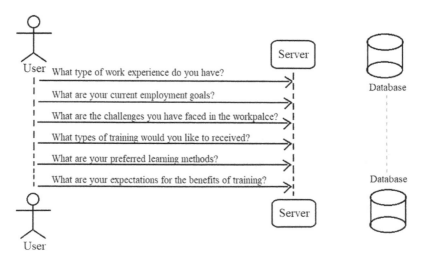

FIGURE 24.8   Use case diagram for designing research questionnaire survey.

of AI in workforce training and development of differently abled human resources. These questions cover a variety of aspects from personal experiences to expectations, potential challenges, and ethical aspects related to AI-based training systems.

1. What are the current challenges you face with existing training systems?
2. How could AI-based training systems improve your learning experience?
3. Can you provide any real-life examples where you've used AI-based systems in your training?
4. What specific features would you like to see in an AI-based training system designed for individuals with physical disabilities?
5. Do you feel that current training methods adequately foster inclusivity?
6. What concerns related to ethics do you have regarding the use of AI in workforce training?
7. Are there any obstacles you foresee in incorporating AI into your training?
8. How important is workplace learning and development in enhancing your employability prospects?
9. What are the potential impacts of AI on your learning and development?
10. How can AI technology empower you in a workplace setting?
11. What are your expectations from AI-based training systems?
12. How familiar are you with existing AI-based training platforms?
13. How do you imagine AI could be used to support individuals with disabilities in the workforce?
14. What kind of support or adjustments would you need to make use of such an AI-based training system?
15. How confident do you feel about using AI-based systems for learning and development purposes?

24.14.2 OBSERVATIONS

Listed are the most important challenges in harnessing AI capabilities in designing inclusive workforce systems that were observed during discussions and analysis (Figure 24.9).

24.14.2.1 Data Availability and Quality

There is a limited amount of data available on the capabilities and needs of physically challenged people. This makes it difficult to train AI models that can accurately assess their skills and potential.

24.14.2.2 Model Complexity

AI models for workforce training are often complex and difficult to interpret. This can make it difficult to explain how the models make decisions, which can be a problem for physically challenged people who may need to understand the rationale behind the decisions to trust the system.

24.14.2.3 Bias

AI models can be biased, which means that they may not accurately reflect the capabilities of all physically challenged people. This can lead to discrimination and

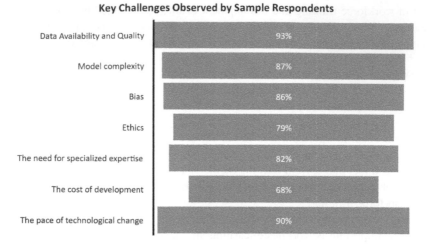

Key Challenges Observed by Sample Respondents

Challenge	Percentage
Data Availability and Quality	93%
Model complexity	87%
Bias	86%
Ethics	79%
The need for specialized expertise	82%
The cost of development	68%
The pace of technological change	90%

FIGURE 24.9   Key challenges observed by sample respondents.

exclusion accessibility. AI systems must be accessible to all physically challenged people, including those with visual, hearing, and mobility impairments. This can be a challenge, as it requires the use of accessible technologies and interfaces.

24.14.2.4   Ethics

There are several ethical issues that need to be considered when building AI-enabled workforce training systems for physically challenged people. These include issues of privacy, discrimination, and fairness.

24.14.2.5   The Need for Specialized Expertise

Developing workforce training systems that incorporate AI technology for individuals with physical disabilities necessitates the acquisition of specialized knowledge in areas such as AI, accessibility, and understanding the unique needs of this population. Given these complex requirements, it can pose a challenge to identify and recruit suitable professionals who possess the precise expertise needed to construct such tailored systems.

24.14.2.6   The Cost of Development

AI-enabled workforce training systems can be expensive to develop. This is due to the need for specialized expertise, the complexity of the systems, and the need to make accessible systems for all physically challenged people.

24.14.2.7   The Pace of Technological Change

The field of AI is constantly changing, which means that AI-enabled workforce training systems need to be regularly updated to keep up with the latest advances. This can be a challenge, as it requires the systems to be designed in a way that makes them easy to update. By effectively addressing the challenges outlined above, organizations can build AI-enabled workforce training systems that can help physically challenged people reach their full potential.

24.14.3 ETHICAL CONSIDERATIONS

There are several key ethical considerations when incorporating AI in workforce training for people with disabilities as shown in Figure 24.10:

1. Privacy and Confidentiality: AI systems need to handle personal data, such as disability specifics, learning patterns, and progress, responsibly. Users need assurance their data will not be misused or shared without consent. Clear policies must be in place.
2. Accessibility: Not all AI technologies may be accessible to people with certain types of disabilities. It's crucial the technology is developed with accessibility in mind so it's inclusive for all users.
3. Bias and Discrimination: AI-enabled workforce systems should be designed in a way that eliminates bias and subjectivity. For example, if a training system is designed with a bias toward a certain kind of disability, it may not be as effective for others.
4. Impersonal Learning Experience: AI systems lack human touch which leads to an impersonal learning experience, affecting the learner's engagement and motivation levels.
5. Dependence on Technology: Over-reliance on AI could potentially hinder the development of certain skills such as peer interaction and group collaboration.
6. Accountability: There needs to be clear accountability for the outcomes of AI-based training systems. If a person feels the training program is ineffective, there should be a method to hold the developers accountable.
7. Job Displacement: AI could potentially displace jobs, creating ethical considerations around the provision of alternative employment or support systems.

These are just a few key considerations that were highlighted by most sample respondents, and there may be others depending on the nature of the technology and learners as shown in Figure 24.11.

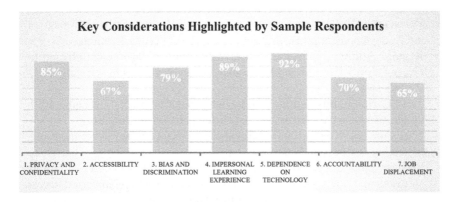

FIGURE 24.10   Key considerations highlighted by sample respondents.

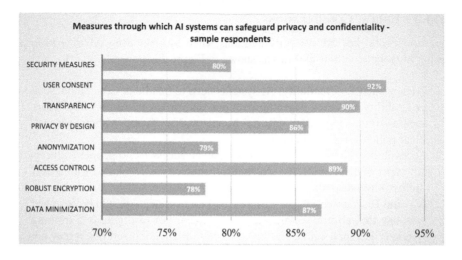

FIGURE 24.11 The nature of the technology and learners.

AI systems can safeguard privacy and confidentiality through several measures as highlighted by the sample respondents:

- **Data Minimization**: Only collect the data necessary for the AI system's operation. The less data collected, the less risk there is of unauthorized access or misuse.
- **Robust Encryption**: Any personal data stored or transmitted should be encrypted. It ensures that even if the data is intercepted or accessed, it remains unintelligible without the decryption key.
- **Access Controls**: Implement strict access controls to ensure that only authorized individuals can access the data. It includes practices like two-factor authentication and layered permissions.
- **Anonymization**: Use anonymization techniques to remove personally identifiable information from data sets. Any data used for training AI models should be anonymized to prevent unauthorized identification of individuals.
- **Privacy by Design**: Incorporate privacy considerations into every stage of the AI development process, starting from the conceptualization.
- **Regular Audits**: Conduct regular privacy audits to detect any vulnerabilities and rectify them promptly. An external auditor could provide an objective assessment.
- **Transparency**: Be clear about how, why, and where data are being used. Users must have easy access to the privacy policy detailing this.
- **User Consent**: Always obtain explicit user consent before collecting, using, or sharing any personal data.
- **Security Measures**: Use firewalls, intrusion detection systems, antivirus software, and other security tools to protect data from unauthorized access or breaches. These strategies, combined with careful adherence to local and global privacy laws, can help AI systems ensure privacy and confidentiality.

There are several AI technologies presently being used in training platforms to enhance educational experiences and personalize learning as highlighted by the sample respondents:

- **Natural Language Processing (NLP)**: This technology enables AI to understand, interpret, and respond to human language, allowing for more natural and intuitive interactions. Platforms can use NLP to assess written assignments, provide feedback, and facilitate chatbot-based Q&A sessions.
- **Machine Learning (ML)**: ML algorithms allow the platform to adapt and personalize content to the learner's pace, style, and understanding. It can track learning progress, identify learner's weak areas, and adjust the curriculum accordingly.
- **Intelligent Tutoring Systems (ITS)**: ITs are AI-based systems that provide personalized instruction and feedback to learners, enhancing the effectiveness of learning.
- **Speech Recognition**: This technology can enhance accessibility for learners with specific needs. It can interpret spoken language and convert it into text, and vice versa, helping learners with visual or motor impairments.
- **Affective Computing**: This involves AI recognizing and responding to human emotions, which can play a role in adaptive learning. A platform could assess a learner's frustration or confusion and adjust the difficulty or style of content accordingly.
- **Recommender Systems**: Similar to Netflix or Amazon recommendations, these AI systems suggest additional resources or course material based on the learner's behavior or preferences.
- **Computer Vision**: This technology powers features like automated proctoring during exams, ensuring academic integrity in remote testing situations.

These examples demonstrate the promise AI holds in transforming and personalizing learning, increasing accessibility, and improving overall educational outcomes as shown in Figures 24.12 and 24.13.

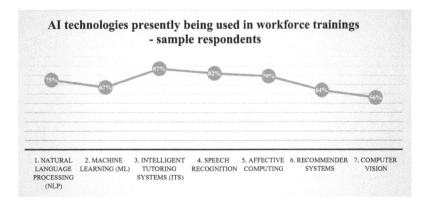

FIGURE 24.12   Transforming and personalizing learning.

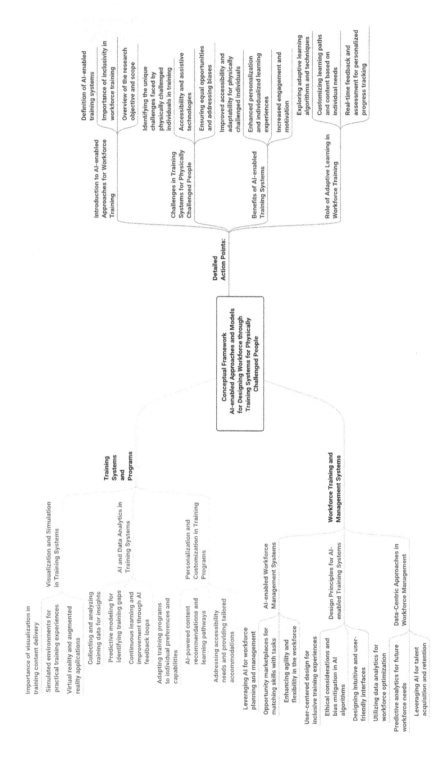

FIGURE 24.13 Designing an inclusive workforce through training systems.

24.15 RECOMMENDATIONS

Analysing AI-based training platforms for people with disabilities requires assessing their effectiveness, accessibility, and inclusivity. Here are some broad aspects to consider:

- **Accessible Design**: Check if the platform is designed with multiple disabilities in mind. It needs to cater to users with various types of disabilities such as motor, visual, hearing, and cognitive impairments. This could manifest in features like screen reader compatibility, speech-to-text or text-to-speech functionality, and easy navigation options.
- **Personalized Learning**: The AI-powered system must adapt to an individual user's learning pace, style, and needs reflecting a personalized approach. It should allow customization for different learning preferences.
- **Interactivity**: Training platforms should have interactive elements to engage learners. This could include VR, AR, or mixed reality (MR) technologies.
- **Analytics and Performance Tracking**: AI platforms should be able to track a user's progress, identify areas of improvement, and suggest learning paths based on these insights.
- **Support and Training**: Good AI-based platforms should offer comprehensive training for users and continuous support to assist with any difficulties in the learning journey.
- **Integrations**: The platform should be compatible with other commonly used software and devices for an integrated learning experience.
- **Ethical Aspects**: The AI platform must ensure the ethical handling of user data, privacy, and any potential bias in its algorithms or design.
- **Cost and Affordability**: Cost can be a major barrier for individuals with disabilities. Analyzing whether these platforms are affordable or have cost-assisted options available is important.

Most platforms will have strengths and weaknesses across these dimensions. It's about identifying which ones best meet the needs of the individuals who will use the system.

24.16 CONCLUSION

This research seeks to investigate the application of AI in developing specialized workforce training systems that cater specifically to individuals with physical disabilities. With the rapid progress and capabilities of AI, there is immense potential for its transformative impact across multiple sectors, including workplace learning and development. In today's rapidly advancing world, there is an increasing demand to leverage the capabilities of AI to develop comprehensive and impactful training programs specifically tailored for individuals facing physical challenges. It has become evident that applying AI technologies can offer a multitude of benefits by addressing accessibility barriers and promoting inclusivity in education and skill development. By harnessing AI's potential, we can overcome various obstacles faced by those with physical disabilities, enabling them to receive personalized training experiences that cater to their unique needs (Shyam & Khang, 2023).

AI holds immense promise when it comes to enhancing the effectiveness of educational initiatives for individuals living with physical impairments. Through advanced algorithms and machine learning techniques, AI systems have the power not only to understand individual requirements but also to adapt instructional content accordingly. This adaptive nature enables customized approaches toward teaching and ensures that each learner receives optimal support based on their specific abilities and limitations. Moreover, incorporating AI into inclusive training programs offers increased autonomy for learners as they navigate through curriculum materials at their own pace while receiving real-time feedback aligned with personal goals. The intelligent analytics provided by these systems further contribute toward tracking progress accurately, thereby allowing trainers or instructors valuable insights into areas needing improvement or additional attention (Pooja et al., 2023).

Incorporation of AI-powered solutions within the training process can bring about significant advantages in addressing the specific needs and requirements of individuals with physical disabilities. The utilization of AI technologies, such as personalized learning algorithms, predictive analytics, and adaptive content delivery systems, enable training programs to be customized based on an individual's unique capabilities, preferences, and accessibility demands. By automating various processes through AI assistance, valuable time can be saved while ensuring a more efficient training experience for those with physical challenges (Muthmainnah et al., 2023).

Furthermore, the integration of AI empowers trainers by providing them with crucial data insights that they can utilize to enhance their teaching methodologies. This insightful information not only assists trainers in better understanding each learner's strengths and weaknesses but also aids in generating tailored learning pathways that cater specifically to an individual's areas requiring improvement. By harnessing these advanced technologies intelligently throughout the training process for physically challenged individuals, the overall effectiveness of educational interventions is enhanced, promoting inclusivity, reducing barriers, and enabling skill development according to each person's unique abilities.

To harness the full potential of AI and embrace this talent-operating model, organizations can optimize their workforce in a way that goes beyond mere task assignment. By effectively aligning skills with specific tasks, organizations can enhance their overall workforce agility and flexibility. Additionally, implementing AI-driven processes allows for the collection of valuable data that serves as a foundation for strategic workforce planning. This approach offers several advantages to organizations. First and foremost, it ensures optimal resource allocation by matching employees' skillsets with appropriate tasks or projects. This precision eliminates inefficiencies caused by misaligned work assignments and helps guarantee high-quality output within shorter time frames (Khang & Muthmainnah et al., 2023).

Furthermore, embracing an AI-savvy talent operating model fosters greater adaptability within the organization's human capital. Employees become better equipped to handle changing requirements and shifting priorities by being exposed to a diverse range of responsibilities based on their unique skillsets. Lastly, leveraging AI not only improves day-to-day operational efficiency but also empowers enterprises with invaluable insights into trends surrounding employee capabilities and workload demand through robust data analysis techniques. Armed with these inputs from strategic workforce planning initiatives enabled by AI technologies, we gain a foresight about future labor demands which leads effective hiring decisions while ensuring business continuity amid dynamic market conditions (Khang & Kali et al., 2023).

24.17 FUTURE SCOPE OF WORK IN INDUSTRY 4.0

While Industry 4.0 presents exciting opportunities for growth and innovation, there are still research problems that need to be addressed to fully realize it's potential. One such problem is the lack of standardization in Internet of Things (IoT) devices and data analytics, which can result in compatibility issues and hinder seamless integration across various systems. Another research problem is the ethical and legal considerations surrounding the use of AI, particularly in decision-making processes. Additionally, there is a need for further research on the social and economic impacts of Industry 4.0, particularly in terms of job displacement and the changing nature of work. Addressing these research problems will be crucial for the continued development and success of Industry 4.0 (Khang & Shah et al., 2023).

REFERENCES

Akerkar, R. (2019). Introduction to artificial intelligence. *Artificial Intelligence for Business*, 2019, 1–18.

American Hospital Association. (2017). Planning for the Workforce of the Future. *Trustee*.

Ayandibu, A. O. & Kaseeram, I. (2019). The future of workforce planning. *Human Capital Formation for the Fourth Industrial Revolution*. IGI Global Press, Hershey, PA. https://doi.org/10.4018/978-1-5225-9810-7.ch006

Brynjolfsson, B. Y. E. & McAfee, A. (2017). The business of artificial intelligence. *Harvard Business Review*, 17, 3–11.

Caner, S. & Bhatti, F. (2020). A conceptual framework on defining businesses strategy for artificial intelligence. *Contemporary Management Research*, 16(3), 175–206. https://doi.org/10.7903/CMR.19970

Dell'Era, M. & Rahova, A. (2019). Policy and artificial intelligence advent. *SSRN Electronic Journal*, 45, 285. https://doi.org/10.2139/ssrn.3463191

Dev, S., Sameki, M., Dhamala, J., & Hsieh, C. J. (2021). Measures and best practices for responsible AI. *Proceedings of the ACM SIGKDD International Conference on Knowledge Discovery and Data Mining*. ACM, New York. https://doi.org/10.1145/3447548.3469458

Eun, S. J., Kim, E. J., & Kim, J. Y. (2022). Development and evaluation of an artificial intelligence-based cognitive exercise game: a pilot study. *Journal of Environmental and Public Health*, 2022, 4403976. https://doi.org/10.1155/2022/4403976

Farrow, E. (2022). Determining the human to AI workforce ratio - exploring future organisational scenarios and the implications for anticipatory workforce planning. *Technology in Society*, 68, 101879. https://doi.org/10.1016/j.techsoc.2022.101879

Gorski, A. T., Gligorea, I., Gorski, H., & Oancea, R. (2022). Workforce and workplace ecosystem - challenges and opportunities in the age of digital transformation and 4IR. *International Conference Knowledge-Based Organization*, 28(1), 372–377. https://doi.org/10.2478/kbo-2022-0028

Healey, M. P. & Hodgkinson, G. P. (2009). Troubling futures: scenarios and scenario planning for organizational decision making. *The Oxford Handbook of Organizational Decision Making*. Oxford, UK. https://doi.org/10.1093/oxfordhb/9780199290468.003.0030

Ishii, E., Ebner, D. K., Kimura, S., Agha-Mir-Salim, L., Uchimido, R., & Celi, L. A. (2020). The advent of medical artificial intelligence: lessons from the Japanese approach. *Journal of Intensive Care*, 8(1), 35. https://doi.org/10.1186/s40560-020-00452-5

Iverson, K. (2000). Managing for effective workforce diversity: Identifying issues that are of concern to employees. *Cornell Hotel and Restaurant Administration Quarterly*, 41(2), 31–38. https://doi.org/10.1016/S0010-8804(00)88895-3

Jiang, Z., Hu, X., & Wang, Z. (2016). Employee-oriented HRM and voice behavior: a moderated mediation model. *Academy of Management Proceedings*, 2016(1), 10731. https://doi.org/10.5465/ambpp.2016.10731abstract

Kahane, A. (2012). Transformative scenario planning: changing the future by exploring alternatives. *Strategy and Leadership*, 40(5), 19–23. https://doi.org/10.1108/10878571211257140

Kalyanaraman, A., Burnett, M., Fern, A., Khot, L., & Viers, J. (2022). Special report: the AgAID AI institute for transforming workforce and decision support in agriculture. *Computers and Electronics in Agriculture*, 197, 106944. https://doi.org/10.1016/j.compag.2022.106944

Karwal, R. & Tandon, S. (2021). Impact of workforce diversity on employee performance: a study on it companies. *Asian Journal of Multidisciplinary Research & Review (AJMRR)*, 3(3), 86–102.

Kaur, A. & Mohammad Aslam, A. (2020). Artificial intelligence in business. *Management of Data in AI Age*. CSMFL Publications, Jagadhri. https://doi.org/10.46679/isbn 978819484834901

Khandelwal, K. & Upadhyay, A. K. (2021). The advent of artificial intelligence-based coaching. *Strategic HR Review*, 20(4), 137–140. https://doi.org/10.1108/shr-03-2021-0013

Khang, A., Kali, C. R., Suresh Kumar, S., Amaresh, K., Sudhansu Ranjan, D., & Manas Ranjan, P. (2023). Enabling the future of manufacturing: integration of robotics and IoT to smart factory infrastructure in industry 4.0. *AI-Based Technologies and Applications in the Era of the Metaverse* (1st ed., pp. 25–50). IGI Global Press, Hershey, PA. https://doi.org/10.4018/978-1-6684-8851-5.ch002

Khang, A., Muthmainnah, M., Seraj, P. M. I., Yakin, A. A., Obaid, A. J., & Panda, M. R. (2023). AI-aided teaching model for the education 5.0 ecosystem. *AI-Based Technologies and Applications in the Era of the Metaverse* (1st ed., pp. 83–104). IGI Global Press, Hershey, PA. https://doi.org/10.4018/978-1-6684-8851-5.ch004

Khang, A., Shah, V., & Rani, S. (2023). *AI-Based Technologies and Applications in the Era of the Metaverse* (1st ed.). IGI Global Press, Hershey, PA. https://doi.org/10.4018/978-1-6684-8851-5

Kim, B. Y. (2006). Managing workforce diversity: developing a learning organization. *Journal of Human Resources in Hospitality and Tourism*, 5(2), 69–90. https://doi.org/10.1300/J171v05n02_05

Kirongo, A. C., Bundi, D. G., Kitaria, D. T., Huka, G. S., & Muketha, G. M. (2022). Implementation of AI-based assistive technologies for learners with physical disabilities in Kenya: a practical design thinking approach. *African Journal of Science, Technology and Social Sciences*, 1(1), 176–196. https://doi.org/10.58506/ajstss.v1i1.71

Kumar, D. & Suresh, B. H. (2018). Workforce diversity and its impact on employee performance. *International Journal of Management Studies*, 4(1), 48. https://doi.org/10.18843/ijms/v5i4(1)/07

Levukova, V. A. (2022). *Artificial Intelligence and Cognitive Computing: A Guide to Artificial Intelligence for Business*. CRC Press, Boca Raton, FL. https://doi.org/10.46554/russian.science-2021.09-1-26/29

Li, Z., Oljaca, M., Firdousi, S. F., & Akram, U. (2021). Managing diversity in the Chinese organizational context: the impact of workforce diversity management on employee job performance. *Frontiers in Psychology*, 12, 733429. https://doi.org/10.3389/fpsyg.2021.733429

Liu, J., Zhu, Y., & Wang, H. (2023). Managing the negative impact of workforce diversity: the important roles of inclusive HRM and employee learning-oriented behaviors. *Frontiers in Psychology*, 14, 1117690. https://doi.org/10.3389/fpsyg.2023.1117690

Muthmainnah, M., Khang, A., Seraj, P. M. I., Yakin, A. A., Oteir, I., & Alotaibi, A. N. (2023). An innovative teaching model - the potential of metaverse for English learning. *AI-Based Technologies and Applications in the Era of the Metaverse* (1st ed., pp. 105–126). IGI Global Press, Hershey, PA. https://doi.org/10.4018/978-1-6684-8851-5.ch005

Pooja, K., Babasaheb, J., Ashish, K., Khang, A., & Sagar, K. (2023). The role of blockchain technology in metaverse ecosystem. *AI-Based Technologies and Applications in the Era of the Metaverse* (1st ed., pp. 228–236). IGI Global Press, Hershey, PA. https://doi.org/10.4018/978-1-6684-8851-5.ch011

Sadiku, M. N. O., Fagbohungbe, O., & Musa, S. M. (2020). Artificial intelligence in business. *International Journal of Engineering Research and Advanced Technology*, 06(07), 62–70. https://doi.org/10.31695/ijerat.2020.3625

Shah, V. & Khang, A. (2023). Metaverse-enabling IOT technology for a futuristic healthcare system. *AI-Based Technologies and Applications in the Era of the Metaverse* (1st ed., pp. 165–173). IGI Global Press, Hershey, PA. https://doi.org/10.4018/978-1-6684-8851-5.ch008

Shyam, R. S. & Khang, A. (2023). Effects of quantum technology on metaverse. *AI-Based Technologies and Applications in the Era of the Metaverse* (1st ed., pp. 104–203). IGI Global Press, Hershey, PA. https://doi.org/10.4018/978-1-6684-8851-5.ch009

Skulimowski, A. M. J. (2013). Exploring the future with anticipatory networks. *AIP Conference Proceedings*, 1510, 224–233. https://doi.org/10.1063/1.4776525

Sooraj, S. K., Sundaravel, E., Shreesh, B., & Sireesha, K. (2020). IoT smart home assistant for physically challenged and elderly people. *Proceedings - International Conference on Smart Electronics and Communication, ICOSEC 2020*. https://doi.org/10.1109/ICOSEC49089.2020.9215389

Southworth, J., Migliaccio, K., Glover, J., Glover, J. N., Reed, D., McCarty, C., Brendemuhl, J., & Thomas, A. (2023). Developing a model for AI Across the curriculum: transforming the higher education landscape via innovation in AI literacy. *Computers and Education: Artificial Intelligence*, 4, 100127. https://doi.org/10.1016/j.caeai.2023.100127

Svetlana, N., Anna, N., Svetlana, M., Tatiana, G., & Olga, M. (2022). Artificial intelligence as a driver of business process transformation. *Procedia Computer Science*, 213(C), 276–284. https://doi.org/10.1016/j.procs.2022.11.067

Vineeta, S. C., Jaydeep, C., & Khang, A. (2023) Smart cities data indicator-based cyber threats detection using bio-inspired artificial algae algorithm. *Handbook of Research on AI-Based Technologies and Applications in the Era of the Metaverse*, pp. 436–447. IGI Global Press, Hershey, PA. https://doi.org/10.4018/978-1-6684-8851-5.ch024

Wilson, H. J. & Daugherty, P. R. (2019). Creating the symbiotic AI workforce of the future. *MIT Sloan Management Review*, 61(1), 1–4.

Xu, J. J. & Babaian, T. (2021). Artificial intelligence in business curriculum: the pedagogy and learning outcomes. *International Journal of Management Education*, 19(3), 100550. https://doi.org/10.1016/j.ijme.2021.100550

Zohuri, B. (2020). From business intelligence to artificial intelligence. *Modern Approaches on Material Science*, 2(3), 1–10. https://doi.org/10.32474/mams.2020.02.000137

25 Human-Centered Approach as a Methodological Tendency of Personnel Management in Workplace

*Alex Khang, Inna Semenets-Orlova,
Hennadiy Dmytrenko, Nataliia Holovach,
Liubov Zgalat-Lozynska, and Roman Shevchuk*

25.1 INTRODUCTION

As the co-founder of the Club of Rome, Aurelio Peccei, writes, "the problem ulti-mately boils down to human qualities and ways to improve them... because only through the development of human qualities and human abilities can we achieve changes in the entire material-oriented civilization and use all its enormous poten-tial for good purposes (Peccei, 2016). We also find confirmation of this opinion in works of outstanding scientist in the field of social sciences: In the 21st century, the socio-economic development of countries will depend not on natural or material and even financial resources but rather on human ones."

The fact of the primacy of person and his or her activity in the appearance of threats to the development of mankind is becoming more and more indisputable (Havrylyshyn, 2016). In the context of globalization and AI-oriented economy, it makes sense to consider the relevance of the problem of improving human capi-tal in three aspects of the social development of mankind: planetary, national, personal.

- **In the planetary aspect**, the solution to the problem lies in two planes: the first case is the theoretical one – in the context of determining the essence of the best human qualities in the direction of a certain human-istic ideology – and the technological one. In the latter case, the prob-lem of systematic formation of the best human qualities in students in all countries of the world (at least in civilized democratic countries),

DOI: 10.1201/9781003440901-25

with the involvement of the students themselves in this process through self-discovery, becomes particularly acute. This is due to the fact that the world desperately needs the establishment of friendly relations between countries both to eliminate hostile confrontations in the peace age and to ensure making joint efforts for building the smart living environment to the humanity of the climatic, informational, medical, and potential pandemic nature (Khang & Rani, 2022).

- **In the national aspect**, the solution to the problem of improving human capital is relevant from the point of view of the need to form a cohesive society without social upheavals, focused on social justice, and to promote the self-realization of each individual, taking into account the interests of green society. This is related to the atmosphere of people's trust in the government, and the government itself is represented in this case by people with the best professional and personal qualities and respect for citizens in workplace (Rani & Khang, 2022).

- **In the personal aspect**, the relevance of this problem is related to the need of each person to find him/herself (first of all, his/her "related work" through the best professional, personal, and social qualities) and the levers of his/her development (Khang & Gujrati et al., 2023). This will contribute to his/her self-realization in all spheres of life not to the detriment of society (those around him/her), that is, according to the principle: "both for oneself and for people."

Such a three-pronged approach to solving the problem of improving human capital in education requires the unification of different ideas on a single ideological basis within the framework of one system-forming idea, and then, the substantiation of appropriate educational technologies for the formation of the best human qualities on a certain fundamental basis, for example, digitalization of the educational process using factor-criterion qualimetry.

25.2 PROBLEM STATEMENT

Today, the fact of the existence of threats to humanity and of powerful involvement of human activities, which are completely dependent on the quality of human capital, in their occurrence, is absolutely proven (Rana & Khang et al., 2021). In this context, it is enough to familiarize oneself with studies of prominent public figures in whom the need for transcendence is developed. We mean, for example, those of Bill Gates regarding a climate catastrophe (Gates, 2021), and Lee Kuan Yew and Yuval Noah Harari regarding a social catastrophe (Kuan, 2016; Harari, 2020).

Klaus Schwab regarding consequences of scientific and technological progress in general and of the revolution after the invention of the Internet in particular (Schwab, 2017); it is the emergence of the pandemic and the postpandemic life, and many others. Of great importance to the understanding of the problem of a planetary scale under consideration are the studies of the Club of Rome, starting from the moment of its foundation in 1968. For example, at the anniversary meeting of the club in 2018, it was recognized that "The old world is doomed. A new world is inevitable."

At the same time, Anders Wijkman (a co-chairman of the meeting) announced the need to urgently adopt the rules of a new unified planetary harmonious civilization, which would be based not on a political, but on a natural and social, spiritual worldview. That is, it is time to use new forms of harmonious lifestyle and spiritual and moral principles of interaction of all social groups of the population (Khanh & Khang, 2021). At the same time, it must be done within the framework of one universal system, both in its territorial and conceptual and ideological format. Such conclusions have ripened in the Club of Rome already at the beginning of its existence. For example, in the late 70s of the last century, it has already expressed some fundamental thoughts.

One of them related to the fact that ultimately the fate of mankind is determined by human qualities and the ways of their improvement. The second opinion was a kind of continuation of the first one, but already with a bias for a path pointer. In particular, after observing the state of permanent crises of a local and regional nature in relations between countries, as well as continuous quarrels within countries, Aurelio came to the conclusion that "the only way to save humanity lies through what I call a human revolution – through a new humanism that leads to the development of the best human qualities."

In this case, "new humanism" is considered as caring for each individual person from the point of view of promoting their self-realization in life in multicultural societies, but necessarily taking into account the interests of other people, society, and even the humanity. The latter can happen if people develop the need for transcendence, that is, the idea of oneself as an organic part of the world (Babasaheb et al., 2023). Therefore, fully sharing the results of the authors determined the goal of their own research, to the justification of the achievement of which this chapter is devoted. The goal is to reveal the content of innovative technologies for the achievement of a "human revolution" through the improvement of human capital through conscious formation of the best human qualities in the education system in the context of new humanism.

25.3 PROPOSED WORK

Let us start with the key concept of "human capital," which, in the conditions of globalization, dynamic information flows, and new humanism, is spreading and even deepening in a certain way. From the very beginning of the introduction of this concept into economic science, its meaning was reduced to "a set of knowledge, abilities and skills accumulated by people, which are implemented in practice as a productive labor force for obtaining income: profits for employers and remuneration for the owners of this force."

Over the years, scientists have added the factor of "human health" to this totality, which is indeed a mandatory organic component of human capital. Then it is determined that the productive power of labor significantly increases if a person is aware of his/her "related work" and finds his/her place in the structure of labor activity. Indeed, his/her work potential increases significantly if his/her professional and personal qualities and abilities of correspond to the position he/she holds and the functions he/she performs (Khang & Rana et al., 2023).

But the concept of "human capital" is broader than the concept of labor potential, "drawing" the latter into itself. Therefore, in the conditions of increased globalization and informatization, as well as accelerated scientific and technical progress (which causes the rapid obsolescence of today's relevant knowledge and the need for its constant updating), such a human quality as the ability of a person to replenish the baggage of his/her knowledge on a reflexive basis acquires particular acuteness.

The modern view of "human capital" is not limited to the content and volume of accumulated knowledge and skills, which provides an opportunity to receive a certain income (added value). This concept also includes a certain set of human qualities, which allows, on the one hand, to strengthen this capital (improve it). On the other hand, these qualities help a person to realize him/herself both as a person and the main productive force of society in the context of personal self-fulfillment (and not only in the labor sphere of life at that). It remains to be determined which combination of qualities is able to strengthen and improve human capital while simultaneously promoting the self-fulfillment in various spheres of life of the "owners" of these qualities themselves.

25.4 THE BEST HUMAN QUALITIES IN THE CONTEXT OF NEW HUMANISM

It has already been mentioned above that a sign of the best human qualities in the context of new humanism is their usefulness for both the personal self-fulfillment of a person and the surrounding social sectors, employers, society, and the state. But the question arises: what exactly are the qualities that correspond to these signs according to the principle "both for themselves and for people"? According to the authors, the definition of these best qualities can be started by receiving answers from a large number of people: what kind of people, based on their qualities, would they like to see around them? And it is desirable for employers and mid-level managers to answer the question: what subordinate specialist do they need (but not more than in terms of basic qualities)? As for the best human qualities required for responsible, high positions, this has already been determined (Snehal et al., 2023).

In this sense, we should remind the reader of the great leader of our time, the author of the economic and social miracle in Singapore, Lee Kuan Yew: "... our government, when selecting personnel, went from 40 qualities to three, which were called "helicopter qualities"... what they consist in? Ability to analyze; accurate logical assessment of facts; concentration on the main points; ability to highlight the main things. However, this is not enough... one must be able to rise above reality..., i.e., have an imagination" (Alison et al., 2018).

If one conducts such a mass survey in different countries of the world, then, after analyzing the information, one can hope to get essentially similar results (with some deviations taking into account different cultures). For example, according to the first question, these qualities can be the following: humanity (ability to come to the aid of others in a difficult situation for them); reliability; responsibility; goodwill; tolerance; sociability; and politeness. As for the best qualities of employees, "normal" employers and managers (answers to the second question) will definitely mention such basic qualities as follows:

- availability of up-to-date professional knowledge and skills to work productively;
- ability to constantly update these knowledge and skills;
- responsibility for one's actions, work, results;
- nonconflict;
- ability to work in a team.

The authors used the term "normal" meaning that, if an employer or a manager adheres to a rigid directive style of management and is excessively ambitious, then there will be additional requirements for flattery, obedience, submission, etc. But the qualities defined above are key. Finally, there is another aspect of determining the best human qualities, but already from the standpoint of the requirements of all humanity on the planet Earth (for the development and even the existence of which there are serious threats). In particular, such qualities include:

- a developed need for transcendence;
- high level of consciousness (the 4th one).

Let us consider the essence of these two interrelated qualities. An outstanding psychological scientist of the 20th century, the author of the well-known pyramid of needs, A. Maslow, already came close to the concept of "transcendence" in the context of actualization (self-fulfillment) of an individual. It was about a person's ability to give preference to much higher, spiritual values of the humanity, rather than the needs of his/her own egoistic self.

S. Kaufman made a concrete contribution to the definition of transcendence (as a higher need within the framework of the self-fulfillment of an individual), deepening the achievements of A. Maslow. He writes: "Our new hierarchy is crowned with the need for transcendence, which goes beyond personal growth (and even happiness and health) and allows rising to the highest level of unity and harmony with oneself and the world."

Transcendence is the point of view that allows us to take a high vantage point and approach our "self" with acceptance, wisdom, and a sense of connection with all humanity. In other words, "transcendence" is a combination of the needs of a personal self with the needs of not only society but also all humanity. The path to transcendence can be characterized by the stages of the formation of human consciousness according to the levels of adopted values. This is covered in sufficient depth in the works of the scientific director of the School of Spiritual Development of a Personality in Nation, doctor of psychological sciences, famous full professor Pomytkin (2015). He substantiated several (5) levels of human consciousness, the fourth of which is the level of universal values, which corresponds to a highly developed need for transcendence.

In particular, he writes: "The fourth level is the level of acceptance and understanding of another person regardless of his/her national characteristics or faith. The acceptance of each person as an individuality, the recognition of the right of everyone to his/her own beliefs, the awareness of the fact that the humanity is a single interconnected organism, in which everyone fulfills his/her important role characterizes the level of universal human values. This level of consciousness frees a person from the closed circle of "mine - someone else's," "friends – enemies." The main principle inherent in the worldview of the fourth level is the so-called "golden rule" of Jesus Christ (and even earlier – that of Confucius (the authors' note)): "Do not do to others what you do not want to be done to yourself" (Pomytkin, 2015). Also, there, all five levels of human consciousness are listed:

1. The level of self-centered consciousness, of values of self-affirmation.
2. The level of family values.
3. The level of social, civic, and national values.
4. The level of universal values.
5. The level of cosmic, expanded consciousness (that is, understanding that the universe is also a single organism, and the Earth is a part of it).

Below, we will take a closer look at why the humanity, society, and man himself need the best human qualities.

25.5 APPLICATION OF THE BEST HUMAN QUALITIES

25.5.1 First, As Regards All Mankind

Bill Gates has proposed a plan to avert a climate catastrophe by eliminating and reducing carbon dioxide emissions to zero by 2050. In this case, recommendations are given on how to act and what to do not only to the governments of all countries, including civilized ones, but also to ordinary citizens (Gates, 2021). There, he also writes: "I also hope that we will unite for the sake of a plan that overcomes political differences.... Everyone can act as a citizen, as a consumer, as an employee or an employer. For example, as a citizen: everyone can take part in constant pressure on politicians to make appropriate decisions. Therefore, he can make phone calls, write letters, attend meetings and rallies, etc., as well as act at both the local and state level through certain public or private organizations, etc."

25.5.2 Role and Responsibility of a Consumer

Reduce emissions of your household by optimizing energy consumption, use solar panels, etc., buy an electric car, sign up for your energy supplier's green tariff program, consume plant-based protein, etc.

25.5.3 Role and Responsibility of an Employee or Employer

Set a domestic carbon tax; prioritize low-carbon innovations; be the first to implement them (without waiting for orders from the authorities); participate in the formation and implementation of a relevant political course; contact budget-funded research; help innovators in the early stages of putting their developments into practice, etc. (Gates, 2021). According to the plan developed by him, Bill Gates provides many more different recommendations in different directions of averting a climate disaster. But there is one but... everything that Bill Gates talks about can be realized under one condition: a developed need for transcendence and a high (at least the fourth) level of consciousness of the critical mass of the population of all countries. And, obviously, first of all, we are talking about management elite which are able to organize and direct in a proper manner scientific and technical progress and the entire education system.

25.5.4 IN RELATION TO ALL OF THE HUMANITY

In relation to all of the humanity, to ward off the threats of the fourth industrial (information) revolution after the invention of the Internet, there is also a need for the best human qualities. The fact is that the consequences of this revolution can be both positive and negative. And, everything will depend on decisions of people and their qualities that are engaged in the implementation of innovations in life, and whose interests are oriented to the benefit of the citizens or to the satisfaction of their egocentric needs, and we hope that the best human qualities are also extremely necessary to develop workplace as shown in Figure 25.1.

In this context, it makes sense to mention the negative impact of social networks on human nature. It writes, "As we approach 2020, we see that social networks are increasingly divisive. It seems that people can no longer talk to each other without getting irritated." He calls the weakening of empathy in people as one of the main reasons for this condition (Arpita et al., 2023). Empathy is the ability to look at life from other people's points of view, to put oneself in others' place.... around 2015, studies began to appear that said that social networks on a biological level destroy the ability of young users to empathize... and already in 2018, former president of Facebook Sean Parker and former head of Facebook. Certain fears regarding the consequences of mass informatization of society are also expressed by other experts who see both positive and simultaneously negative nature of these consequences.

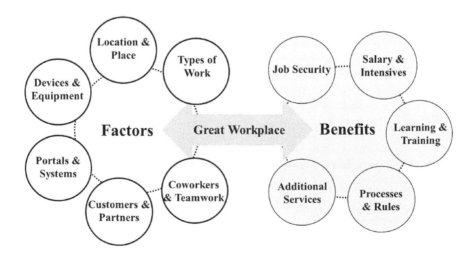

FIGURE 25.1 Factors and benefits of a great workplace.

In particular, Klaus Schwab (the founder and president of the World Economic Forum in Geneva) writes: The fourth industrial revolution has the potential to robotize the humanity and threaten our traditional sources of sense, such as work, society, family, and personality. It is in our power not to allow such a development of the scenario but to use this revolution for an upward movement of the humanity toward a new collective and moral consciousness based on a unified vision of destiny. It is up to all of us to try to make it happen exactly like this... in the end, it all comes down to people, to culture, and to values. Cooperation is the only thing that will save the humanity (Schwab, 2017). So what human qualities are needed to create a "unified vision of destiny and cooperation for the sake of the humanity's movement toward a new collective and moral consciousness"? Undoubtedly, it is a developed need for transcendence and a high level of consciousness based on universal morality.

The essence of the latter is the respect of each person for others, regardless of different views, skin color, religious preferences, etc., as well as the respect for nature through a careful attitude to the lithosphere, hydrosphere, and atmosphere, and finally, the last thing about all of the humanity and each individual. To remove the threats of pandemics, one should pay attention to one of the best human qualities – good health (Vrushank et al., 2023). To solve this problem, it is necessary to realize that the primary role is not the development of vaccines after the detection of the signs of a pandemic but the prevention of diseases through the immune defenses of people. The primary role is played by immunology and the general indicator of the immune defense of each person. And, it is this indicator that should be determined as one of the main guidelines for the activities of teachers with the involvement of students themselves and their parents in the context of promoting the preservation of people's health and improving human capital (Vrushank & Khang, 2023).

25.6 PERSONNEL MANAGEMENT IN ORGANIZATIONS IN THE CONTEXT OF THE DIGITAL ECONOMY

In the context of globalization processes, which are characterized by the dynamic development of the digital economy, the use of innovative digital tools in the field of HR management in organizations has become relevant. Using artificial intelligence, automated data analytics systems, and other digital tools for HR management, it is possible to optimize the HR management system, improve the efficiency of work with personnel, make informed decisions based on objective data, etc.

Here are the main advantages of using digital tools in HR management:

- developing a strategy for digital transformation of HR management involves analyzing the current state, defining goals and objectives of HR management strategy and policy, selecting and implementing digital initiatives and tools that can be used in the HR management process (HR platforms, learning management systems, analytical tools, etc.), developing a plan for implementing digital tools and processes, and involving personnel in the digital transformation process;

- the use of data analytics will provide valuable insights into the efficiency of personnel utilization, forecast personnel losses, and develop strategies to attract and retain competitive personnel;
- the use of cloud platforms and mobile applications will simplify and automate HR management processes;
- developing digital skills in the effective use of digital tools and technologies in the workforce will optimize the composition and functions of the workforce.

In addition, digital tools such as data analytics systems and predictive models can be used to forecast workforce needs, identify internal team talent, and also analyze organizational challenges and trends to help make informed management decisions. One of the key aspects of effective human resource management in the digital economy is understanding the impact of digital technologies and artificial intelligence on the workforce and the organization as a whole (Verina & Titko, 2019). The introduction of digital tools such as virtual communication platforms and collaborative workspaces will change the way employees communicate and collaborate, particularly in virtual team structures.

The use of artificial intelligence in personnel management opens up quite a wide range of opportunities to improve the efficiency of this process, in particular:

- The process of automating standard HR management tasks, such as document processing, personnel record keeping, work time allocation, etc., frees personnel from monotonous work and allows them to concentrate on strategic aspects of management.
- Recruitment and personnel selection – artificial intelligence can analyze resumes, can assess the skills and experience of candidates, using a set of algorithms, and can qualitatively select the most suitable candidates for specific vacancies.
- Analytics and forecasting – artificial intelligence can analyze large amounts of HR data, including productivity metrics, talent pools, employee satisfaction, and many other factors. With the help of algorithms, it can make analytical predictions about personnel retention, identify the risk of losing key employees, and recommend strategies to optimize HR processes.
- Learning and development management – artificial intelligence can accurately identify individual training and development needs of employees, analyze employee data, their competency levels, and potential capabilities. Given the individual needs of workers, artificial intelligence can suggest training materials, courses, or trainings that best meet the needs and abilities of each worker, and can develop personalized training programs.

Today's digital transformation requires executives and HR managers to rethink the HR functions in an organization. Traditional tasks such as recruiting and administration are now increasingly automated, enabling HR to focus more on strategic aspects such as talent development, corporate culture, and innovation. The adoption of digital technologies and the digital transformation of organizations are accompanied by significant changes in work processes, functions, and employee attitudes. Managers must actively work to engage and support personnel in the change process, ensuring readiness for new requirements and rapid adaptation to change. As the digital economy develops, the role of developing digital skills in employees is increasing, including such skills as the ability to work with digital tools, understanding of cyber security and data protection, adaptability to change, and readiness for continuous development (Koval et al., 2021). Self-sufficient organizations today are already actively investing in the development of digital competencies of their personnel, organizing training to improve their digital literacy (Semenets-Orlova et al., 2021).

The modern labor market offers enough IT specialists and specialists of various professions with developed digital competencies with the development of the digital economy, in particular, such as the ability to work in the digital environment, search and analyze information, interact and communicate with various actors in cyberspace, work with digital content and others. Companies that actively implement digital tools in their activities attract the attention of a younger generation of employees who appreciate innovativeness and development opportunities. The use of digital tools in the recruiting process, creation of favorable working conditions, and development of online learning is now becoming popular in many fast-growing companies (Semenets-Orlova et al., 2020).

Today, the ethical issues of using digital technologies in HR management are becoming increasingly acute. Due to the increase in personal data and the use of algorithms to make management decisions, risks related to privacy, confidentiality, and discrimination may arise. Therefore, users of digital technologies need to comply with ethical standards and policies that govern the use of digital technologies and protect the rights and interests of personnel. An important aspect of successful HR management in the digital economy is a strategic partnership with senior management.

The digital transformation of an organization requires significant investments and changes in business strategy. Therefore, HR managers must work in close partnership with senior management to understand the strategic goals of the organization and identify the necessary transformations in HR management. Consequently, HR management in the digital economy requires an orientation toward innovation and the use of digital tools to attract, evaluate, and develop personnel. Data analysis and digital skills are becoming key competencies for managers, HR managers, HR, and labor economics specialists. Adoption of digital technologies allows organizations to make more objective assessments of workforce needs forecasting, attract and retain talent, and generally, optimize the HR management process. Successful implementation of digital HR management tools contributes to the competitiveness of organizations and their adaptation to changes in the digital economy.

25.7   RESULTS AND DISCUSSION

So, having considered certain nuances of the application of the best human qualities in general in the planetary aspect, let us return to the personal aspect in order to pay serious attention to an innovative strategy of conscious formation of these qualities as the results of educational activities in national education systems.

25.7.1 THE TECHNOLOGICAL ESSENCE OF A NEW STRATEGY OF HUMAN CAPITAL IMPROVEMENT

One of the features of this innovative strategy and the first stage is the identification of key human qualities from the complex of the best ones. This is necessary in order to concentrate the efforts of all pedagogical personnel on them (with the involvement of the students themselves and their parents) with the aim of their conscious formation in an end-to-end manner at institutions of all levels of education. For example, of the eleven best qualities mentioned above (which are useful for both oneself for one's own self-fulfillment and others for their self-fulfillment, including employers), only two are identified as key ones, and the rest are derived from the key qualities.

We are talking, firstly, about universal morality (UM) and, secondly, about analytical and cognitive activity (ACA). With regard to UM, we will try to prove it as follows, through answering the question: "What are these based on: humanity based on empathy; reliability; goodwill; tolerance; sociability (not to be confused with talkativeness); politeness; non-conflict; responsibility?"

Deep consideration of this issue will definitely lead us to one thing: the foundation of all these qualities is respect for other people (Khang, Shah et al., 2023). And it is "respect" that is the essence of universal morality (this "respect" applies not only to people but also to nature: preservation of fauna and flora, nonpollution of the earth's surface, as well as the atmosphere and water resources). Of course, for some time, an egocentric person with a low level of consciousness can pretend to be tolerant, benevolent, polite, humane, etc. But this is only for a certain time, and also, usually, it refers to certain persons on whom something depends. But "life not in one's body" by itself does not contribute to self-realization and does not bring happiness.

In this sense, such a quality as "responsibility" is interesting. It can be either natural (that is, genetically determined), or instilled by the parents from the childhood, or the result of effective management. But in any case, this quality is highly valued primarily by employers in employees as self-respect and to facilitate management. As for ACA, it is at the basis of such qualities as the presence of relevant professional knowledge and skills, the ability to constantly update this knowledge and these skills, as well as erudition.

The fact is that a highly developed ACA conditions the person's assimilation of knowledge and skills at a reflexive level (and not based on memorization), which ensures their creative application and the internal need for constantly updating this knowledge and these skills. Also, it is the ACA indicator that becomes an indicator of a school's fulfillment of its primary purpose: "teaching children to learn throughout life." The UM indicator serves as a similar indicator of a school's fulfillment of its second mission regarding the socialization of an individual: "teaching children to live together in a multicultural society." The second feature and, accordingly, the second stage of the formation of this New Strategy for Improving Human Capital is the modernization of the education system in the context of determining the final results of the activities of teachers, students and their parents, and finally, the educational institutions themselves.

25.7.2   Digitization of the Educational Process Based on Factor-Criterion Qualimetry (FCQ)

In general, each person is an extremely complex biosocial phenomenon that includes many different qualities and character traits and is to some extent a whole microcosm. But from this set, it is possible to single out the main significant qualities that are critical in the context of influencing the behavior of a person as a social being and a productive force. It is on the diagnosis of these qualities for the purpose of their further development (or inhibition, if they are negative) that the educational activities of teachers should begin to concentrate. The modern world has created many opportunities to use digital tools in education. For example, there are possibilities of using a chatbot for improving learning, given in Table 25.1, which shows that a chatbot is extremely useful as an additional tool for students' learning.

A student's communication with a teacher and his/her communication with a chatbot differs in many factors (Table 25.2).

At the same time, the traditional methods of diagnosis with the help of qualitative characteristics or quantitative scoring have already proven their inefficiency from the point of view of motivational influence on the self-knowledge and self-improvement of students.

TABLE 25.1

Possibilities of Using Chatbots to Improve Student Learning (by Authors)

Type	Possibilities
Feedback	Enables a teacher to keep in touch with a group of students without much difficulty
Personal mentor	Using as a program for individual development
Interactivity in learning	Acquisition and consolidation of knowledge in the form of tests, quizzes, surveys, providing instant feedback
Student assessment	Automatic check of performed works
Source of information	Exchanging information about consultations, meetings, useful links, work samples, etc.

TABLE 25.2

Advantages and Disadvantages of Students' Communication with Chatbots (by Authors)

Advantages	Disadvantages
Effective collection of information	Difficulty of implementation
Request processing speed	Ability to process simple requests
Increase in the level of satisfaction	Lack of creative approach to solving problems
Processing a large number of requests simultaneously	Lack of emotions from live communication

- The diagnosis of the best human qualities at EU schools (taking Germany as an example) can serve as an evidence of the presence of this inefficiency. There, information about social work and the behavior of students are usually highlighted by diagnosing the following qualities: readiness for learning and achieving success; reliability; independence; readiness for responsibility; readiness to work in a team; the ability not to conflict; compliance with accepted rules; and culture of interaction.
- The diagnosis is carried out according to the following criteria: pronounced; expressed; partially expressed; weakly expressed (which teachers identify through certain signs). Such cards are provided to students and parents every 6 months in parallel with the report cards.
- But the final result of such diagnosis is very weak. The absolute majority of the students and parents remain indifferent to such criteria, and the teachers make their observations formally, only as reports on the execution of orders from above. That is, it is not possible to achieve the main thing, namely, to "draw" the students into the process of self-discovery and self-improvement.

The situation can change significantly if factor-criterion qualimetry from the field of fuzzy mathematics is used. Then the "pronounced" criterion, for example, as the optimal norm (or a standard, or an ideal) will be equal to 1.0, "expressed" -0.75, "partially expressed" -0.50, "weakly expressed" -0.25. There can be other criteria, and their number can be greater, but the essence is the same: from a maximum of 1.0–0.0, if the presence of one or another quality is not observed at all. In this chapter, it is not possible to cover in detail the methodology of forming factor-criterion qualimetric models (FCQMs), but the main thing is different.

Thanks to FCQM, it is possible to express the results of diagnosis according to the innovative methodology in an index form (e.g., 0.36, 0.45, 0.68, etc.). In this case, there are great advantages of such a digital interpretation of the results of diagnosing the qualities of students as personality parameters, which is the basis for their purposeful formation with the involvement of adolescent students themselves in this process.

- Firstly, this purposefulness is determined by the emergence of clear benchmark indices of educators' activities, positive dynamics of which can be considered as the final results of their educational activities. And this will stimulate creative searches for certain methods to increase these indices.
- Secondly, students clearly see their lag behind the optimal level -1.0. Then they see that their every step in the direction of improvement is reflected in a change in the index. And, finally, with the correct use of rating comparisons, an additional and sufficiently powerful motivation appears in an adolescent student: competitiveness.

There is another interesting aspect of the digitalization of the educational process on a factor-criterion qualimetric basis. We are talking about an innovative way to the formation at educational institutions (and later at management structures of the educational sector) of effective management, with which the educational sector has never been associated. The fact is that the key condition of effective management according to Khang and Gujrati et al. (2023) is a clear definition of goals and measurement of final results. Until recently, this could not be achieved in education due to the complexity of a person as an object of research and the lack of tools capable of absolutely comparative measurement of human qualities as personality parameters (Khang & Muthmainnah et al., 2023).

The application of FCQMs means the emergence of such a toolkit, which determines the possibility of using effective management in the education system, as well as forming a system management culture (SMC) at each educational institution based on final results (Semenets-Orlova, 2020). Risk management of forming enterprises in the latter are precisely indices of the best human qualities, including, first of all, the indices of UM (universal morality) and analytical cognitive activity (ACA), as well as others characterizing, for example, a person's suitability for a certain professional activity.

25.8   CONCLUSION

The need to increase the level of development of human capital (HC) is connected with solving problems of various scales: at the planetary level (of all humanity); national levels of different states; at the personal level of citizens for successful self-fulfillment of the latter within the framework of new humanism in the workplace (Khang & Gupta et al., 2023). The essence of the New Strategy for Improving Human Capital is specifically related to the formation of the best human qualities in the education system and includes the justification of the three main areas reflected in this chapter.

- First, the definition of the essence of the best human qualities in the conditions of the modern civilization from the point of view of an organic combination of a person's own self with the interests of other people, society, and the humanity.
- Second, the development of an innovative technique of the formation of the best human qualities in the education system based on the digitization of the educational process with the help of factor-criterion qualimetry (FCQ) from the field of fuzzy mathematics.
- Third, the gradual spread of effective management in the educational field. This is done through the transition of educational institutions of all levels (starting with preschool) and management structures to a system management culture (SMC) based on final results, displacing the current bureaucratic culture of administrative pressure (CAP) without feedback.

The authors link the continuation of these studies with the development of the New Strategy for the Modernization of Public Administration in Nation, which is clearly focused on improving the quality of life of citizens (QLC) in the conditions of the modern civilization (Luke et al., 2024). The role of a human-centered smart education system in improving the quality of life of citizens is ultimately extraordinary.

The relevance of the problem of improving human capital is covered in three aspects: at the planetary, national, and personal levels. This is connected, first of all, with the need to divert from the humanity the threats of climatic, social, and informational catastrophes (as well as pandemics), the occurrence and approach of which depend on human activity (Subhashini & Khang, 2024).

The usefulness of any activity, in turn, determines the corresponding human qualities, among which the best ones are singled out. The authors attribute the peculiarity of the latter to two organically interdependent consequences of each person's behavior: in their own interests and at the same time in the interests of society and even the humanity (Khang & Kali et al., 2023). In this case, "their own interests" are considered as the satisfaction of a higher need: the need for the self-fulfillment of the individual in all spheres of life (but with the obligatory consideration of the interests of other people). To determine the specific best human qualities, the technique of sociological survey from the standpoint of "consumers" at different levels – personal, national, and planetary – was used. Two key qualities are singled out from all the best ones, which include analytical and cognitive activity (ACA) and universal morality (UM).

- The first one ensures the maturity of the need for continuous development through continuous updating and assimilation of knowledge and skills at the reflexive level.
- The second one is the basis for other best human qualities, including humanity, benevolence, loyalty, tolerance (non-conflict), responsibility, ability to work in a team, etc.

The technique essence of the formation of the best human qualities in the education system based on the digitization of the educational process at educational institutions with the help of factor-criterion qualimetry from the field of fuzzy mathematics is revealed. It is emphasized that all of the above in its totality determine the New Strategy for the Improvement of Human Capital in the interests of both an individual and society and all of the humanity in the age of smart working ecosystem.

This strategy is implemented, thanks to the implementation of effective management in the education system by displacing the current culture of administrative pressure (CAP). Its gradual replacement with the system management culture (SMC) according to the final results is carried out through measurements of these results and their interpretation in the form of certain indices (Khang & Shah et al., 2023).

REFERENCES

Alison, H., Blackwill, R. D., Vine, E., & Lee Kuan, Y. (2018). *Reflections of a Great Leader on the Future of China, the United States and the World*. KM group – BUKS, China. https://wires.onlinelibrary.wiley.com/doi/abs/10.1002/wcc.354

Arpita, N., Satpathy, I., Patnaik, B. C. M., Sukanta Kumar, B., & Khang, A. (2023). Impact of artificial intelligence (AI) on talent management (TM): a futuristic overview. *Designing Workforce Management Systems for Industry 4.0: Data-Centric and AI-Enabled Approaches* (1st ed., pp. 32–50). CRC Press, Boca Raton, FL. https://doi.org/10.1201/9781003357070-9

Babasaheb, J., Sphurti, B., & Khang, A. (2023). Industry revolution 4.0: workforce competency models and designs. *Designing Workforce Management Systems for Industry 4.0: Data-Centric and AI-Enabled Approaches* (1st ed., pp. 14–31). CRC Press, Boca Raton, FL. https://doi.org/10.1201/9781003357070-2

Gates, B. (2021). How to avert a climate catastrophe. *Where We are now and What should We do Next*. Laboratory, Kyiv. https://www.nejm.org/doi/full/10.1056/NEJMp1502918

Harari, Y. N. (2020). Man is intelligent. *History of mankind from the past to the future*. Family Leisure Club, Kharkiv. https://www.ajol.info/index.php/hts/article/view/213074

Havrylyshyn, B. (2016). A nation has a future if there is a plan for tomorrow. *Viche*, 2016, 1–2. https://link.springer.com/article/10.2307/4621690

Khang, A., Gupta, S. K., Rani, S., & Karras, D. A. (2023). Smart cities: IOT technologies, big data solutions, cloud platforms, and cybersecurity techniques (1st ed.). CRC Press, Boca Raton, FL. https://doi.org/10.1201/9781003376064

Khang, A., Kali, C. R., Suresh Kumar, S., Amaresh, K., Das, S. R., & Panda, M. R. (2023). Enabling the future of manufacturing: integration of robotics and IoT to smart factory infrastructure in industry 4.0. *AI-Based Technologies and Applications in the Era of the Metaverse* (1st ed., pp. 25–50). IGI Global Press, Hershey, PA. https://doi.org/10.4018/978-1-6684-8851-5.ch002

Khang, A., Muthmainnah, M., Seraj, P. M. I., Yakin, A. A., Obaid, A. J., & Panda, M. R. (2023). AI-aided teaching model for the education 5.0 ecosystem. *AI-Based Technologies and Applications in the Era of the Metaverse* (1st ed., pp. 83–104). IGI Global Press, Hershey, PA. https://doi.org/10.4018/978-1-6684-8851-5.ch004

Khang, A., Rana, G., Tailor, R. K., & Hajimahmud, V. A. (2023). *Data-centric AI solutions and emerging technologies in the healthcare ecosystem* (1st ed.). CRC Press, Boca Raton, FL. https://doi.org/10.1201/9781003356189

Khang, A., Rani, S., Gujrati, R., Uygun, H., & Gupta, S. K. (2023). *Designing Workforce Management Systems for Industry 4.0: Data-Centric and AI-Enabled Approaches* (1st ed.). CRC Press, Boca Raton, FL. https://doi.org/10.1201/9781003357070

Khang, A., Rani, S., & Sivaraman, A. K. (2022). *AI-Centric Smart City Ecosystems: Technologies, Design and Implementation* (1st ed.). CRC Press, Boca Raton, FL. https://doi.org/10.1201/9781003252542

Khang, A., Shah, V., & Rani, S. (2023). *AI-Based Technologies and Applications in the Era of the Metaverse* (1st ed.). IGI Global Press, Hershey, PA. https://doi.org/10.4018/978-1-6684-8851-5

Khanh, H. H. & Khang, A. (2021). The role of artificial intelligence in blockchain applications. *Reinventing Manufacturing and Business Processes through Artificial Intelligence*, pp. 20–40. CRC Press, Boca Raton, FL. https://doi.org/10.1201/9781003145011-2

Koval, V., Mikhno, I., Udovychenko, I., Gordiichuk, Y., & Kalina, I. (2021). Sustainable natural resource management to ensure strategic environmental development. *TEM Journal*, 10(3), 1022–1030.

Lee Kuan, Y. (2016). From the third world to the first. *History of Singapore: 1965–2000*. BUKS Publishing Group, Kyiv. https://www.tandfonline.com/doi/abs/10.1080/10357823.2019.1625864

Luke, J., Khang, A., Vadivelraju, C., Antony Richard, P., & Kumar, S. (2024). Smart city concepts, models, technologies and applications. *Smart Cities: IoT Technologies, Big Data Solutions, Cloud Platforms, and Cybersecurity Techniques* (1st ed.). CRC Press, Boca Raton, FL. https://doi.org/10.1201/9781003376064-1

Peccei, A. (2016). The Human Quality, Pergamon. https://books.google.com/books?hl=en&lr=&id=2UUvBQAAQBAJ&oi=fnd&pg=PP1&dq=Peccei+The+Human+Quality,+Pergamon&ots=Z8ugj5ucfM&sig=9uwE4xHx9hPeQOCxLKI6yOceyiA

Pomytkin, E. (2015). Spiritual Potential of the Individual: Psychological Diagnosis, Actualization and Development, Inner World. https://search.proquest.com/openview/07 36e2163c3ba65d939c7a0061d94e7a/1?pq-origsite=gscholar&cbl=18750&diss=y

Rana, G., Khang, A., Sharma, R., Goel, A. K., & Dubey, A. K. (2021). *Reinventing Manufacturing and Business Processes through Artificial Intelligence*. CRC Press, Boca Raton, FL. https://doi.org/10.1201/9781003145011

Schwab, K. (2017). The Fourth Industrial Revolution, Penguin. https://library.oapen.org/bitstream/handle/20.500.12657/23279/1/1006877.pdf#page=216

Semenets-Orlova, I., Klochko, A., Shkoda, T., Marusina, O., & Tepliuk, M. (2021). Emotional intelligence as the basis for the development of organizational leadership during the covid period (educational institution case). *Estudios De Economia Aplicada*, 39(5), 5074. https://doi.org/10.25115/eea.v39i5.5074

Semenets-Orlova, I., Klochko, A., Tolubyak, V., Sebalo, L., & Rudina, M. (2020). Functional and role-playing positions in modern management teams: an educational institution case study. *Problems and Perspectives in Management*, 18(3), 129–140.

Snehal, M., Babasaheb, J., & Khang A. (2023). Workforce management system: concepts, definitions, principles, and implementation. *Designing Workforce Management Systems for Industry 4.0: Data-Centric and AI-Enabled Approaches* (1st ed., pp. 1–13). CRC Press, Boca Raton, FL. https://doi.org/10.1201/9781003357070-1

Subhashini, R. & Khang, A. (2024). The role of internet of things (IoT) in smart city framework. *Smart Cities: IoT Technologies, Big Data Solutions, Cloud Platforms, and Cybersecurity Techniques* (1st ed.). CRC Press, Boca Raton, FL. https://doi.org/10.1201/9781003376064-3

Verina, N. & Titko, J. (2019). Digital Transformation: Conceptual Framework. https://doi.org/10.3846/cibmee.2019.073

Vrushank, S. & Khang, A. (2023). Internet of medical things (IoMT) driving the digital transformation of the healthcare sector. *Data-Centric AI Solutions and Emerging Technologies in the Healthcare Ecosystem* (1st ed., pp. 1). CRC Press, Boca Raton, FL. https://doi.org/10.1201/9781003356189-2

Vrushank, S., Vidhi, T., & Khang, A. (2023). Electronic health records security and privacy enhancement using blockchain technology. *Data-Centric AI Solutions and Emerging Technologies in the Healthcare Ecosystem* (1st ed., pp. 1). CRC Press, Boca Raton, FL. https://doi.org/10.1201/9781003356189-1

Index

Milton Keynes UK
Ingram Content Group UK Ltd.
UKHW031125141024
449569UK00006B/438